D0153598

FUNDAMENTAL
CONCEPTS
OF GEOMETRY

FUNDAMENTAL CONCEPTS OF GEOMETRY

by

BRUCE E. MESERVE

DOVER PUBLICATIONS, INC., NEW YORK

Copyright © 1955, 1983 by Bruce E. Meserve.
All rights reserved under Pan American and International Copyright Conventions.

Published in Canada by General Publishing Company, Ltd., 30 Lesmill Road, Don Mills, Toronto, Ontario.
Published in the United Kingdom by Constable and Company, Ltd.

This Dover edition, first published in 1983, is an unabridged and slightly corrected republication of the second printing (1959) of the work originally published by Addison-Wesley Publishing Company, Inc., Reading, Mass., in 1955.

Manufactured in the United States of America
Dover Publications, Inc., 31 East 2nd Street, Mineola N.Y. 11501

Library of Congress Cataloging in Publication Data

Meserve, Bruce Elwyn, 1917–
  Fundamental concepts of geometry.

  Reprint. Originally published: Reading, Mass. : Addison-Wesley Pub. Co., 1959. (Addison-Wesley mathematics series) With slight corrections.
  Bibliography: p.
  Includes index.
  1. Geometry. I. Title. II. Series: Addison-Wesley mathematics series.
[QA445.M45 1983]   516   82-18309
ISBN 0-486-63415-9

PREFACE

This book and its companion volume * on algebra and analysis are based upon a course entitled "Fundamental Concepts of Mathematics" as it has evolved at the University of Illinois during the past forty years. These two books provide a broad mathematical perspective for readers with a maturity equivalent to at least one year and preferably two years of college mathematics. Neither book is a prerequisite for the other. When both texts are to be used, the author prefers to cover the algebra before the geometry, but many of his students take the geometry first. Both texts have been used in mimeographed form for several years at the University of Illinois. There is ample material in each text for a course having 45 class hours. Both texts reflect the recognition of a basic need for a knowledge of fundamental concepts of mathematics apart from what is gained in the specialized courses in each of its many subdivisions. Prospective teachers of secondary mathematics, students preparing for specialized advanced courses in mathematics, and those desiring a broad liberal education have a particular need for the mathematical perspective gained from an elementary treatment of the fundamental concepts of mathematics.

The primary purpose of this book is to help the reader

(i) to discover how euclidean plane geometry is related to, and often a special case of, many other geometries,

(ii) to obtain a practical understanding of "proof,"

(iii) to obtain the concept of a geometry as a logical system based upon postulates and undefined elements, and

(iv) to appreciate the historical evolution of our geometrical concepts and the relation of euclidean geometry to the space in which we live.

These goals are sought through an introduction to the foundations of geometry (Chapter 1) and a consideration of the following hierarchy of geometries.

---

* Meserve, B. E., *Fundamental Concepts of Algebra*. Addison-Wesley Publishing Company, Inc., 1953.

v

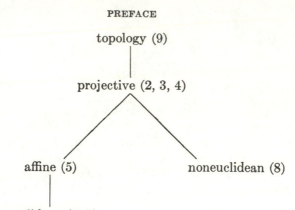

The numbers in parentheses indicate the chapters in which the geometries are discussed. The use of both synthetic and algebraic methods enables the reader to develop also an appreciation for each of these important methods.

The fundamental concepts of euclidean geometry are treated in the specialization of projective geometry to obtain euclidean geometry (Chapters 2 through 6). Chapters 7 (the evolution of geometry), 8 (noneuclidean geometry), and 9 (topology) are introductions to topics that will increase the reader's understanding of euclidean geometry. Unfortunately these topics can not be treated thoroughly in the space available.

Chapters 1 through 6 form a convenient unit for short courses in which the discussion of conics (Sections 2–11, 2–12, 2–13, 4–9, 4–10, and 4–11) and angles (Sections 6–6 and 6–7) may be omitted if necessary. Chapters 7 and 9 may be read independently either before or after Chapters 1 through 6. Chapter 8 is based upon Chapters 1 through 7.

The author is deeply indebted to Professors J. W. Young, E. B. Lytle, Echo Pepper, and J. H. Chanler for their part in the evolution of the course on which this book is based; to the constructive criticisms of many students; to his wife, who typed the manuscript; and to the publishers for their cooperation and very efficient service. To each and all the author is sincerely grateful.

B. E. M.

July, 1954

CONTENTS

CHAPTER 1

FOUNDATIONS OF GEOMETRY

The word "geometry" is derived from the Greek words for "earth measure." Since the earth was assumed to be flat, early geometers considered measurements of line segments, angles, and other figures on a plane. Gradually, the meaning of "geometry" was extended to include the study of lines and planes in the ordinary space of solids, and the study of spaces based upon systems of coordinates, as in analytic plane geometry, where points are represented by sets of numbers (coordinates) and lines by sets of points whose coordinates satisfy linear equations. During the last century, geometry has been still further extended to include the study of abstract spaces in which points, lines, and planes may be represented in many ways. We shall be primarily concerned with the fundamental concepts of the ordinary high-school geometry — euclidean plane geometry. Our discussion of these concepts is divided into three parts: the study of the foundations of mathematics (Chapter 1), the development of euclidean plane geometry from the assumption of a few fundamental properties of points and lines (Chapters 2 through 6), and a comparison of euclidean plane geometry with some other plane geometries (Chapters 7 through 9). The treatment of the second part — the development of euclidean plane geometry — forms the core of this text and emphasizes the significance of the assumptions underlying euclidean geometry. Together, the three parts of our study enable us to develop an understanding of and appreciation for many fundamental concepts of geometry.

1–1 Logical systems. We shall consider geometries as logical systems. That is, we shall start with certain elements (points, lines, . . .) and relations (two points determine a line, . . .) and try to deduce the properties of the geometry. In other words, we shall assume certain properties and try to deduce other properties that are implied by these assumptions.

In a geometry or any other logical system, some undefined elements or terms are necessary in order to avoid a "circle" of definitions. For example, one makes little progress by defining a point

1

to be the intersection of two distinct lines and defining a line to be the join of two distinct points.   A nonmathematical example would be obtained by defining a child to be a young adult and an adult to be a full-grown child.   Thus, in any geometry, some of the elements must be accepted without formal definition; all other elements may be defined in terms of these *undefined elements*.*   Similarly, in any geometry some of the relations among the elements must be accepted without formal proof.   These assumed relations are often called assumptions, axioms, or *postulates*.   Other relations, which may be proved or deduced, are called *theorems*.

Definitions enable us to associate names with elements and relations that may be expressed in terms of the undefined elements, postulates, previously defined elements, and previously proved relations.   The postulates and definitions are combined according to the rules of logic (Sections 1–2 and 1–3) to obtain statements of properties of the geometry.   Necessary and desirable properties of postulates are considered in Sections 1–4 through 1–6.   Definitions should be concisely stated, should give the distinguishing characteristics of the element or relation being defined with reference to the element or relation most similar to it, should be reversible, and should not contain any new elements or relations.

A *reversible* statement may be expressed in the form of "if and only if."   For example, the statement: *A triangle is equilateral if and only if its three sides are equal* is reversible.   This statement is commonly expressed as a definition: *An equilateral triangle is a triangle having three equal sides.*   That is, we interpret the definition to mean that all equilateral triangles have three equal sides *and* that all triangles having three equal sides are equilateral.   In other words, we usually assume that definitions are reversible.

A definition of a new term is not acceptable if it involves terms that have not been previously defined.   A proof of a theorem is not acceptable if it involves relationships that have not been previously proved, postulated, or stated as assumptions in the hypothesis of the theorem.   Such faulty definitions and proofs often involve "circular reasoning."   If a statement such as: *A zig is a zag* and its converse statement (Section 1–2): *A zag is a zig* are both taken

---

* Throughout this text new terms will be italicized when they are defined or first identified.

as definitions, we have a circle of definitions and have not improved our understanding of either zigs or zags. In general, whenever the definitions of two elements, say A and B, are related such that the definition of A depends upon the element B and the definition of B depends upon the element A, we have an example of reasoning in a circle in setting up our definitions. Whenever the conclusion of a theorem is used as a basis for a step in the proof of the theorem, we have an example of reasoning in a circle in the proof of a theorem. To avoid such reasoning in a circle we must have undefined elements and unproved relations (postulates) among these elements. Throughout this text we shall be concerned with definitions and theorems based upon sets of undefined elements and unproved postulates.

Exercises

1. Discuss the reasoning in the following story:

Long ago some Christian monks heard that in a certain medieval village there lived a holy man who talked with angels. In order to verify this report the monks traveled to the village and talked with some of the local people who knew the holy man. These people repeated the story of the holy man and, when asked how they knew that he talked with angels, said that he had told them of his experiences. The monks then asked the people how they knew that the holy man was telling the truth. The local people were astounded at the question and replied "What! A man lie who talks with angels?"

2. Look through current newspapers and magazines for 20th century examples of reasoning similar to that illustrated in Exercise 1.

3. Give a nonmathematical example of reasoning in a circle.

4. Give a mathematical example of reasoning in a circle.

5. In Euclid's *Elements*, a point is that which has no parts, or which has no magnitude; a line is length without breadth. Discuss the effectiveness of these two definitions.

6. Identify the elements upon which each of the following definitions of Euclid is based: a straight line is that which lies evenly between its extreme points; a surface is that which has only length and breadth; a plane surface is that in which any two points being taken, the straight line between them lies wholly in that surface.

7. Indicate which of the following statements are reversible.

(a) A duck is a bird.

(b) A line is on a plane if and only if at least two points of the line are on the plane.

(c) If a circle is on a plane, then at least two points of the circle are on the plane.

(d) The boy is a McCoy if he has red hair.

(e) The equality of the lengths of two sides of a quadrilateral is necessary and sufficient for the quadrilateral to be a square.

(f) *Rat* is *tar* if and only if *r* is *t*.

(g) A circle is a square if and only if two radii are equal.

(h) A necessary and sufficient condition for a rectangle to be a square is that the diagonals be equal.

1–2 Logical notation. In any system, logical thinking may be hampered by the language used in stating the propositions under consideration. Sometimes the statements become so complex that they are difficult to comprehend. At other times lack of preciseness in the language (English is a good example) is a handicap. These difficulties may be minimized by using special symbols. We shall consider a few such symbols and their applications to the concepts that have just been introduced. Further information regarding these symbols may be found in [11; 1–119]* and introductory treatments of symbolic logic.

All formal proofs are based upon implications of the form "$p \rightarrow q$," that is, "the statement p implies the statement q." The logical processes underlying geometric proofs may be most easily represented in terms of the symbols

$\rightarrow$	implies,
$\leftrightarrow$	is equivalent to,
$\sim$	not,
$\wedge$	and,
$\vee$	or.

For any given statements p, q we use the symbols as follows:

p	p is valid,
$\sim p$	p is not valid,
$p \wedge q$	both p and q are valid,
$p \vee q$	at least one of the statements p, q is valid,

---

* The notation [11; 1–119] is used to refer to pages 1 to 119 of reference number 11 in the bibliography at the end of this text.

where a statement is *valid* if it holds in the logical system under consideration.

Two statements p, q are *equivalent* if $p \to q$ and $q \to p$, that is, $p \leftrightarrow q$. A reversible definition asserts that two elements or statements are equivalent. For example, the definition of an equilateral triangle may be considered in the form $p \leftrightarrow q$, where the statements are

$p:$ a triangle has three equal sides,

and

$q:$ a triangle is equilateral.

Given any statement of the form $p \to q$, the *converse* statement is $q \to p$, the *inverse* (or opposite) statement is $(\sim p) \to (\sim q)$, and the *contrapositive* (opposite of converse) statement is $(\sim q) \to (\sim p)$. Briefly then, we have

statement $p \to q$,   converse $q \to p$,
inverse $(\sim p) \to (\sim q)$,   contrapositive $(\sim q) \to (\sim p)$.

Two statements p, q are *contradictory* if
$$[p \to (\sim q)] \wedge [(\sim p) \to q]$$
For example, the statement

$p:$ the apple is all red,

and the statement

$q:$ the apple is not all red

are contradictory statements. If the statement, "The apple is all red," is valid, then the statement, "The apple is not all red," must be invalid. In other words, if the first statement holds, then the second statement does not hold; i.e., if p, then $\sim q$. Similarly, if the first statement does not hold, then the second statement does hold; i.e., if $\sim p$, then q.

The word "contradictory" should not be confused with the word "contrary." Two statements p and q are *contrary* if they cannot both hold; i.e., if $\sim (p \wedge q)$. The following statements are contrary: The apple is all red. The apple is all green. Notice that two contrary statements may both be false. In the above example, the apple may be yellow. All contradictory statements are contrary, but many contrary statements are not contradictory. The above distinction between contrary and contradictory statements is often overlooked in popular discussions.

We shall not attempt to give a complete discussion of the logical processes that may be used to derive or deduce theorems from a given set of undefined elements and unproved postulates. Instead we shall consider very briefly the following laws of logic.

ARISTOTLE'S LAWS OF LOGIC

(i) $p \leftrightarrow p$, *law of identity,*

(ii) $\sim [p \wedge (\sim p)]$, *law of noncontradiction — a statement and its contradictory cannot both be valid,*

(iii) $p \vee (\sim p)$, *law of the excluded middle — at least one of any two contradictory statements must be valid.*

Throughout this text we shall assume that Aristotle's Laws of Logic apply to the development of any geometry from its postulates. In particular, these laws provide a basis for the indirect method of proof (Exercise 4) and for the inverse and contrapositive statements of any given statement.

EXERCISES

1. Indicate which of the following pairs of statements are contrary.
   (a) That is Bill.   That is Jim.
   (b) It is a citrus fruit.   It is an orange.
   (c) The car is a Ford.   The car is not a Chevrolet.
   (d) x is positive. x is negative.
   (e) $x < 0$. $x \geq 0$.   (f) $x < 4$. $x = 2$.
   (g) $x < 4$. $x = 5$.   (h) $x^2 = 4$. $x \neq 2$.

2. Indicate which of the pairs of statements in Exercise 1 are contradictory.

3. Bring in current examples of the confusion of contrary and contradictory statements.

4. Discuss the dependence of the method of indirect proof upon Aristotle's Laws of Logic.

5. State and discuss the validity of the converse, inverse, and contrapositive of each of the following statements in euclidean geometry:
   (a) If two lines are parallel, they do not intersect.
   (b) If the angles A and B of a triangle ABC are equal, the triangle is isosceles.
   (c) If a triangle is on a euclidean plane, its angle sum is $180°$.

6. Describe three advertisements in current periodicals or newspapers in which the advertiser hopes that the reader will assume the converse or inverse of the fact or situation presented.

7. Compare the following statements:
    (a) $p \leftrightarrow q$.   (b) p if and only if q.
    (c) p is a necessary and sufficient condition for q.
    (d) p is equivalent to q.

*8. Discuss the basis for and the validity of each of the following state-ments. Give examples of each statement.
    (a) A statement does not necessarily imply its converse.
    (b) A statement does not necessarily imply its inverse.
    (c) Any statement implies its contrapositive.
    (d) The contrapositive of the contrapositive of a statement is the statement.
    (e) The contrapositive of a statement implies the statement.
    (f) Any statement is equivalent to its contrapositive.
    (g) A direct proof of the contrapositive of a statement is an indirect proof of the statement.
    (h) The converse of any statement is equivalent to the inverse of the statement.

9. Compare the following statements:
    (a) The statement $p \rightarrow q$ is reversible.
    (b) The statement $p \rightarrow q$ implies its inverse.
    (c) The statement $p \rightarrow q$ implies its converse.
    (d) $p \leftrightarrow q$.

10. Give an example of at least one statement in each of the following forms:

    (a) $p \rightarrow q$.   (b) $(\smallsmile p) \rightarrow q$.
    (c) $(p \wedge q) \rightarrow r$.   (d) $p \rightarrow (r \vee q)$.
    (e) $(p \wedge q) \rightarrow (r \vee s)$.   (f) $p \vee (\smallsmile q)$.

11. Is $\leftrightarrow$ an *equivalence relation?* In other words, is it reflexive $(p \leftrightarrow p)$ symmetric $(p \leftrightarrow q$ implies $q \leftrightarrow p)$, and transitive $(p \leftrightarrow q$ and $q \leftrightarrow r$ imply $p \leftrightarrow r)$? Explain.

12. Consult an appropriate reference and discuss briefly the use of Venn diagrams in any two-valued logical system.

1–3 Inductive and deductive reasoning.

1–3 Inductive and deductive reasoning. Before endeavoring to derive properties of a particular geometry, let us consider two types of reasoning that are often used in the development of a geometry. We have seen that no terms or elements can be defined until at least one element is accepted and allowed to remain undefined.   In plane geometry it is customary to accept points and lines as undefined

---

   * Throughout this text the exercises designated by an asterisk lead to results that will be used in the development of the text.

elements. In any science, the elements to be left undefined are generally chosen for their simplicity and fundamental relation to other elements that are to be considered.

Similarly, not all statements or theorems can be proved. Some statements must be accepted as postulates or "rules of the game." These statements often arise as generalizations based upon observations of specific cases. It is not known how specific cases suggest general rules or laws in the minds of men, but most scientific progress originates in this manner. Specific cases and happenings lead men to wonder and finally to devise theories, principles, or laws. This process of formulating a generalization based upon the observation of specific cases is called inductive reasoning or *induction*.

The statements that are used as postulates are frequently devised by induction. Each postulate must be accepted without proof. If there are several postulates, it is desirable (Section 1–4) that no two postulates be contrary and, furthermore, that no two theorems obtainable from them be contrary. The logical process that we use to obtain a proposition or theorem from a set of undefined terms and unproved postulates is called deductive reasoning or *deduction*.

The *proof* or deduction of a theorem of the form $p \rightarrow q$ from a given set of undefined elements, unproved statements of relations (postulates), defined elements, and previously proved relations (theorems) may be direct or indirect (Exercise 8g, Section 1–2). A direct proof consists of a finite sequence of statements having the form

$$p \rightarrow p_1, \quad p_1 \rightarrow p_2, \quad \cdots, \quad p_{n-1} \rightarrow p_n, \quad p_n \rightarrow q$$

and such that each statement is valid in the logical system under consideration. When all the undefined elements and postulates of a science have been accepted and the proved theorems arranged in an order such that each theorem may be proved without using any succeeding (later) theorems, the science is said to be *well established* as a deductive science.

Throughout this text we shall be primarily concerned with various geometries as deductive sciences; i.e., we shall accept various sets of postulates without proof and shall consider the properties that may be deduced from them. We shall be concerned with the implications of sets of postulates rather than with interpretations of the space in which we live. The emphasis of this approach is quite different from that of an experimental or inductive science, in which the emphasis is upon the determination of a set of postulates (generalizations of observable phenomena) from which the science may be deduced.

EXERCISES

1. State a common nonmathematical postulate.
2. Indicate in what sense physical laws such as Newton's are postulates.
3. Give an example of inductive reasoning.
4. Give an example of deductive reasoning.
5. Discuss briefly a science that is well established.
6. Mention at least one science that is not yet well established.

In each of the following exercises find one or more conclusions implied by the given propositions.

7. (a) Jim likes to hunt.
   (b) Only a country boy enjoys hunting.
8. (a) Students from different countries enjoy talking to each other.
   (b) John and Arthur are students from Cuba.
   (c) Jane and Alice are students from Brazil.
9. (a) No lazy student has an eight o'clock class.
   (b) Robert studies hard.
   (c) Donald has an eight o'clock class.
   (d) Francis gets up at seven.
10. (a) All peace-loving people should work together.
    (b) All people want peace.
11. (a) Books in which at least half the words contain at most three letters are easy reading.
    (b) Symbols are one-letter words.
    (c) Over half the words are symbols in some mathematics books.
12. (a) People who are overweight eat too much.
    (b) Overeating is an extravagance.
    (c) Thrifty people are not extravagant.
    (d) Most Americans are overweight.
13. (a) All flowers are beautiful.
    (b) Beautiful objects make good decorations.
    (c) Some plants do not make good decorations.

1–4 Postulates.   The postulates of any science must satisfy certain conditions before being generally accepted.   It is desirable that they be easily understood, be few in number, and involve only a few accepted undefined terms, and that the set be complete (categorical, Section 1–6) in the sense that any additional statement either may be proved as a theorem or will lead to a special case of the science.   If there are several postulates, it is desirable that no two be contrary and, furthermore, that no two theorems obtainable from them be contrary.   Formally, it is always required that a set of postulates be consistent.   It is also often required that each of the postulates in the set be independent and that the set of postulates be categorical.   These terms will now be defined.

A set of postulates and the science based upon it are each said to be *consistent* if no two statements (postulates and theorems) of the science are contrary; i.e., a science is consistent if any two (all possible pairs) of its statements are consistent.   Thus, by definition, a science may be proved to be consistent by deducing all possible theorems of the science and proving that all possible pairs of statements in the science are consistent.   However, this procedure raises two problems — the determination of all the theorems of a science and the determination of the consistency of pairs of statements. Especially in the physical sciences, it is practically impossible to deduce all the theorems of a science.   Also, we do not yet have a test for the consistency of a pair of statements.   Thus we still need a test for the consistency of a science.   Since an irrational universe would invalidate all the physical sciences, it is generally assumed that *the universe is subject to the laws that govern the reasoning of men.* Under this assumption, a set of postulates and the associated science may be proved to be consistent by finding a single concrete representation of the postulates, i.e., by finding in the physical universe an interpretation of the undefined elements and relations of the science such that the statements of the science are valid for this interpretation.   We shall apply this test for consistency to specific sets of postulates in the next section.

A set of postulates and the science based upon it are each consistent if they are self-consistent.   The question as to the *truth* of the statements of the science is another matter entirely; i.e., does the science present a valid picture of the universe?   This question is often very difficult to answer, since the validity of a picture can be verified only by comparison with the original.   Furthermore, since this comparison must be made by man, it can be accomplished only within the margins of error in man's observations.   Thus, as with the theory of relativity, sciences may be confirmed only to within the error of man's observations.   Whenever several distinct, consistent, and categorical sets of postulates may be made to fit the same physical phenomenon within the error of man's observations, a choice is usually made on the basis of usefulness, simplicity, and productivity in the sense of interesting theorems. The choice must remain a matter of convenience and usefulness until man's observations can be sharpened sufficiently to make a scientific selection.

These considerations of validity and truth illustrate an important aspect of logic; namely, a logical consequence of a postulate or set of postulates need not be true. For example, the postulate *all fruits are red* implies that *all oranges are red*. The reasoning is valid even though the conclusion is not true in the sense of presenting an accurate picture of our universe. In general, a statement is said to be *valid* if it is a logical consequence of the set of postulates under consideration; a statement is said to be *true* if it appears to be valid in our universe. We say that the geometric properties of the physical space in which we live are *true*, whereas geometric properties deduced from a set of postulates are *valid* relative to that set of postulates. The properties deduced from a set of postulates may or may not be true when considered in our physical universe. Geometry may be considered either as a study of physical space or as a deductive system. Mathematicians usually think of a geometry as a deductive system based upon a consistent and categorical (Section 1-6) set of postulates.

Bertrand Russell once said, "Mathematics is the subject in which we never know what we are talking about nor whether what we say is true." From a very similar point of view, mathematics is a study of implications (consequences of sets of postulates). The postulates then serve simply as the rules of the game — just as baseball is played according to certain accepted rules. The pure mathematicians often play their games with considerable enthusiasm and without any concern for concrete representations. Sometimes practical applications are found. At other times the games appear to have no possible application, and yet they, too, may at any time have a practical application in one of the physical sciences.

EXERCISES

1. Give an example of a consistent set of postulates.

2. Give a set of postulates that is consistent but does not present a true picture of the universe.

3. Discuss the above comment of Bertrand Russell.

4. Give a nonmathematical example of valid reasoning that gives a false (in our universe) conclusion.

5. Give a mathematical example of valid reasoning that gives a false (in our universe) conclusion.

6. Discuss the distinction between validity and truth.

1–5 Independent postulates. The postulates of a given set are said to be *independent* in that set if no one of them is a consequence of the others. A given postulate is shown to be independent of the others in a particular set by exhibiting a concrete representation in which the others are valid but the given postulate does not apply. It is desirable, but not necessary, that the postulates of a given set be independent. The desirability of independent postulates is based upon the desirability of a minimum number of postulates and the isolation of the fundamental concepts underlying the geometry. It is not necessary that the postulates be independent, since any dependent postulate or part of a postulate may be proved as a theorem. Indeed, it may at times be convenient in elementary courses to consider systems in which the postulates are not all independent. Thus the independence of postulates is theoretically desirable but often forfeited in order to avoid difficult steps of reasoning or proof.

Let us consider the following set of postulates, in which the undefined elements are points and lines, the undefined relation is the incidence of points and lines [a point is (is not) on a line, distinct points (lines) do not coincide, . . .], and the undefined operations are the determination of a line by two points and the determination of a point by two lines.

P–1.1: There exist exactly three distinct points.

P–1.2: Two distinct points determine a unique line.

P–1.3: Not all points are on the same line.

P–1.4: Two distinct lines determine at least one point.

P–1.5: Two distinct lines determine at most one point.

In this set, Postulate P–1.5 is a consequence of P–1.2 and may be dropped from the set (Exercise 1). The set P–1.1 through P–1.4 is consistent, since it is valid in the geometry of the three sides (lines) and three vertices (points) of a plane triangle (Fig. 1–1). In the set P–1.1 through P–1.4, Postulate P–1.1 can be proved to be

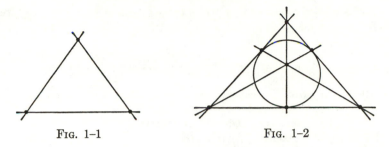

FIG. 1-1   FIG. 1-2

independent by using Fig. 1-2, in which the circle is interpreted as a line and only the indicated points exist. In this figure there are seven points and seven lines, three points on each line, and three lines on each point. Postulate P-1.2 is independent of P-1.1, P-1.3, and P-1.4, since these three hold in Fig. 1-3, where P-1.2 does not apply. Postulate P-1.3 can be proved to be independent by using a figure composed of one line with three points on it. Postulate P-1.4 can be proved to be independent of P-1.1, P-1.2, and P-1.3 by using Fig. 1-4, in which two particular lines are assumed to have no points in common. We have now indicated that Postulates P-1.1, P-1.2, P-1.3, and P-1.4 form a consistent set and that each postulate is independent of the others in the set.

EXERCISES

1. Prove that Postulate P-1.5 is a consequence of P-1.2 in the set P-1.1 through P-1.5.

*2. Prove that Postulates P-2.1, P-2.2, P-2.5, and P-2.6 are independent in the following set, where A and B refer to any two points of a set S of undefined elements called points (i.e., in the statements of the postulates, A and B may refer to any two points of S, irrespective of the particular letters

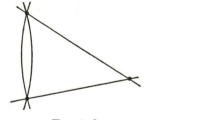

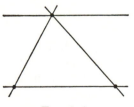

FIG. 1-3   FIG. 1-4

used to designate the points). (*Note:* Postulates P–2.3, P–2.4, and P–2.7 are also independent.)

P–2.1: If A and B are points of S, there is a line AB on them.

P–2.2: If A and B are distinct points of S, there is at most one line AB on them.

P–2.3: Any two lines have at least one point of S in common.

P–2.4: There exists at least one line.

P–2.5: Every line is on at least three points of S.

P–2.6: Not all points of S are on the same line.

P–2.7: No line is on more than three points of S.

3. Prove that the set of postulates in Exercise 2 is consistent.

4. Give a second representation of the set of postulates in Exercise 2.

1–6 Categorical sets of postulates. A set of postulates is said to be *categorical* if there is essentially only one system or representation for which the postulates are valid. In other words, for any two representations of a categorical set of postulates there must be a one-to-one correspondence between the elements, between the relations ($=$, $<$, ...), and between the operations ($+$, $-$, $\cdot$, ...) in the two representations such that the properties of the relations and operations among the elements are preserved under the correspondence. For example, if a, b, c, $+$, and $=$ correspond respectively to A, B, C, plus, and equal in two such representations and $a + b = c$, then A plus B equals C. We may formally describe the above correspondence by stating that the two representations are *isomorphic*. A set of postulates is categorical if all representations of the set are isomorphic.

Postulates P–1.1 through P–1.4 stated in Section 1–5 are valid in Fig. 1–1. They are also valid when the undefined points are taken as symbols A, B, C, and the lines are taken as columns in the following array:

$$\begin{array}{ccc} A & B & C \\ B & C & A \end{array}$$

In this representation there exist exactly three distinct symbols, two distinct symbols determine a unique column, not all symbols are on the same column, and two distinct columns have at least one symbol in common. Thus Postulates P–1.1 through P–1.4 are satisfied when the undefined elements are represented as symbols and columns. The one-to-one correspondence between the representation in Fig. 1–1 and the representation in the above array is indicated in Fig. 1–5, where each symbol is associated with a point, and each column with a line.

We now consider a second *finite geometry* (i.e., another geometry involving only a finite number of elements) as determined by its undefined elements and postulates. We shall again find that the various interpretations given the undefined elements invoke very different representations of the geometry.

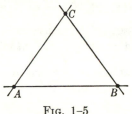

Fig. 1–5

The postulates are so chosen that these different-appearing geometries are equivalent (isomorphic); i.e., a categorical set of postulates is used.

We first consider as our undefined elements the students and student committees in a modernistic secondary-school class. There is one undefined relationship, namely, serving on a committee. This relationship may be conveniently indicated for a given student and a given committee by saying that the student is a member of the committee. In this sense the word "student" is used interchangeably with the word "member" in the following postulates. We make seven postulates and seek conclusions about the class that can be deduced from them.

P–2.1′: If A and B are students, there exists a committee on which they serve together.

P–2.2′: If A and B are different students, there exists at most one committee on which they serve together.

P–2.3′: Any two committees have at least one member in common.

P–2.4′: There exists at least one committee.

P–2.5′: Every committee has at least three members.

P–2.6′: Not all students serve on the same committee.

P–2.7′: No committee has more than three members.

From these postulates we may deduce the following theorems, where the postulates specified in parentheses are given to suggest a method of proof of the theorem. A list of postulates does not constitute a proof of a theorem.

THEOREM 1. *Any two distinct students serve together on exactly one committee* (P–2.1′, P–2.2′).

THEOREM 2. *Any two distinct committees have exactly one member in common* (P–2.2′, P–2.3′).

THEOREM 3. *There exist three students who do not serve together on a committee* (*Exercise 2*).

THEOREM 4. *Every committee has exactly three members* (*Exercise 3*).

We now endeavor to determine the number of students in the class and the number of committees on which they serve. There exists at least one committee (P–2.4′) having exactly three members (Theorem 4). Let us designate students by the symbols A, B, C, D, . . . and the above committee by ABC. Since not all students serve on the same committee (P–2.6′), there exists a student D who is not on committee ABC. Now D must serve with A (P–2.1′) on a committee of three students (Theorem 4), and neither B nor C can serve on that committee (Theorem 1). Thus there exists a committee ADE and, similarly, a committee BDF, where E and F are other students in the class.

At present we have three committees (ABC, ADE, BDF) and six students. However, by Theorem 1 every pair of students must serve together on exactly one committee. Student A now serves with B, C, D, and E. Thus (P–2.2′) A and F cannot serve with B, C, D, or E, and there must be a seventh student G in the class who serves with A and F on the committee AFG. If there were an eighth student H, then the committee containing A and H could not (P–2.2′) have a member in common (P–2.3′) with BDF, since A already serves with B, D, and F. Thus the class consists of exactly seven students. By continuing this type of reasoning, committees CEF, BEG, and CDG can be found. There are then exactly seven committees, as indicated by the following seven columns:

$$
\begin{array}{ccccccc}
A & A & B & A & C & B & C \\
B & D & D & F & E & E & D \\
C & E & F & G & F & G & G
\end{array}
$$

The class thus consists of seven pupils who form seven committees. Each committee has three members, and each student serves on three committees.

An entirely different representation of the above geometry is obtained by taking points and lines of a set S as undefined elements, and taking incidence (a point is on a line; a line is on a point) as the undefined relation. Postulates P–2.1′ through P–2.7′ may then be expressed in the form of Postulates P–2.1 through P–2.7 stated in Exercise 2, Section 1–5, with points and lines as undefined elements. We may deduce the following statements (theorems) from Postulates P–2.1 through P–2.7 in exactly the same manner as the corresponding theorems were obtained from Postulates P–2.1′ through P–2.7′:

(i) Any two distinct points of S are on a unique line.

(ii) Any two distinct lines are on a unique point of S.

(iii) There exist three points of S that are not on the same line.

(iv) Every line is on exactly three points of S.

We may also prove, exactly as before, that S consists of seven points that lie on seven lines, each line being on three points of S and each point on three lines. A graphical representation of this geometry is given in Fig. 1–2. The equivalence (isomorphism) between the point-line representation and the student-committee

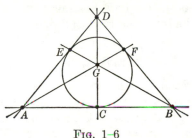

Fig. 1–6

representation indicated in the above array of seven students (A, B, ..., G) on seven committees (columns), can be indicated by identifying the points in Fig. 1–2 with students, as in Fig. 1–6.

Veblen and Young [16; 2–3] state postulates similar to the set P–2.1 through P–2.7 using "element of S" and "m-class" as undefined terms and "belonging to a class" as an undefined relation. The geometry thereby obtained is again isomorphic to those considered above. Still another geometry isomorphic to those above may be obtained by using number triples as points (Section 1–7).

EXERCISES

1. Discuss the representation of Postulates P–1.1 through P–1.4 when the undefined lines are taken as symbols A, B, C and the points as columns in an array.

2. Prove Theorem 3.

3. Prove Theorem 4.

4. Restate Postulates P–2.1′ through P–2.7′, using "element of S" and "m-class" as undefined terms and "belonging to a class" as the undefined relation.

5. Repeat Exercise 4, using "telephone" and "line" as undefined terms and "being on a line" as the undefined relation.

6. Prove that the representations obtained in Exercises 3 and 4 of Section 1–5 are isomorphic.

1–7 A geometry of number triples.   We are accustomed to representing points on a plane by pairs of real numbers (x,y) and lines by equations of the form $ax + by + c = 0$. These representations will be obtained after coordinates are introduced in Section 3–7. For the present, we may dismiss visual geometric concepts and consider points abstractly as number triples (x_1,x_2,x_3), subject to the following two conditions:

(i) $(kx_1,kx_2,kx_3) = (x_1,x_2,x_3)$ when $k \neq 0$,

(ii) there is no point corresponding to $(0,0,0)$.

When $x_3 \neq 0$, we may choose $k = 1/x_3$ and associate with each point a number triple $(x,y,1)$ where $x = x_1/x_3$ and $y = x_2/x_3$. Then for real values of the coordinates x, y we have, essentially, the points (x,y) on an ordinary plane. The two representations (x,y) and (x_1,x_2,x_3) of the points on an ordinary plane are called, respectively, *nonhomogeneous* and *homogeneous coordinates* of the points (Sections 3–7 and 3–8). When $x_3 = 0$, there exist number triples (points) that cannot be represented in the form $(x,y,1)$ and therefore cannot be represented using nonhomogeneous coordinates. For real values of the coordinates, the totality of points (x_1,x_2,x_3) will be called the real projective plane (Chapter 4).

The totality of points (x_1,x_2,x_3) such that $u_1x_1 + u_2x_2 + u_3x_3 = 0$, where at least one u_j $(j = 1,2,3)$ is different from zero, is called a line and is indicated by $[u_1,u_2,u_3]$. When $k \neq 0$, we write

$$[u_1,u_2,u_3] = [ku_1,ku_2,ku_3]$$

to indicate that the equations

$$u_1x_1 + u_2x_2 + u_3x_3 = 0$$

and

$$(ku_1)x_1 + (ku_2)x_2 + (ku_3)x_3 = 0$$

represent the same line. Thus the number triples representing lines satisfy the same two conditions as the number triples representing points. The symbol (x_1,x_2,x_3) represents a point in terms of its homogeneous point coordinates; the symbol $[u_1,u_2,u_3]$ represents a line in terms of its homogeneous line coordinates. On any line the points with real point coordinates form a real line; those with rational coordinates form a rational line or a line of rational points.

We now consider the points and lines with coordinates from a set of numbers S, where S consists of two numbers: 0 and 1. Such a system S may be obtained from the set of integers by replacing each even number by 0 and each odd number by 1, that is, by replacing each integer by the remainder when it is divided by 2. In the symbol (x_1,x_2,x_3) there are then 2 possibilities for each coordinate and $2^3 = 8$ possible number triples. However, $(0,0,0)$ does not represent a point. Thus there are exactly seven points in this geometry, namely

$(1,0,0)$, $(0,1,0)$, $(0,0,1)$, $(1,1,0)$, $(1,0,1)$, $(0,1,1)$, $(1,1,1)$.

Similarly, there are exactly seven lines:

$[1,0,0]$, $[0,1,0]$, $[0,0,1]$, $[1,1,0]$, $[1,0,1]$, $[0,1,1]$, $[1,1,1]$.

Under the convention that each even number is equivalent to zero, a point (x_1,x_2,x_3) and a line $[u_1,u_2,u_3]$ are incident if and only if $u_1x_1 + u_2x_2 + u_3x_3$ is an even number. The points listed in each column below are on a line which is indicated at the base of the column.

$(1,0,0)$	$(1,0,0)$	$(0,1,0)$	$(1,0,0)$	$(1,1,0)$	$(0,1,0)$	$(1,1,0)$
$(0,1,0)$	$(0,0,1)$	$(0,0,1)$	$(0,1,1)$	$(1,0,1)$	$(1,0,1)$	$(0,0,1)$
$(1,1,0)$	$(1,0,1)$	$(0,1,1)$	$(1,1,1)$	$(0,1,1)$	$(1,1,1)$	$(1,1,1)$
$[0,0,1]$	$[0,1,0]$	$[1,0,0]$	$[0,1,1]$	$[1,1,1]$	$[1,0,1]$	$[1,1,0]$

This array can be made equivalent to the seven-column array of student committees (Section 1–6) by the substitution of A for $(1,0,0)$, B for $(0,1,0)$, ... The above geometry of triples (x_1,x_2,x_3) and

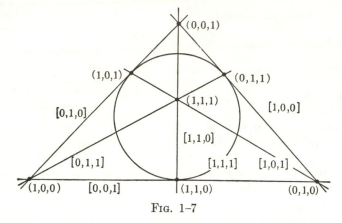

FIG. 1-7

$[u_1, u_2, u_3]$ is represented in Fig. 1-7 and can be readily shown to be isomorphic to that represented by Fig. 1-6 under the same substitution as that used above.

A geometry of thirteen points and thirteen lines with four points on each line and four lines on each point may be obtained by taking points and lines with coordinates from the set of numbers 0, 1, 2, obtained by replacing each integer by the remainder when it is divided by 3. Thus 3 is replaced by 0, 4 is replaced by 1, and so forth. As in the above geometry of points and lines with coordinates from the set of numbers 0 and 1, the symbol (0,0,0) does not represent a point. Also, two symbols related in the manner of (kx_1, kx_2, kx_3) and (x_1, x_2, x_3) where $k \neq 0$ represent the same point. For example, (1,0,1) and (2,0,2) represent the same point. Also (1,0,2) and (2,0,4) or (2,0,1) represent the same point. We shall develop this geometry of thirteen points in the following exercises.

EXERCISES

Perform the following exercises in the geometry of thirteen points, i.e., the geometry of number triples obtained using the numbers 0, 1, and 2.

1. Write down the thirteen number triples that represent distinct points.
2. Write down the thirteen number triples that represent distinct lines.
3. Find the four points on the line [1,0,0].
4. Find the four points on the line [1,2,1].
5. Make an array indicating the four points on each of the thirteen lines of this geometry.

6. Give a set of postulates for this geometry of thirteen points.

7. State and prove four theorems in the geometry postulated in Exercise 6.

8. Find a second representation of the geometry postulated in Exercise 6 and show that this representation is isomorphic to the one above [16; 6-7].

1-8 Geometric invariants. We have now set the stage for the consideration of a geometry as a logical system based upon undefined elements and unproved relations (postulates). Given any consistent and categorical set of postulates expressing relations among a set of undefined elements, we may deduce properties (theorems) of the geometry, define new elements and operations, deduce more theorems, and gain an understanding of many of the consequences of the given set of postulates. In this sense a geometry is a deductive system. We have also mentioned that, especially in the early days, geometry was a study of properties of the physical space in which we live. These two concepts of geometry are closely related in the minds of most of us, since we have previously been primarily concerned with euclidean geometry, a deductive system that usually appears to be concerned with properties of the physical space in which we live.

From an entirely different point of view, any geometry is a study of properties (expressed by postulates, definitions, and theorems) that are left invariant (unchanged) under a group of transformations. This view was proposed by Felix Klein in 1872 and has received considerable attention. Although the word "group" is used here in a technical sense, we may gain some insight into Klein's approach through a few examples involving concepts that have not yet been introduced in this book but which are familiar to most readers.

Consider the points on a number line (Fig. 1-8) in euclidean geometry. The isomorphism between the points P_x and their coordinates x enables us to express transformations of the points in terms of their coordinates. For example, when each point in Fig. 1-8 is moved 2 units to the right, the coordinate x' of the point corresponding to the point P_x is related to the coordinate x by the equation

$$x' = x + 2.$$

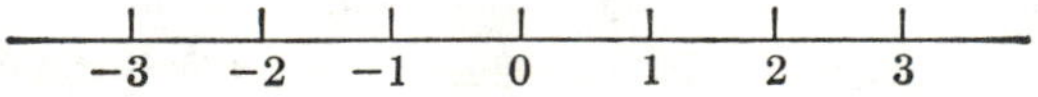

$$\begin{array}{cccccccc} & & & & & & & \\ -3 & -2 & -1 & 0 & 1 & 2 & 3 \end{array}$$

Fig. 1-8

Under this transformation (a translation), all points are changed; i.e., there are no invariant points. However, distances are invariant, since $x_1 - x_2 = x_1' - x_2'$. Similarly, direction or sense along the line is invariant, since $x_1' < x_2'$ if and only if $x_1 < x_2$. In general, each equation of the form

$$x' = x + a$$

represents a translation of the points P_x to the points $P_{x'} = P_{x+a}$, and every translation may be represented by an equation of the above form. Under each translation, distance and direction along the line are invariant. We say that distance and direction are invariants under the group of translations on the line.

Now let us consider a uniform stretching of the line about the origin. For example, the equation

$$x' = 2x$$

represents a transformation of each point P_x at a directed distance x from the origin into the point P_{2x} at a directed distance $2x$ from the origin. Under this transformation, the origin is an invariant point, distances are not invariant, and directions are invariant. In general, the origin is invariant and directions are invariant under the group of transformations

$$x' = ax,$$

where a is positive.

Finally, let us consider a group of transformations on a coordinate plane in euclidean geometry. For any real number $a > 0$, the equations

$$x' = ax, \qquad y' = ay$$

represent a transformation of the points (x,y) into the points $(x',y') = (ax,ay)$. Under such transformations, the origin is an invariant point, lines through the origin are invariant lines, the property of being a line is invariant (i.e., lines correspond to lines), and measures of angles are invariant. Thus under such transformations, triangles correspond to triangles and any two corresponding triangles are similar but not necessarily of the same area, since distances are not invariant when $a \neq 1$.

The present text is primarily concerned with the development of euclidean geometry as a special case of projective geometry — an inherently simple geometry. We shall first develop projective geometry as a deductive system.   Then, after developing analytic expressions for general projective transformations, we shall adopt Klein's viewpoint and consider a geometry as a study of properties that are invariant under groups of transformations.   The advantage of this point of view lies in the fact that each invariant of a geometry is also an invariant of every special case of the geometry.   Thus we shall consider the invariant properties of projective geometry and gradually increase the number of invariant properties by specializing the transformations under consideration, until we have the invariant properties of euclidean geometry.

EXERCISES

1. Give five transformations of a line onto itself and discuss the properties left invariant under each transformation.

2. Repeat Exercise 1 for transformations of a plane onto itself.

3. Enumerate some invariants in euclidean plane geometry.

4. Use your previous knowledge of euclidean geometry and give a geometric basis for the invariance of each of the following in euclidean geometry:

(a) The determinant $\begin{vmatrix} x_1 & y_1 & 1 \\ x_2 & y_2 & 1 \\ x_3 & y_3 & 1 \end{vmatrix}$,

where each (x_j, y_j) is a point on the plane.

(b) The sign of $B^2 - 4AC$ for conic sections

$$Ax^2 + Bxy + Cy^2 + Dx + Ey + F = 0$$

(c) The vanishing of the determinant

$$\begin{vmatrix} 2A & B & D \\ B & 2C & E \\ D & E & 2F \end{vmatrix}$$

for the above conic sections.

5. With the goal of increasing your understanding of the procedures and viewpoint of this text, make precise, but not necessarily brief, statements of your personal answers to each of the following:

(a) What is meant by "geometry"?

(b) What is a "proof"?

(c) How do we understand any body of knowledge?

(d) How can we best understand, appreciate, learn to use, and teach euclidean geometry?

(e) How are we endeavoring to understand euclidean geometry in this text?

REVIEW EXERCISES

1. Distinguish between
   (a) defined and undefined elements,
   (b) postulates and theorems,
   (c) deductive and inductive reasoning,
   (d) contrary and contradictory statements,
   (e) true statements and valid statements.
2. Discuss the desired characteristics of a set of postulates.
3. (a) Give a pair of statements that are contrary but not contradictory.
   (b) Use logical notation and define contradictory statements.
4. State Aristotle's Laws of Logic and give one basis for their importance.
5. Indicate which of the following statements are valid.
   (a) Any theorem implies its converse.
   (b) Any theorem implies its inverse.
   (c) Any theorem implies its contrapositive.
   (d) The converse of any theorem implies the inverse of the theorem.
   (e) The contrapositive of any theorem implies the theorem.
6. Consider a theorem of the form $p \rightarrow q$ and rewrite the statements in Exercise 5 in logical notation.
7. Make a statement that is (a) reversible, (b) nonreversible.
8. Select a transformation of a euclidean plane onto itself and state four properties that are invariant under this transformation.
9. Prove that the following set of postulates is consistent.
   (a) There is at least one line on any two points.
   (b) There is at most one line on any two points.
   (c) There exists at least one line.
   (d) There are exactly two points on each line.
   (e) Corresponding to each line there is exactly one line that does not intersect it.
10. Prove that the last postulate is independent in the set of postulates stated in Exercise 9.

CHAPTER 2

SYNTHETIC PROJECTIVE GEOMETRY

Projective geometry has been mentioned as an inherently simple and elegant geometry having euclidean geometry as a special case. We shall first consider projective geometry as a deductive system based upon a set of postulates having points and lines as undefined elements. The resulting geometry is often called synthetic projective geometry to distinguish it from analytic projective geometry, in which points are represented by sets of numbers. We shall develop a coordinate system in synthetic projective geometry (Chapter 3) and then consider analytic projective geometry (Chapter 4). Both synthetic projective geometry and analytic projective geometry may be considered as deductive systems.

After considering projective geometry as a deductive system and developing algebraic representations for projective transformations, we shall change our point of view and consider projective geometry as a study of properties that are invariant under projective transformations. Then the set of projective transformations under consideration will be restricted systematically to obtain the transformations of affine and euclidean geometries. Thus these geometries are special cases of projective geometry.

Our use of both synthetic and algebraic methods enables the reader to see the beauty and the advantages of each method. Also, our shift in point of view from a geometry as a deductive system to a geometry as a study of properties that are invariant under a group of transformations adds breadth to the reader's concept of geometry. Although we shall endeavor to give a survey of the fundamental properties of projective geometry, we shall find it necessary to omit completely many interesting theorems and merely to mention others, in order to have time to attain our principal goal, the development of euclidean plane geometry from projective geometry. This goal will provide a basis for selecting most of the topics that we shall consider.

The organization of this chapter, and indeed our complete development of euclidean geometry from projective geometry, was originally obtained by selecting, modifying, and expanding sequences of topics from the two volumes of *Projective Geometry*, [16] and [17], by Oswald

Veblen and J. W. Young.   Accordingly, several of the postulates, definitions, and theorems, as well as parts of the organization of the present text, are closely related to the development in the two texts by Veblen and Young.   These texts provide an excellent source of more advanced concepts and problems related to those discussed in the present text.

2–1 Postulates of incidence and existence.   We have seen that some undefined elements are necessary in any logical system.   In synthetic projective geometry the undefined elements are usually taken as points and lines.   A plane and a three-space may then be defined using the postulates of incidence and existence.   In the n-space notation a point is a zero-space, a line is a one-space, and a plane is a two-space.   The postulates of existence stated below assure us that these spaces exist.   Postulate P–8 is called a *postulate of closure*, since it restricts all points to three-space.   We shall find that for any positive integer n, the following set of postulates may be modified and extended such that the totality of points forms an n-space.   The term *incidence* in the phrase "postulates of incidence" is used to indicate the property that a line is on a point or that a point is on a line.

The following two sets of postulates may be found in [16; 16, 24], where they are called assumptions of alignment and assumptions of existence, respectively.

POSTULATES OF INCIDENCE

P–1: *If A and B are distinct points, there is at least one line on both A and B.*

P–2: *If A and B are distinct points, there is not more than one line on both A and B.*

P–3: *If A, B, and C are points not all on the same line, and D and E are distinct points such that B, C, and D are on a line and C, A, and E are on a line, there is a point F such that A, B, and F are on a line and also D, E, and F are on a line* (Fig. 2–1).

POSTULATES OF EXISTENCE

P–4: *There exists at least one line.*

P–5: *There are at least three distinct points on every line.*

P–6: *Not all points are on the same line.*

P–7: *Not all points are on the same plane.*

P–8: *If S_3 is a three-space, every point is on S_3.*

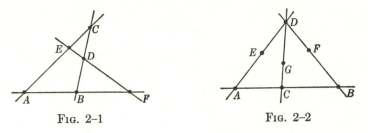

FIG. 2–1   FIG. 2–2

Let us now endeavor to build up a representation for Postulates P–1 through P–8 just as we built up a representation for Postulates P–2.1′ through P–2.7′ in Section 1–6. We shall represent a point by a single capital letter A, B, C, $\ldots$; a line by two capital letters representing distinct points on the line AB, AC, BC, $\ldots$ or by a single small letter a, b, c, $\ldots$; and a plane by three capital letters representing three noncollinear points of the plane ABC, BCF, $\ldots$ or by a single small Greek letter α, β, $\ldots$, π, $\ldots$.

There exists a line, say m, by Postulate P–4, and by P–5 this line contains at least three distinct points, say A, B, C. Also by P–6 there exists a point D that is not on the line m. Then by P–1 there exist lines DA, DB, DC, and possibly others joining the point D to the points of m. Furthermore, by P–5 each of these lines contains at least one other point, say E, F, and G as in Fig. 2–2. In general, for any line AB and any point D that is not on AB, the totality of points on lines joining D to points of AB is called the *projective plane* ABD. The projective plane in Fig. 2–2 may contain additional points or may contain only seven points as indicated in Fig. 1–6 (Section 1–6). Postulates P–1 through P–6 are satisfied by the points and lines indicated in Fig. 1–6 (Exercise 1).

Postulate P–7 indicates that there exists a point H that is not on the plane ABD. By P–5 each of the lines joining H to points on ABD contains at least three distinct points as in Fig. 2–3. In general, for any projective plane ABD and any point H that is not a point on ABD, the totality of points on lines joining H to points on ABD is called the *projective three-space* $ABDH$. A projective three-space may contain only a finite number of points (Exercise 5) or infinitely many points. In particular Postulates P–1 through P–8 are satisfied by the points, lines, and planes of the three-space whose points are indicated in Fig. 2–3. This three-space contains exactly fifteen points. There are also many very different finite

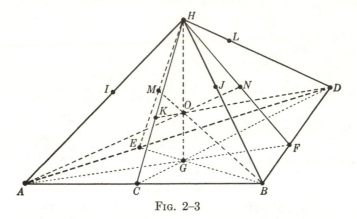

FIG. 2-3

projective three-spaces on which Postulates P-1 through P-8 are satisfied.

Projective plane geometries are obtained by replacing Postulates P-7 and P-8 by P-7': *If S_2 is a plane, every point is on S_2.* Projective three-dimensional geometries satisfy Postulates P-1 through P-8. In general, for any positive integer n, a geometry on an n-space is called an *n-dimensional geometry.* A four-dimensional projective geometry may be obtained by replacing P-8 by P-8': *Not all points are on the same three-space* and by a postulate of closure P-8'': *If S_4 is a four-space, every point is on S_4.* Then if $ABDH$ is a three-space, there exists by P-8' a point, say T, that is not on the three-space, and the totality of points on lines joining T to points of the three-space is called a *projective four-space $ABDHT$.* By P-8'' every point is on the four-space $ABDHT$. In general, an n-dimensional projective geometry ($n = 4,5,...$) may be obtained by replacing P-8 by postulates stating that

(i) not all points are on the same $S_3, S_4, ..., S_{n-1}$,

(ii) if S_n is an n-space, every point is on S_n.

Then there exist n points $A_1, A_2, ..., A_n$ determining an S_{n-1} and a point A_{n+1} that is not on the S_{n-1}. The totality of points on lines joining A_{n+1} to points of S_{n-1} is called a *projective n-space S_n.* The study of these higher-dimensional spaces ($n > 3$) has many important applications in advanced mathematical theories. However, we shall leave the study of these spaces for more specialized courses and return to the properties of a projective plane.

Exercises

*1. Verify that Postulates P-1 through P-6 are satisfied by the points and lines indicated in Figure 1-6.

2. Using the notation in the above discussion of n-dimensional projective geometry, identify or state the postulates needed for the corresponding n-dimensional geometry, and define a projective n-space for $n = 4$, 5, and 6.

*3. Prove that in any projective space satisfying Postulates P-1 through P-6, every projective plane contains at least seven points. (*Note:* This result is independent of the number or form of any additional postulates for the projective geometry.)

*4. Prove that a line may be determined by any two of its distinct points.

5. Consider the projective three-space in which each line contains exactly three points (Fig. 2-3), and

    (a) list all points (fifteen),
    (b) list all planes (fifteen),
    (c) list the points (seven) on each plane, and
    (d) list the points (three) on each line (thirty-five) where lines are determined as intersections of distinct planes.

(*Hint:* The projective plane geometry in which every line contains exactly three points is isomorphic to the geometry of number triples (Section 1-7) with coordinates 0 and 1. Any projective three-dimensional geometry in which each line contains exactly three points is isomorphic to the geometry of number quadruples with coordinates 0 and 1. Points and planes in this geometry on a three-space have a relationship similar to that between points and lines on the corresponding plane geometry.)

6. Verify that Postulates P-1 through P-8 are satisfied by the points, lines, and planes identified in Exercise 5.

2-2 Properties of a projective plane. We now consider properties of a plane in projective geometry, i.e., properties of a projective plane. We shall assume that every projective plane satisfies at least Postulates P-1 through P-6. By Postulates P-1, P-2, P-4, and P-5, there exist distinct points A and B and a unique line AB. Also, by P-6 there exists a point D that is not on the line AB. The plane ABD has been defined to consist of the totality of points on lines joining D to points of AB. The three noncollinear points A, B, and D and the three lines AB, BD, and AD determined by them form a *triangle* ABD. The points are called *vertices* of the triangle; the lines, *sides* of the triangle. Any two sides of the triangle *intersect*, i.e., have a point in common. These definitions may be used to restate

Postulate P–3 as follows:

> *If a line m intersects two lines of a triangle, then it intersects the third line.*

Given any point C on a line AB and a point D that is not on AB, every point on the line DC is by definition on the plane ABD. Also by definition each point on the plane ABD is on some line DC where C is a point on the line AB. Thus every line FG on the plane ABD must satisfy one of three conditions. It must coincide with AB, pass through D and a point on AB, or be determined by two points such as F and G, as in Fig. 2–4. In the last case J may coincide with any of the points on the line AB, and the points C, H on AB are not necessarily distinct from the original points A, B. If FG coincides with AB or passes through D, every point on the line FG is on the plane ABD by Postulates P–1 and P–2 and the definition of the plane ABD. In the remaining case, we apply Postulate P–3 to triangle DFG and obtain a point J on FG such that J is collinear with H considered as a point of DF and C considered as a point of DG. Since by P–2 there is at most one line on H and C, the line HC coincides with AB and the point J is also on AB. When P–3 is applied to triangle GCJ (or to triangle FHJ if the points G, C, and J coincide), we find that every point K on FG is collinear with D and some point K' on AB. Thus we have proved that every line FG determined by two points on the plane ABD intersects the line AB and every point K on FG is on the plane ABD; that is,

THEOREM 2–1. *If two points of a line are on a given plane, then every point of the line is on that plane.*

Thus a line is on a plane if and only if two points of the line are on the plane. We next wish to prove that any two distinct lines FG and RS on a plane ABD intersect in a unique point P. Since the lines are assumed to be distinct, they intersect at most in one point by P–2. It remains to show that they intersect in at least one point. If $F = S$, the lines intersect at F. If either of the lines coincides with the line AB, they intersect, as proved in the proof of Theorem 2–1. Thus we may assume that $F \neq S$ and that the lines FG and RS are both distinct from AB. Then, since the line FG has at most one point in common with the line AB, is determined by any two of its points (Exercise 4, Section 2–1), and contains at least three points, we may assume that the points F and G are not on AB.

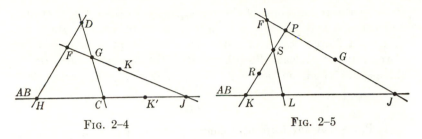

FIG. 2-4   FIG. 2-5

Similarly, we may assume that the points R and S are not on AB. Under these assumptions we may designate the intersection of FG and AB by J, the intersection of RS and AB by K, and the intersection of FS and AB by L (Fig. 2–5). If $J = K$, the given lines intersect at K; if $J = L$, they intersect at S; and if $L = K$, they intersect at F. Finally, if J, K, and L are distinct points on the line AB, we apply Postulate P–3 to triangle JLF, with K considered as a point on JL and S considered as a point on LF, to obtain a point P collinear with $KS = RS$ and $FJ = FG$. We have now proved

THEOREM 2–2. *Any two distinct coplanar lines intersect in a unique point.*

The following properties of projective planes and three-spaces can be proved in a manner very similar to the properties above. The proofs will be assigned as exercises. If C, E, and F are three noncollinear points on the plane ABD, then the plane CEF coincides with the plane ABD; that is, any three noncollinear points on a plane determine that plane (Exercise 1). If F and G are distinct points of a three-space $ABCD$, then every point on the line FG is on the three-space $ABCD$ (Exercise 2). If F, G, and H are noncollinear points of a three-space $ABCD$, then every point on the plane FGH is on the three-space $ABCD$ (Exercise 3). Any four noncoplanar points of a three-space determine that three-space (Exercise 4). A line and a plane on a three-space must have at least one point in common (Exercise 5). Two distinct planes on a three-space have a unique line in common (Exercise 6). Three distinct planes on a three-space have at least one point in common (Exercise 7). These properties of projective planes and spaces will be used throughout our study of projective geometry.

We have illustrated, by means of the above proofs and statements,

the importance and some of the consequences of the set of postulates
P–1 through P–8.   It is also important to recognize that there may
be only a finite number of points on a line in projective geometry
(Exercise 8).   Indeed, Postulates P–1 through P–8 hold on the
three-space of fifteen points in Fig. 2–3 (Exercise 6, Section 2–1).

EXERCISES

*1. Prove that any three noncollinear points of a plane determine that
plane.

*2. Use Postulates P–1 through P–7 and prove that if two points of a line
are on a three-space, then every point of the line is on the three-space.

*3. Use Postulates P–1 through P–7 and prove that if three noncollinear
points of a plane are on a three-space, then every point of that plane is on
the three-space.

*4. Use Postulates P–1 through P–7 and prove that any four noncoplanar
points of a three-space determine-that three-space.

*5. Prove that a line and a plane on a three-space must have at least one
point in common.

*6. Prove that any two distinct planes on a three-space have a unique
line in common.

*7. Prove that any three distinct planes on a three-space have at least
one point in common.

*8. Use Postulates P–1 through P–8 and the assumption that there exists
at least one line m containing exactly $N + 1$ points to prove each of the
following:

   (a) Every line coplanar with m has exactly $N + 1$ points.
   (b) Every line has exactly $N + 1$ points.
   (c) Every plane containing the line m has exactly $N^2 + N + 1$ points.
   (d) Every plane has exactly $N^2 + N + 1$ points.
   (e) Every plane has exactly $N^2 + N + 1$ lines.
   (f) Every three-space has exactly $N^3 + N^2 + N + 1$ points.

2–3 Figures.   You have all studied geometric figures.   Can you
define a figure in euclidean geometry (ordinary high-school geometry)?
In any geometry the definition of a figure must depend upon the
elements of the geometry.   Point and line are undefined elements in
projective geometry.   A plane may be defined in terms of points
and lines.   We now define any set of points, lines, and planes in
space to be a *space figure* or simply a *figure*.   Similarly, any set of
points and lines on the same plane is called a *plane figure;* any set of
planes and lines on the same point is called a *point figure;* any set of
points and planes on the same line is called a *line figure*.

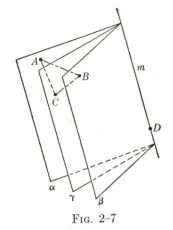

Fig. 2-6   Fig. 2-7

Given any space figure F and any point P that is not a point of F, the points and lines of F determine with P a set of lines and planes called the *point section* of the space figure F by P. For example, if the figure F is taken as triangle ABC and if P is not on the plane ABC (Fig. 2–6), then the point section of F by P consists of the lines PA, PB, PC and the planes PAB, PAC, PBC. Given any space figure F and any line m that is not a line of F, the points, lines, and planes of F determine with m a set of planes and points called the *line section* of the space figure F by m. For example, if the figure F is taken as triangle ABC and if m is not on the plane ABC (Fig. 2–7), then the line section of F by m consists of the planes $mA = \alpha$, $mB = \beta$, and $mC = \gamma$, and the point D of intersection of m with the plane ABC. Given any space figure F and any plane π that is not a plane of F, the planes and lines of F determine with π a set of lines and points that is called the *plane section* of F by π.

These definitions of sections of space figures may be rephrased as follows to obtain definitions for sections of plane figures: Given any plane figure F and any point P that is on the plane of F but not a point of F, the points of F determine with P a set of lines that is called the *point section* of the plane figure F by P. For example, if the figure F is taken as triangle ABC in Fig. 2–6, then the point section of F by P consists of the lines PA, PB, and PC. Given any plane figure F and any line m that is on the plane of F but not a line of F, the lines of F determine with m a set of points that is called the *line section* of the plane figure F by m. For example, if the figure F

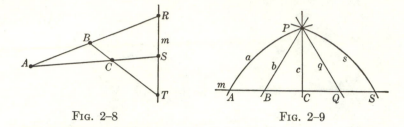

FIG. 2–8   FIG. 2–9

is taken as triangle ABC in Fig. 2–8, then the line section of F by m consists of the points $R = m \cdot AB$, $S = m \cdot AC$, and $T = m \cdot BC$, where the notation $m \cdot AB$ indicates the point of intersection of the line m and the line AB. Similarly, we shall use the notation $AB \cdot ST$ to indicate the point R of intersection of AB and ST.

We may use vertical columns to indicate incidence relations; assume that the point, line, or plane used in the section is not an element of a space figure F; and indicate the elements of each of the above sections by means of the following array:

	F:	points	lines	planes
point section:		lines	planes	
line section:		planes		points
plane section:			points	lines.

Similarly, for a plane figure F we have the following elements:

	F:	points	lines
point section:		lines	
line section:			points.

In the above definitions and arrays, the "points" of F may be interpreted either as the totality of points on elements of F (i.e., as all isolated points of F, all points on lines of F, and all points on planes of F) or as the isolated points and the vertices (intersections of distinct lines) of F. Unless otherwise specified, we shall use the second interpretation throughout our discussions of figures.

The following three figures occur frequently in our study of synthetic projective geometry and are given special names. The set of all planes on a line m is called a *pencil of planes* with *axis* m. The set of all points on a line m is called a *pencil of points* with *axis* m. On a plane, the set of all lines on a point P is called a *pencil of lines* with *center* P.

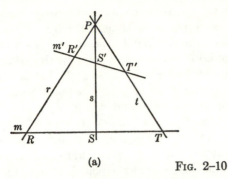
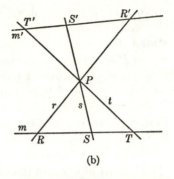

(a)   Fig. 2-10   (b)

On any projective plane, each line of a given pencil of lines with center P is on a unique point (Theorem 2-2) of any given pencil of points with axis m, where m is not on P (Fig. 2-9). Similarly, on any plane each point of a pencil of points with axis m is on a unique line of any pencil of lines with center P, where P is not on m (Fig. 2-9). The points Q and Q' of any two coplanar pencils of points with axes m and m' may be associated by considering the pairs of points on each line of a pencil of lines with center P, where P is not on m or m', that is, R and R' on r, S and S' on s, ... as in the figures in Fig. 2-10. Similarly, the lines q and q' of any two coplanar pencils of lines with centers P and P' may be associated by considering the pairs of lines on each point of a pencil of points with axis m, where m is not on P or P', that is, r and r' on R, s and s' on S, ... as in the figures in Fig. 2-11. In both Figs. 2-10 and 2-11, two figures are given having the same property, in order to illustrate different-appearing figures having this property.

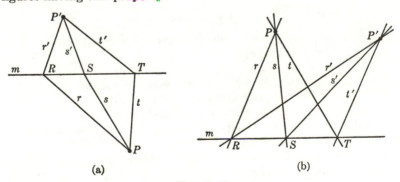

(a)   (b)

Fig. 2-11

The above paragraph may be stated much more concisely using the concepts of point, line, and plane sections. Briefly, it was stated that on a projective plane the section of a pencil of lines with center P by a line m that is not on P is a pencil of points with axis m; the section of a pencil of points with axis m by a point P that is not on m is a pencil of lines with center P (Fig. 2–9). A point section of two pencils of points on coplanar axes m and m' by any point P that is on the plane of the pencils of points but not on m or m' may be used to associate the points of the two pencils of points in pairs (Fig. 2–10). A line section of two coplanar pencils of lines on centers P and P' by any line m that is on the plane of the pencils of lines but not on P or P' may be used to associate the lines of the two pencils of lines in pairs (Fig. 2–11).

The pencils of points, lines (on a plane), and planes are interrelated as follows: A plane section of a pencil of planes on a line m by a plane π which is not on m is a pencil of lines. A point section of a pencil of points on a line m by a point P that is not on m is a pencil of lines. On a projective plane, a line section of a pencil of lines on a point M by a line p that is not on M is a pencil of points. These relationships among pencils of planes, lines, and points provide the basis for perspective and projective correspondences (Sections 2–5 and 2–6).

Throughout this section the words "point" and "plane" have been interchanged in statements regarding points and planes to obtain new statements. Also the words "point" and "line" have been interchanged in statements regarding points and lines on a projective plane. The resulting pairs of statements illustrate the principles of duality (Section 2–4).

Exercises

1. Let F be a plane triangle and draw plane figures indicating
    (a) a point section of F,
    (b) a line section of F,
    (c) a point section F_1 of F and a line section of F_1, and
    (d) a line section F_2 of F and a point section of F_2.
2. Repeat Exercise 1 where F is a plane quadrilateral.
3. Any four points that are not on the same plane determine a *tetrahedron*. Repeat Exercise 1 for space figures where F is a tetrahedron.
4. Let F be a tetrahedron and draw figures indicating
    (a) a plane section of F,
    (b) a plane section F_1 of F and a line section of F_1, and
    (c) a plane section F_2 of F and a point section of F_2.

5. Name the figure obtained and draw a section by a line m of
  (a) a pencil of lines coplanar with m where m is not a line of the pencil,
  (b) a pencil of points that is not coplanar with m, and
  (c) a pencil of planes where m is not on any one of the planes.

2–4 Duality. There are two principles of duality — the principle of planar duality and the principle of space duality.

> PRINCIPLE OF PLANAR DUALITY: *Any properly worded valid statement concerning incidences of points and lines on a projective plane gives rise to a second valid statement when the words "point" and "line" are interchanged.*

It is assumed here that either the statement is expressed in terms of the "on" terminology (two distinct points are on a unique line; two distinct lines are on a unique point) or that suitable interchanges of phrases such as "point of intersection" and "line joining" are made.

The principle of planar duality has been illustrated in Section 2–3 by the definitions on a projective plane of

point section   and   line section,
pencil of points   and   pencil of lines.

It is also illustrated by the following pair of statements:

  (i) a point section of a pencil of points with axis m by a point P that is not on m is a pencil of lines with center P;
  (ii) a line section of a pencil of lines with center P by a line m that is not on P is a pencil of points with axis m.

A triangle has been defined as a plane figure composed of three points that are not on the same line and the three lines determined by them. The plane dual of this definition is: A triangle (or trilateral) may be defined as a plane figure composed of three lines that are not on the same point and the three points determined by them. Thus a triangle is its own plane dual; i.e., a triangle is a *self-dual* plane figure.

> PRINCIPLE OF SPACE DUALITY: *Any properly worded valid statement concerning incidences of points, lines, and planes in projective space gives rise to a second valid statement when the words "point" and "plane" are interchanged.*

The assumptions on the wording are the same here as in the case of planar duality. The word "line" is left unchanged in the case of space duals, since a line may be considered either as the join of two points or as the intersection of two planes, i.e., since a line is a self-dual space figure. The word "space" is used to mean "projective three-space" in the principle of space duality. There also exists a principle of duality for projective n-spaces (Exercise 15).

The principle of space duality has been illustrated in Section 2–3 by the definitions of

| point section | and | plane section, |
| pencil of points | and | pencil of planes. |

It is also illustrated by the following pair of statements:

(i) a point section of a pencil of points with axis m by a point P that is not on m is a pencil of lines on P;

(ii) a plane section of a pencil of planes with axis m by a plane π that is not on m is a pencil of lines on π.

A tetrahedron is a self-dual space figure.

The principles of planar and space duality are based upon the validity of the plane and space duals of the postulates. The plane dual of Postulate P–1 is

If a and b are distinct lines (on a plane), there is at least one point on both a and b.

The inserted phrase "on a plane" is appropriate, since any two points are necessarily on a plane and planar duality is concerned only with figures on a plane. The plane dual of P–1 is a consequence of Theorem 2–2. The space dual of P–1 is

If α and β are distinct planes, there is at least one line on both α and β.

Notice that this statement is not the space dual of the above plane dual of P–1. In general, the space dual of a statement is distinct from the space dual of the plane dual of the statement (Exercise 10). The space dual of P–1 is a consequence of Exercise 6, Section 2–2. The statements and proofs of the plane and space duals of Postulates P–2 through P–8 are given as Exercises 1 through 8 of this section.

Any given theorem concerning incidence of points and lines on a plane has a corresponding plane dual theorem. Each step in the proof of the given theorem has a plane dual that may be used as a step in the proof of the plane dual theorem. Thus the complete proof of the given theorem has a corresponding plane dual which forms a complete proof of the plane dual theorem. In general, all valid theorems regarding incidences of points and lines on a plane have valid plane duals, since all theorems ultimately depend upon the postulates and the plane duals of the postulates are valid. Similar statements may be made for space duals. Thus in projective geometry the proof of one theorem gives rise to three other theorems (not necessarily distinct). For example, Theorem 2–2 may be stated in the form:

Any two distinct lines on the same plane determine a unique point.

The plane dual of Theorem 2–2 is

Any two distinct points on the same plane determine a unique line.

The space dual of Theorem 2–2 is

Any two distinct lines on the same point determine a unique plane

The space dual of the plane dual of Theorem 2–2 is

Any two distinct planes on the same point determine a unique line.

The principles of duality and the proof of Theorem 2–2 are sufficient to establish the validity of all four of these statements. Thus the principles of duality often make it possible to prove four theorems at once. This procedure has an obvious practical advantage. The principles of duality also have other advantages. They may be used to find new theorems by considering duals of known theorems. They may be used to organize the theorems of the geometry systematically. We shall use these principles freely and frequently in our discussion of projective geometry. Indeed, we have already used them in our discussion of pencils of points, lines, and planes.

Exercises

*1. State and prove the plane dual of Postulate P–2.
*2. Repeat Exercise 1 for P–3.
*3. Repeat Exercise 1 for P–4.
*4. Repeat Exercise 1 for P–5.
*5. Repeat Exercise 1 for P–6.
*6. Repeat Exercise 1 for P–7.

*7. Repeat Exercise 1 for P–8.

*8. Repeat Exercises 1 through 7 for space duals.

9. State plane duals for Postulates P–2.1 through P–2.7 given in Exercise 2, Section 1–5.

10. State necessary and sufficient conditions upon a statement in order that the space dual of the statement coincide with the space dual of the plane dual of the statement.

11. Draw a figure illustrating the fact that a section of a pencil of points with axis m by a point that is not on m is a pencil of lines.

12. State the plane dual of Exercise 11 and draw the figure.

13. State the space dual of Exercise 11.

14. State and discuss the space dual of the statement obtained in Exercise 12.

15. Formulate a principle of duality for projective n-spaces.

2–5 Perspective figures. Each of us makes use of the basic properties of perspective figures in his daily living. For example, we use these properties whenever we use mirrors. Consider an object O as a set of points. The points of its apparent image M in the plane of the mirror and the corresponding points of its virtual image V behind the mirror are perspective with respect to the eye of the observer (Fig. 2–12). Similarly, the corresponding points of the apparent image M and the object O are perspective with respect to a point P on the other side of the mirror from the observer. In a study of optics, we find that the point P and the eye E of the observer are equidistant from the plane of the mirror and that the line PE is perpendicular to the plane of the mirror.

Perspective figures are also involved in many other common situations. The image I on the negative in a pinhole camera and the object photographed, O, are perspective with respect to the

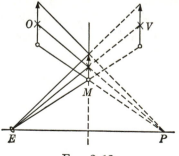

Fig. 2–12

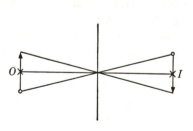

Fig. 2–13

pinhole when the photograph is made (Fig. 2–13). Any circle C on a right circular cone is perspective with respect to the vertex V of the cone with any circle C_1 or any ellipse E on the cone (Fig. 2–14).

The perspectivity between a circle and an ellipse (Fig. 2–14) indicates that perspective figures do not necessarily have the same shape. However, since a landscape or an individual may often be recognized by a photograph, perspective figures must have some invariant properties. Projective geometry may be considered as a study of these properties. Size, shape, distance, and angles are not invariant. Incidence relations, existence, and a few other properties are invariant, and they suffice to enable us to recognize photographs, to make practical as well as vain use of mirrors, and to find many other applications of properties of perspective figures.

The pencils of points on m and m' in Fig. 2–10 are said to be perspective from the point P, since the lines joining corresponding points of m and m' all pass through P. In general, any two pencils of points on a plane are said to be *perspective* from a point P if the lines joining corresponding points form a pencil of lines with center P. The point P is called the *center of perspectivity;* the lines through P are called the *projectors* of the points of the pencils of points. In accordance with the principle of planar duality, any two pencils of lines on a plane are said to be perspective from a line m if the points of intersection of corresponding lines form a pencil of points with axis m (Fig. 2–11). The line m is called the *axis of the perspectivity;* the points on m are called the *traces* of the lines of the pencils of lines. Finally, in accordance with the principle of space duality, any two pencils of planes on a point (i.e., any two pencils of planes whose axes intersect at the point) are said to be perspective from a plane π if the lines of intersection of corresponding planes form a pencil of lines on π.

The term "perspective" is also used to indicate a relationship between two figures that are of different types. A pencil of lines with center P is perspective with a pencil of points with axis m

$E \quad C \qquad\qquad C_1$

FIG. 2–14

(where m is not on P) if the corresponding elements of the two pencils are incident (Fig. 2–9) This definition is self-dual on the plane. The space dual of the definition is: A pencil of lines on a plane π is perspective with a pencil of planes with axis m (where m is not on π) if the corresponding elements of the two pencils are incident. These definitions and those considered above illustrate two underlying properties of perspective figures:

> (i) a one-to-one correspondence of corresponding elements;
> (ii) an incidence relation shared by corresponding elements.

We now use these properties to define figures perspective from a point, a line, or a plane.

Any two figures F and F' are said to be *perspective from a point O* if there is a one-to-one correspondence between the elements of F and F' and if the lines joining corresponding points of F and F' all pass through O. For example, triangles ABC and $A'B'C'$ are perspective from O in Fig. 2–15. The point O is called the *center of perspectivity;* the lines through O joining corresponding points of F and F' are called *projectors.*

In the case of plane figures, we define figures perspective from a line by taking the plane dual of the above paragraph. Any two plane figures F and F' are said to be *perspective from a line m* if there is a one-to-one correspondence between the elements of F and F' and if the points of intersection of corresponding lines of F and F' all lie on m. For example, triangles ABC and $A'B'C'$ are perspective from m in Fig. 2–16. The line m is called the *axis of perspectivity;* the points of intersection on m of corresponding lines of F and F' are called *traces.*

In the case of space figures we define figures perspective from a plane by taking the space dual of the definition of figures perspective from a point. Any two figures F and F' are said to be *perspective from a plane α* if there is a one-to-one correspondence between the elements of F and F' and if the lines of intersection of corresponding planes of F and F' all lie on α. The plane α is called the *plane of perspectivity;* the lines of intersection of corresponding planes of F and F' are called *traces.*

The correspondence between the elements of the two figures in each of the above cases is called a *perspective correspondence.* When two figures are related by a perspective correspondence, each figure

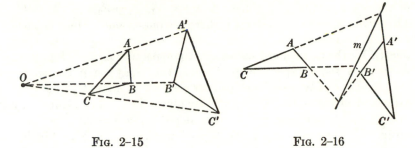

Fig. 2–15   Fig. 2–16

is said to be obtained from the other by a *perspective transformation* and the figures are called *perspective figures*. The Theorem of Desargues (Section 2–10) states that whenever two triangles are perspective from a point, as in Fig. 2–15, they are also perspective from a line, as in Fig. 2–16. This is only one of the many examples that we shall find of the importance of perspective correspondences. In the next section we shall use perspective transformations to define projective transformations.

EXERCISES

1. Sketch several common pairs of figures that are perspective from a point.

2. Draw a figure representing two pencils of points perspective from a point.

3. State and draw the plane dual of Exercise 2.

4. State the space dual of Exercise 2.

5. State the space dual of the statement obtained in Exercise 3.

6. Use space duality and define a pencil of lines perspective with a pencil of planes.

2–6 Projective transformations. The concept of a projective transformation is simply an extension of that of a perspective transformation (Section 2–5). We shall use the notation $F \overset{}{\underset{\wedge}{=}} F'$ to indicate that the figures F and F' are perspective. Figures perspective from a point O may be indicated by $F \overset{O}{\underset{\wedge}{=}} F'$; figures perspective from a line m by $F \overset{m}{\underset{\wedge}{=}} F'$; and figures perspective from a plane α by $F \overset{\alpha}{\underset{\wedge}{=}} F'$.

Two figures F and F' are *projective* and each may be obtained from the other by a *projective transformation* if there exists a finite sequence of figures $F_1, F_2, \ldots, F_k$ such that $F \overset{\sim}{=} F_1, F_1 \overset{\sim}{=} F_2, \ldots, F_k \overset{\sim}{=} F'$. Thus two figures are projective if there is a one-to-one correspondence between their elements, and all pairs of corresponding elements are related by the same finite sequence of perspectivities. In Fig. 2–12 we have a sequence of perspectivities that may be indicated by $O \overset{P}{\overline{\wedge}} M \overset{E}{\overline{\wedge}} V$. The relationship between the figures O and V is expressed in words by "O is projective with V" and in symbols by "$O \overline{\wedge} V$." A perspectivity is considered as a special case of a projectivity and we may write $F \overline{\wedge} F'$ whenever $F \overset{\sim}{=} F'$. Finally, since $F \overset{\sim}{=} F'$ implies $F' \overset{\sim}{=} F$, we may write $F' \overline{\wedge} F$ whenever $F \overline{\wedge} F'$. In other words, if a figure F' may be obtained by a projective transformation of a figure F, then F may be obtained by a projective transformation of F'.

THEOREM 2–3. *Any three distinct points on a line m are projective with any three distinct points on a line m'.*

The complete proof of this important theorem is given as Exercise 1 of this section. We shall indicate the proof for the special case in which the lines m and m' are distinct. The proof is valid whether m and m' are on the same or different planes.

We are given three distinct points A, B, C on a line m and three distinct points A', B', C' on a line $m' \neq m$. At most one of the given points on m' is also on m, and at most one of the given points on m is also on m'. Suppose C' is not on m and C is not on m'. Draw the line CC' and let O_1 be a third point (Postulate P–5) on this line (Fig. 2–17). On the plane ACC' let m_1 be any line through C' that is distinct from m' and from CC'. Then, since m and m_1 are on the plane ACC', m and m_1 are coplanar. Also, since m' and m_1 intersect at C', m' and m_1 are coplanar. On the plane ACC', let $m_1 \cdot O_1B = B_1$ and $m_1 \cdot O_1A = A_1$. Then

$$ABC \overset{O_1}{\overline{\wedge}} A_1B_1C'.$$

On the plane of m_1 and m', let $A_1A' \cdot B_1B' = O_2$. Then

$$A_1B_1C' \overset{O_2}{\overline{\wedge}} A'B'C'.$$

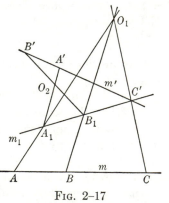

FIG. 2–17

These two perspectivities may be combined in the form

$$ABC \overset{O_1}{\doublebarwedge} A_1B_1C' \overset{O_2}{\doublebarwedge} A'B'C'$$

to indicate the projectivity $ABC \barwedge A'B'C'$. If $m = m'$, a perspectivity is used first to transform the points A, B, C onto points A'', B'', C'' on another line and then the above procedure is used for the points A'', B'', C'' and A', B', C' (Exercise 1).

Theorem 2–3 states that there exists a projective transformation $ABC \barwedge A'B'C'$ whenever A, B, C are three distinct points on a line m and A', B', C' are three distinct points on a line m'. The corresponding theorem for points on planes is

THEOREM 2–4. *Any four points on a plane π such that no three are collinear are projective with any four points on a plane π' such that no three are collinear.*

The corresponding theorem for points in space may be stated as follows:

Any five points such that no four are coplanar are projective with any five points such that no four are coplanar.

Since we are primarily interested in the development of euclidean plane geometry, we shall

(i) consider projective planes as sets of points on projective three-space;

(ii) use the incidence and existence properties on the three-space to obtain properties of projective planes;

(iii) give a detailed development of projective transformations on a line and on a plane;

(iv) mention (usually without formal proof) many of the corresponding relations for projective transformations in space.

Accordingly, we shall prove Theorem 2–4 only for the special case when the theorem involves a projective transformation of a plane onto itself, i.e., when $\pi = \pi'$.

Let A, B, C, D be four points on a plane π, no three collinear, and let A', B', C', D' be four points on the same plane π, no three collinear. Theorem 2–4 states that there exists at least one projectivity $ABCD \barwedge A'B'C'D'$. The following proof of the existence of such a projectivity is based upon the determination of four coplanar points A^*, B^*, C^*, D^* (Fig. 2–19) such that

$$ABCD \barwedge A^*B^*C^*D^* \qquad \text{and} \qquad A^*B^*C^*D^* \barwedge A'B'C'D'.$$

We shall make use of the assumption (Postulate P–7) that planes are sets of points in space and seek two points O, O_1 such that

$$AB \overset{O_1}{\barwedge} A^*B^* \overset{O}{\barwedge} A'B'$$

and the points A, B, C, D are perspective from O_1 with four distinct points on the plane $OA'B'$. In order to satisfy these conditions, we need two pairs of corresponding points, say A, A' and B, B', such that there exists a point F satisfying the three conditions:

  (i) A, A', F are collinear;
  (ii) B, B', F are collinear;
  (iii) A', B', F are not collinear.

Thus we must first prove the existence of such a point F.

  If the corresponding points of any two pairs are identical (i.e., for example, if $A = A'$ and $B = B'$), the point F may be taken as any point that is not on the line $A'B'$. If there do not exist two pairs of identical corresponding points, then there are at least three pairs of distinct corresponding points. If, among these three pairs of distinct corresponding points there exist two pairs of points $A \neq A'$ and $B \neq B'$ such that the intersection of AA' and BB' does not contain A' or B', then

$$\begin{array}{ll} A' \text{ is not on } BB', & AA' \neq BB', \\ AA' \cdot BB' \text{ is a single point}, & B' \text{ is not on } AA', \end{array}$$

and the above three conditions are satisfied when $F = AA' \cdot BB'$. Finally, we must prove that there exists a point F satisfying the above three conditions whenever both of the following conditions hold:

  (i) there are at least three pairs of distinct corresponding points;
  (ii) for every two pairs of distinct points A, A' and B, B' the intersection $AA' \cdot BB'$ contains either A' or B'.

In this case we have $B \neq B'$, $C \neq C'$, and $D \neq D'$ from (i). From (ii) we have $BB' \cdot CC' = B'$ or C', $BB' \cdot DD' = B'$ or D', $CC' \cdot DD' = C'$ or D'. Suppose $BB' \cdot CC' = B'$. Then $BB' \cdot DD' \neq B'$, since this would imply that $CC' \cdot DD' = B' \neq C'$ or D'. Accordingly, $BB' \cdot DD' = D'$, $CC' \cdot DD' = C'$, and the three lines BB', CC', DD' are distinct (Exercise 5). We now have $BB' = B'D'$, $CC' = B'C'$, and $DD' = C'D'$. This implies that A' is not on any of the lines BB', CC', DD', since no three of the points A', B', C', D' are collinear. If $A = A'$, we may take $F = D'$. If $A \neq A'$, we may assume (as a matter of notation) that AA' does not contain B' (since the line AA' contains at most one of the points B', C', D') and take $F = AA' \cdot BB'$.

We have proved that for any four points A, B, C, D on a plane π, no three collinear, and any four points A', B', C', D' on the same plane π, no three collinear, there always exist two pairs of corresponding points which (as a matter of notation) we shall indicate by A, A' and B, B', such that there exists a point F satisfying the above three conditions. The remainder of the proof of Theorem 2–4 for the case $\pi = \pi'$ is concerned with projective transformations.

Let O be any point that is not on the plane π and let F be a point on π satisfying the three conditions considered above. Let O_1 be a point on OF distinct from O and F. Let π_1 be the plane $OA'B'$. The planes π_1 and π' are distinct, since O is on π_1 and is not on π. The point O_1 is not on π_1 since $\pi \cdot \pi_1 = A'B'$, F is not on $A'B'$, the line OF intersects the plane π_1 in only one point O, and $O_1 \neq O$ is on OF. The fact that O_1 is not on π_1 assures us that the points A, B, C, D on π are perspective with respect to O_1, with four distinct points on π_1. Let $AB \cdot CD = E$ and define (Fig. 2–18) the points A^*, B^*, C_1, D_1, E_1 on π_1 by the perspectivity

$$ABCDE \overset{O_1}{\doublewedge} A^*B^*C_1D_1E_1.$$

Then on the plane OFA', the intersection of O_1A and OA' must be A^*, since A^* is the only point on the line O_1A that is also a point on the plane $OA'B'$. Similarly, $O_1B \cdot OB' = B^*$ on the plane OFB'.

The point E_1 is on A^*B^* since E is on AB. Let $A'B' \cdot C'D' = E'$ and $OE' \cdot A^*B^* = E^*$. If $E_1 = E^*$, we take $C_1 = C_2$ and $D_1 = D_2$. If $E_1 \neq E^*$, there exists a projective transformation

$$A^*B^*E_1 \overline{\wedge} A^*B^*E^*$$

of the line A^*B^* onto itself (Theorem 2–3). There also exists (Exercise 7) at least one projective transformation

$$A^*B^*C_1D_1E_1 \overline{\wedge} A^*B^*C_2D_2E^*$$

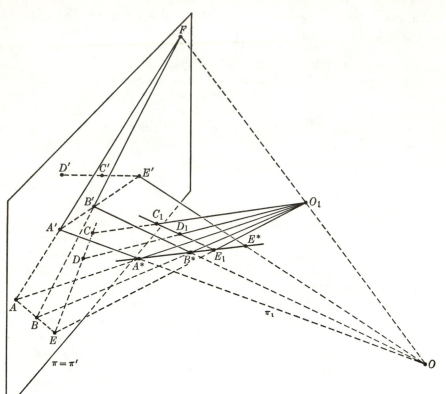

FIG. 2–18

of the plane $\pi_1 = OA^*B^*$ onto itself, where C_2 and D_2 are determined by the transformation.   We now have

$$ABCDE \mathrel{\overset{}{\underset{\wedge}{-}}} A^*B^*C_2D_2E^*,$$

where $A^*B^*E^* \mathrel{\overset{O}{\underset{\wedge}{=}}} A'B'E'$.   We shall complete our proof of the special case $\pi = \pi'$ of Theorem 2–4 by determining two perspectivities as follows:

$$(2\text{–}1) \qquad A^*B^*C_2D_2E^* \mathrel{\overset{}{\underset{\wedge}{=}}} A^*B^*C^*D^*E^* \mathrel{\overset{}{\underset{\wedge}{=}}} A'B'C'D'E'.$$

Let C^* be a point on OC' that is distinct from O and C' (Fig. 2–19). On the plane $OC'E'$ let $E^*C^* \cdot OD' = D^*$, and on the plane $C^*E^*C_2$

let $C_2C^* \cdot D_2D^* = O_3$. The first
perspectivity in formula (2–1)
now has center O_3 and represents a perspective transformation of the plane $\pi_1 = OA^*B^*$
onto the plane $\pi^* = A^*B^*C^*$.
The points on the line A^*B^*
are invariant (unchanged) under this perspective transformation, since the planes π_1
and π^* intersect on A^*B^*. The
second perspectivity of formula
(2–1) has center O and represents a perspective transformation of the plane π^* onto the
plane π'.

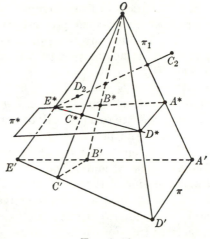

Fig. 2–19

We have proved that there must exist at least one finite sequence
of perspective transformations (i.e., at least one projective transfor-
mation) of the plane π onto itself such that $ABCD \mathrel{\overset{\wedge}{}} A'B'C'D'$.
In the next section we shall prove that even though there may be
many such sequences of perspectivities, there is essentially only
one such projectivity; i.e., all such projectivities involve the same
correspondence of points of π.

EXERCISES

*1. Give a complete proof of Theorem 2–3 and prove that at most three
perspectivities are necessary.

2. Draw a figure showing two triangles related by a sequence of four
perspectivities. Letter the figures and indicate the sequence of perspectivities.

3. Repeat Exercise 2 for quadrilaterals.

4. Repeat Exercise 2 for tetrahedrons related by two perspectivities.

*5. Prove that the three lines BB', CC', DD' are distinct whenever the
points B', C', D' are noncollinear; $B \neq B', C \neq C'. D \neq D', BB' \cdot CC' = B'$
or C', $BB' \cdot DD' = B'$ or D', and $CC' \cdot DD' = C'$ or D'.

6. Prove that $ABCD \mathrel{\overset{\wedge}{}} ABCD$ for any four coplanar points A, B, C, D.

*7. Rephrase the proof of Theorem 2–3 and prove that there exists at
least one projective transformation of a plane π onto itself such that any
three distinct points A, B, C on a line m on the plane π are projective with
any three distinct points A', B', C' on m.

8. Prove that $\mathrel{\overset{\wedge}{}}$ is an equivalence relation (See Exercise 11, Section 1–2).

2-7 Postulate of Projectivity. We have seen (Theorem 2-3) that there exists at least one projectivity taking any three distinct points on a line m onto any three distinct points on a line m'. The projectivity obtained may be used to associate each point D on m with a point D' on m' by taking (in the notation of Fig. 2-17) $DO_1 \cdot m_1 = D_1$ and $D_1O_2 \cdot m' = D'$. Thus any projective transformation of three distinct points on a line m onto three distinct points on a line m' gives rise to a projective transformation of the pencil of points on m onto the pencil of points on m'. Similarly, any projective transformation of four points on a plane π, no three collinear, onto four points on a plane π', no three collinear, gives rise to a projective transformation of the set of all points on π onto the set of all points on π'. Also in space, a projective transformation of five points, no four coplanar, onto five points, no four coplanar, gives rise to a projective transformation of space onto itself.

We now postulate that the projective transformation obtained in Theorem 2-3 is unique. This complete determination or uniqueness of the projectivity does *not* imply that there is only one possible sequence of perspectivities that may be used to obtain the projectivity. Indeed, the line m_1 in Fig. 2-17 may be chosen in many ways. Instead, the uniqueness of the projectivity implies that if $ABCD \overset{\wedge}{-} A'B'C'D_1'$ using one sequence of perspectivities, and if $ABCD \overset{\wedge}{-} A'B'C'D_2'$ using the same or a different sequence of perspectivities, then $D_1' = D_2'$. That is, there is a unique projective correspondence under which three distinct points on m are associated with three distinct points on m', and this projectivity associates a unique point P' on m' with each point P of the pencil of points on m. In other words, the projectivity between three distinct points on m and three distinct points on m' determines a unique projective transformation of the pencil of points on m onto the pencil of points on m'.

POSTULATE OF PROJECTIVITY

P-9: *A projectivity between two pencils of points is completely determined by three distinct pairs of corresponding points.*

This postulate may also be stated in several other ways, such as:

(i) A projective transformation of one pencil of points onto another is completely determined by three distinct pairs of corresponding points.

(ii) A projective transformation of a pencil of points onto itself is completely determined by three distinct pairs of corresponding points.

(iii) A projective transformation of a pencil of points onto itself that leaves three distinct points invariant leaves every point of the pencil of points invariant.

After a coordinate system has been introduced, we may state the postulate in the following form:

(iv) Any projective transformation upon the points of a line is completely determined by the transformation upon the rational points of the line.

When projective geometry is considered from an algebraic point of view, it is customary to use postulates in terms of the coordinates of points. In such cases the above postulate may be proved as a theorem and is called the *Fundamental Theorem of Projective Geometry* (Exercise 22, Section 4–4).

Postulate P–9 implies that the projective transformations in Theorems 2–3 and 2–4, and the corresponding statement in space, are all unique. When this result is combined with the results considered previously, we have the following theorems and statement.

THEOREM 2–5. *There exists a unique projective transformation of a line m onto a line m' such that $ABC \barwedge A'B'C'$ where A, B, C are any three distinct points on m and A', B', C' are any three distinct points on m'.*

THEOREM 2–6. *There exists a unique projective transformation of a plane π onto a plane π' such that $ABCD \barwedge A'B'C'D'$ where A, B, C, D are any four points on π, no three collinear, and A', B', C', D' are any four points on π', no three collinear.*

The corresponding statement for three-space is:

There exists a unique projective transformation of projective three-space onto itself such that $ABCDE \barwedge A'B'C'D'E'$ where A, B, C, D, E are any five points, no four on the same plane, and A', B', C', D', E' are any five points, no four on the same plane.

Theorem 2–5 may be proved as follows: By Theorem 2–3 there exists at least one projective transformation of the desired form.

By Postulate P–9 there exists at most one such projectivity. Thus there exists a unique projectivity such as stated in Theorem 2–5.

The proof of Theorem 2–6 for the special case $\pi = \pi'$ is nearly as short as that for Theorem 2–5. By Theorem 2–4 there exists at least one projective transformation of the desired form. Under this transformation $AB \cdot CD = E$ corresponds to $A'B' \cdot C'D' = E'$, whence the projectivity is uniquely determined on the lines AB and CD. Similarly, it is uniquely determined on the lines AC, AD, BC, and BD. Finally, since every line m on the plane π intersects these six lines in at least three distinct points, the projective transformation is uniquely determined on every line and therefore for all points (intersections of lines) on π. This completes the proof of Theorem 2–6 when $\pi = \pi'$. As indicated in the last section, we shall not take the time to prove the general case of Theorem 2–6 or the corresponding statement regarding transformations in three-space.

EXERCISES

*1. Prove that two projective pencils of points on different axes are perspective if and only if their common point is invariant (self-corresponding).

*2. State and prove the plane dual of Postulate P–9.

3. State and prove the plane dual of Exercise 1.

4. State the plane dual of Theorem 2–5.

5. State the space dual of Theorem 2–5.

6. State the space dual of Theorem 2–6.

2–8 Quadrangles. Since the postulates of any geometry involve relations among the elements of the geometry, it is often desirable to define certain elements and relations in order to state the postulates in a convenient form. For example, pencils of points and projective

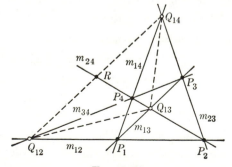

FIG. 2–20

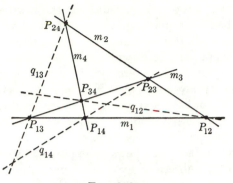

Fig. 2–21

transformations were needed before Postulate P–9 could be stated, as in Section 2–7. In the present section, we shall define a complete quadrangle and introduce an additional postulate that will imply that each line contains at least four distinct points. There still may be only a finite number of points on a line, but there will be at least four points. All projective geometries satisfy at least Postulates P–1 through P–9.

The figure consisting of four coplanar points (vertices), no three of which are collinear, and the six lines (sides) determined by them is called a *complete quadrangle* on the plane (Fig. 2–20). The figure consisting of four coplanar lines (sides), no three of which are concurrent, and the six points (vertices) determined by them is called a *complete quadrilateral* on the plane (Fig. 2–21). Note that a complete quadrangle and a complete quadrilateral are plane dual figures. The figure consisting of four coplanar points (vertices) P_1, P_2, P_3, P_4, no three of which are collinear, and the four lines (sides) P_1P_2, P_2P_3, P_3P_4, P_4P_1 obtained by taking the vertices in cyclic order is called a *simple plane quadrangle* (Fig. 2–22). The plane dual of a simple plane quadrangle is a *simple plane quadrilateral* (Fig. 2–23) (Exercise 1).

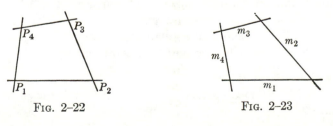

Fig. 2–22   Fig. 2–23

We now construct the figure needed for the next postulate. Let P_1, P_2, P_3, P_4 be four coplanar points, no three of which are on the same line. We shall use the notation m_{12} to designate the line P_1P_2 and, in general, m_{jk} to designate the line P_jP_k. The six sides of the complete quadrangle determined by the given four points are then designated by m_{12}, m_{13}, m_{14}, m_{23}, m_{24}, m_{34}, as in Fig. 2–20. Two sides that do not have a vertex in common are said to be *opposite sides*. Thus m_{12} and m_{34} are opposite sides, m_{13} and m_{24} are opposite sides, and m_{14} and m_{23} are opposite sides. One advantage of our notation lies in the fact that pairs of opposite sides may be easily identified by their subscripts. Since in projective geometry any two coplanar lines intersect (Theorem 2–2), the three pairs of opposite sides determine three points called the *diagonal points* of the complete quadrangle. In Fig. 2–20 we have designated the diagonal points as $Q_{12} = m_{12} \cdot m_{34}$, $Q_{13} = m_{13} \cdot m_{24}$, and $Q_{14} = m_{14} \cdot m_{23}$, using the subscripts of the lines through P_1 that are on the diagonal point. This notation emphasizes the fact that for any vertex P_1 of the complete quadrangle, the lines joining this vertex to the other three vertices pass through the three diagonal points. If the diagonal points are not collinear, the triangle $Q_{12}Q_{13}Q_{14}$ indicated by dotted lines in Fig. 2–20 is called the *diagonal triangle* of the complete quadrangle. The plane dual of this paragraph may be used to define *opposite vertices, diagonal lines*, and the *diagonal triangle* of any complete quadrilateral (Exercise 2).

Before postulating that the diagonal points of a complete quadrangle are not collinear and thus that the diagonal triangle exists, let us observe the need for this additional postulate by considering a finite projective geometry in which the diagonal points of any complete quadrangle are collinear. This situation arises whenever each line contains exactly three points, as in Fig. 1–6. When designated in the notation of this section, the seven points of Fig. 1–6 may be taken as P_1, P_2, P_3, P_4, Q_{12}, Q_{13}, Q_{14} (Fig. 2–24), where the last three points are considered to be collinear, as in Fig. 1–6. Thus there does exist a finite projective geometry in which the diagonal points of any complete plane quadrangle are collinear, and the independence of the following postulate is established.

Postulate on Quadrangles

P–10: *The diagonal points of any complete quadrangle are not collinear.*

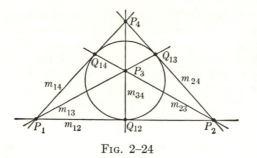

FIG. 2–24

A complete quadrangle may be visualized as in Fig. 2–20 on any plane satisfying Postulates P–1 through P–8 and P–10. In this figure it is now known that Q_{12}, Q_{13}, and Q_{14} are not collinear. Thus the line $Q_{12}Q_{14}$ intersects the line m_{24} in a point $R \neq Q_{13}$. Furthermore, the assumption that no three of the vertices of the quadrangle are collinear implies that $R \neq P_2$ and $R \neq P_4$ (Exercise 3). Thus R is a fourth point on the line m_{24}, and every line of a projective plane satisfying Postulates P–1 through P–8 and P–10 contains at least four points [Exercise 5(a)]. It is still possible that every line contains exactly four points. When taken with Postulates P–1 through P–10, the postulates of separation (Section 3–4) will imply that every line contains infinitely many points. Postulates P–1 through P–10 suffice to obtain the set of points (Section 2–9) needed for the Theorem of Desargues (Section 2–10).

EXERCISES

*1. Define a simple plane quadrilateral.

*2. Define the opposite vertices, diagonal lines, and diagonal triangle of a complete quadrilateral by writing the plane dual of the third paragraph of this section.

*3. Prove that $R = Q_{12}Q_{14} \cdot m_{24}$ in Figure 2–20 is distinct from Q_{13}, P_2, and P_4.

*4. State and prove the plane dual of Postulate P–10.

*5. Prove that in any projective geometry satisfying Postulates P–1 through P–8 and P–10

    (a) every line contains at least four distinct points;

    (b) every point is on at least four distinct coplanar lines;

    (c) every line is on at least four distinct planes;

    (d) every plane contains at least thirteen distinct points;

(e) every plane contains at least thirteen distinct lines;
(f) every point is on at least thirteen distinct planes;
(g) every point is on at least thirteen distinct lines;
(h) projective space contains at least forty points.

6. Use number quadruples with coordinates 0, 1, 2 to represent points and to obtain a representation of a projective geometry in which every line contains exactly four points. (*Hint:* See the exercises of Section 1–7 and Exercise 5, Section 2–1.)

2–9 Complete and simple *n*-points. In the *n*-point and *n*-line terminology, a triangle may be considered either as a three-point or as a three-line. Also a triangle may be considered either as a simple three-point (each point is joined by a line to the next point in the order given) or as a complete three-point (each point is joined by a line to each of the other given points). A simple plane quadrangle (Fig. 2–22) is a simple plane four-point, a complete plane quadrangle (Fig. 2–20) is a complete plane four-point, a simple plane quadrilateral (Fig. 2–23) is a simple plane four-line, and a complete plane quadrilateral (Fig. 2–21) is a complete plane four-line. In general, recognizing that in finite geometries the definitions for large values of *n* may be meaningless, the definitions of simple and complete quadrangles and quadrilaterals may be extended to obtain definitions of simple and complete *n*-points and *n*-lines.

The figure consisting of five coplanar points (vertices), no three of which are on the same line, and the ten lines (sides) determined by them is called a *complete plane five-point*. The figure consisting of five coplanar lines (sides), no three of which are on the same point, and the ten points (vertices) determined by them is called a *complete plane five-line*. The figure consisting of five coplanar points R, S, T, U, V, no three of which are on the same line, and the five lines RS, ST, TU, UV, VR obtained by taking the given points in cyclic order is called a *simple plane five-point*. The figure consisting of five coplanar lines r, s, t, u, v, no three of which are on the same point, and the five points rs, st, tu, uv, vr obtained by taking the given lines in cyclic order is called a *simple plane five-line*. The figure consisting of *n* coplanar points (vertices), no three of which are on the same line, and the $n(n-1)/2$ lines (sides) determined by them is called a *complete plane n-point*. The figure consisting of *n* coplanar lines (sides), no three of which are on the same point, and the $n(n-1)/2$ points (vertices) determined by them is called a *complete plane n-line*.

The statements of the definitions of *simple plane n-points* and *simple plane n-lines* are given as an exercise (Exercise 6).

The definitions of complete and simple space n-points are similar to the above definitions of complete and simple plane n-points. We shall state the general definition of a complete space n-point and leave the statements of the definitions for specific values of n as exercises (Exercises 8 through 11). The figure consisting of n-points, no four of which are on the same plane, together with the $n(n-1)/2$ lines and the $n(n-1)(n-2)/6$ planes determined by them is called a *complete space n-point*. The space dual of this definition may be used to define a *complete space n-plane* (Exercise 7).

We conclude this section with a brief discussion of a complete space five-point, since one of the figures used for the Theorem of Desargues (Section 2–10) may be obtained as a plane section of a complete space five-point.

Let P_1, P_2, P_3, P_4, P_5 be five points, no four of which are on the same plane. Then, as indicated by the subscripts, these points determine ten lines m_{12}, m_{13}, m_{14}, m_{15}, m_{23}, m_{24}, m_{25}, m_{34}, m_{35}, m_{45} and ten planes π_{123}, π_{124}, π_{125}, π_{134}, π_{135}, π_{145}, π_{234}, π_{235}, π_{245}, π_{345}. We next consider a plane π which does not contain any of the five given points. We have already seen that Postulate P–10 is a necessary condition for the existence of a projective plane containing more than seven points. It is a necessary condition for the existence of the plane π, since each of the above ten lines m_{jk} must intersect the plane in a point P_{jk}; that is, the plane must contain at the least the ten points P_{12}, P_{13}, P_{14}, P_{15}, P_{23}, P_{24}, P_{25}, P_{34}, P_{35}, and P_{45}. The plane π must also contain ten lines m_{ijk} corresponding to the above ten planes π_{ijk}. All incidence relations are easily obtained from a consideration of the complete space five-point or from the notation. For example, in space the planes π_{123}, π_{124}, and π_{125} are on the line m_{12} and the lines m_{12}, m_{13}, and m_{23} are on the plane π_{123}. Therefore, on the plane section the lines m_{123}, m_{124}, and m_{125} are on the point P_{12} and the points P_{12}, P_{13}, and P_{23} are on the line m_{123}. In general, any plane section of a complete space five-point by a plane that does not contain any of the five given points consists of ten points and ten lines, where there are three points on every line and three lines on every point (Fig. 2–25). This is the figure that we shall use in our discussion of the Theorem of Desargues in the next section.

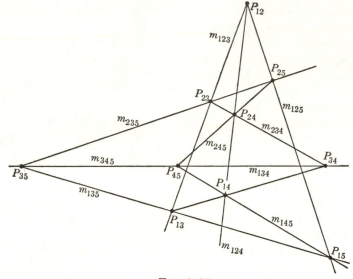

Fig. 2–25

Exercises

1. Draw a complete plane five-point.
2. Draw a complete plane five-line.
3. Draw a simple plane five-point and a simple plane five-line.
*4. Define each of the following: complete plane six-point, complete plane six-line, simple plane six-point, simple plane six-line.
5. Draw figures illustrating each of the figures defined in Exercise 4.
6. Define a simple plane n-point and a simple plane n-line.
7. Define a complete space n-plane.
8. Define a complete space four-point.
9. Define a complete space four-plane.
10. Define a complete space five-point.
11. Define a complete space five-plane.

2–10 Theorem of Desargues.

The Theorem of Desargues, in-volving a relationship between two triangles, has many applications in projective geometry.

THEOREM 2–7. THEOREM OF DESARGUES. *If two triangles are perspective from a point, they are perspective from a line, and conversely.*

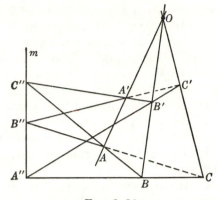

FIG. 2–26

This theorem can be proved whether the triangles are on the same or different planes. Thus the theorem is equivalent to the following four special cases of it:

    (i) If two triangles on distinct planes are perspective from a point, they are perspective from a line.

    (ii) If two triangles on distinct planes are perspective from a line, they are perspective from a point.

    (iii) If two triangles on the same plane are perspective from a point, they are perspective from a line.

    (iv) If two triangles on the same plane are perspective from a line, they are perspective from a point.

Since cases (iii) and (iv) are plane duals, the proof of either one of these two cases will imply the proof of the other. We shall prove case (i), give the proof of case (ii) as an exercise (Exercise 1), and outline the proof of case (iii).

We first prove case (i). Given triangle ABC on a plane π and triangle $A'B'C'$ on a plane π' distinct from π such that the lines AA', BB', and CC' all pass through a point O, let the line of intersection of the two planes be m. We shall prove that corresponding sides of the two triangles intersect upon the line m. The figure needed for this proof (Fig. 2–26) may be visualized by thinking of a tetrahedron with base ABC and the fourth vertex O. The plane of the base ABC is intersected along the line m by another plane $A'B'C'$ where A' is on AO, B' is on BO, and C' is on CO. The proof involves several applications of Theorems 2–1 and 2–2 (Section 2–2). Let AC and $A'C'$ on the plane ACO intersect at B'', AB and $A'B'$ on the plane ABO intersect at C'', and BC and $B'C'$ on the plane BCO intersect at A''. The points A'', B'', and C'' are on the plane ABC and also on the plane $A'B'C'$. Therefore, since these two distinct planes can have only the points of the line m in common, A'', B'', and C'' are on the line m; that is, the intersections of corresponding sides of the two given triangles are collinear and the triangles are perspective from a line. We note that when the triangles are on distinct planes, the projectors OA, OB, and OC are necessarily distinct lines.

The proof of case (ii) is very similar to that just considered for case (i) and is given as an exercise (Exercise 1). In this case the intersections of pairs of lines joining corresponding points are considered, and the result of Exercise 7, Section 2–2 is used.

We have now proved the Theorem of Desargues for triangles on different planes and are ready to consider the case of two coplanar triangles that are perspective from a point. As mentioned above, it is sufficient to consider only one of the cases (iii) and (iv). We shall consider case (iii), first when the projectors are distinct, and then when the projectors are not distinct. Given two triangles $P_{23}P_{24}P_{25}$ and $P_{13}P_{14}P_{15}$ perspective from a point P_{12} (Fig. 2–25), we wish to prove that the points $P_{34} = P_{13}P_{14} \cdot P_{23}P_{24}$, $P_{35} = P_{13}P_{15} \cdot P_{23}P_{25}$, and $P_{45} = P_{14}P_{15} \cdot P_{24}P_{25}$ are collinear. The proof may be considered as consisting of the construction of a complete space five-point to obtain a single triangle $P_3P_4P_5$ that is not on the plane of the given triangles and that is perspective with each of the given triangles. In other words, a complete space five-point $P_1P_2P_3P_4P_5$ is to be constructed such that $P_{13}P_{14}P_{15} \overset{P_1}{\wedge} P_3P_4P_5$ and $P_{23}P_{24}P_{25} \overset{P_2}{\wedge} P_3P_4P_5$. The remainder of the proof consists of the application of Theorem 2–7 for triangles in different planes to each of these pairs of triangles and the resulting recognition that the given triangles are perspective from the line of intersection of their plane with the plane $P_3P_4P_5$. The details of the proof when the projectors are distinct are left as an exercise (Exercise 4). When the projectors are not all distinct, the triangles appear as in Fig. 2–27. In all such cases the proof is trivial, since the point $C'' = AB \cdot A'B'$ may be taken as any point on the line $AB = A'B'$. In particular, C'' may be taken as the point of intersection of AB and $A''B''$.

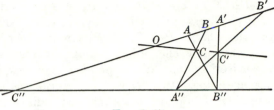

FIG. 2–27

We have now considered all possible cases of the Theorem of Desargues. We next consider the Theorem of Pappus, which may be used to prove the Theorem of Desargues when all points are on the same projective plane.

EXERCISES

*1. Prove case (ii) of Theorem 2–7.

2. Draw a figure for Theorem 2–7 when triangles ABC and $A'B'C'$ are on different planes and $A = A'$. Are the proofs of cases (i) and (ii) still valid?

3. Repeat Exercise 2 when $A = A'$ and $B = B'$.

*4. Prove Theorem 2–7 when the triangles are on the same plane.

2–11 Theorem of Pappus. We first consider a few definitions that will make possible a concise statement of the Theorem of Pappus. A *hexagon* may be defined as a simple plane six-line, i.e., as a figure consisting of six coplanar lines, no three of which are on the same point, and the six points obtained by taking the intersections of the given lines in cyclic order (Section 2–9). The lines are called *sides* and the points are called *vertices* of the hexagon. As in Fig. 2–28, it is intuitively evident that the three pairs of opposite sides of the

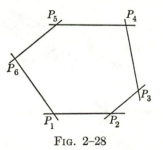

FIG. 2–28

hexagon $P_1P_2P_3P_4P_5P_6$ are P_1P_2 and P_4P_5, P_2P_3 and P_5P_6, and P_3P_4 and P_6P_1. These pairs of sides may also be determined directly from the symbol $P_1P_2P_3P_4P_5P_6$ for the hexagon by considering the points in cyclic order, taking the line through two consecutive vertices as one side, skipping one vertex, and taking the line through the next two consecutive vertices as the opposite side. This pairing of opposite sides may be indicated by placing bars on the symbol for the hexagon as follows:

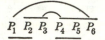

$$P_1 \ \overline{P_2 \ P_3} \ \overline{P_4 \ P_5} \ P_6$$

We define the points of intersection of the pairs of opposite sides of a hexagon to be *diagonal points* of the hexagon. Finally, we say that the hexagon is *inscribed* in a plane figure F if each vertex of the hexagon is on a line or point of F.

THEOREM 2–8. **THEOREM OF PAPPUS.** *If a hexagon is inscribed in two lines that are not sides of the hexagon, the diagonal points of the hexagon are collinear.*

We shall follow the method of proof in [16; 98]. Consider the hexagon $P_1P_2P_3P_4P_5P_6$ inscribed in the lines m and m' (Fig. 2–29), with diagonal points $S = P_1P_2 \cdot P_4P_5$, $T = P_2P_3 \cdot P_5P_6$, and $U = P_3P_4 \cdot P_6P_1$. Let $P_1P_6 \cdot P_4P_5 = A$, and $P_1P_2 \cdot P_4P_3 = B$. Then $P_4SAP_5 \overset{P_1}{\wedge} P_4BUP_3 \overset{P_2}{\wedge} CSUD$, where $SU \cdot P_2P_4 = C$ and $SU \cdot P_2P_3 = D$. Since S is a self-corresponding point of the projectivity $P_4SAP_5 \overline{\wedge} CSUD$, this projectivity is a perspectivity (Exercise 1, Section 2–7). Thus the lines $P_4C = m'$, $AU = P_1P_6$, and P_5D all

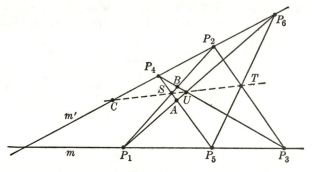

Fig. 2–29

pass through P_6, the center of the perspectivity. Then D is on P_5P_6, P_2P_3, and SU, whence $D = T$ (since $P_2P_3 \cdot P_5P_6$ is a unique point) and the diagonal points of the hexagon are collinear. This completes the proof of Theorem 2–8.

The proof of the Theorem of Desargues in Section 2–10 involved Postulate P–7, that is, the assumption that not all points are on the same plane. When Postulates P–7 and P–8 are replaced by the single postulate, say P–7′, stating that all points are on the same plane, there exist geometries satisfying Postulates P–1 through P–6 and P–7′ in which the Theorem of Desargues does not hold. In this case the Theorem of Pappus is often taken as a postulate, since on a plane the Theorem of Desargues is a consequence of Postulates P–1 through P–6, P–7′, and the Theorem of Pappus (Exercise 4).

EXERCISES

1. Draw a hexagon inscribed in two lines and find the line through the diagonal points.

2. Repeat Exercise 1 in euclidean geometry when the two given lines are parallel.

3. State the plane dual of the Theorem of Pappus and draw the corresponding figure.

4. Use the Theorem of Pappus to prove the Theorem of Desargues on a plane. (*Hint:* In Fig. 2–25, assume triangles $P_{13}P_{14}P_{15}$ and $P_{23}P_{24}P_{25}$ are perspective from the point P_{12}; define $A = P_{13}P_{15} \cdot P_{24}P_{25}$, $B = P_{13}P_{24}$ $P_{15}P_{25}$, $C = P_{13}P_{14} \cdot P_{12}A$, and $D = P_{23}P_{24} \cdot P_{12}A$; and apply the Theorem of Pappus to the three hexagons $P_{14}P_{15}P_{12}AP_{24}P_{13}$, $P_{23}P_{25}P_{12}AP_{13}P_{24}$, and $P_{24}AP_{13}CBD$.)

2–12 Conics. A conic is frequently defined in euclidean geometry as a plane section of a right circular cone. The nondegenerate conics, i.e., parabola, circle, ellipse, and hyperbola, are studied in analytic geometry. A single point and any two lines (coincident, intersecting, or parallel) may be considered as degenerate conics. In projective geometry we shall use the word conic to mean nondegenerate conic. This does not cause any difficulty since degenerate conics may be readily identified as above in terms of points and lines.

Let m and m' be distinct coplanar lines and suppose that there is given a projective transformation of the points of m onto the points of m'. By the definition of a perspective transformation, all lines determined by the pairs of corresponding points on m and m'

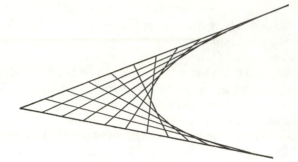

FIG. 2–30

pass through a single point P if and only if the projectivity is a perspectivity. By Exercise 1, Section 2–7, the projectivity is a perspectivity if and only if the point of intersection of the two lines is a self-corresponding point. This gives a very simple test for determining whether or not a projective correspondence between two lines is a perspectivity. If the correspondence between the points of m and m' is projective and not perspective, (i.e., a projective nonperspective correspondence), the lines joining corresponding points of m and m' form a line conic. Formally, we have the following definitions.

The set of all lines joining corresponding points of two projective nonperspective pencils of points that are on the same plane but not on the same line is called a *line conic* (Fig. 2–30). The set of all points of intersection of corresponding lines of two projective nonperspective pencils of lines on the same plane but not on the same point is called a *point conic* (Fig. 2–31). A line having exactly one point in common with a point conic is called a *tangent* of the conic. A *conic section* is the figure formed by a point conic and its tangents. A plane figure F is said to be *inscribed* in a point conic if its vertices (points) are points of the conic, *circumscribed* about the conic if its lines are tangents of the conic.

We shall, for convenience, refer to a point conic simply as a *conic*. Then a conic is defined, as above, as the set of all points of intersection of two projective nonperspective pencils of lines on the same plane but not on the same point. When the centers of the pencils of lines are A and B, the conic passes through both A and B as in Fig. 2–31. When the line AB is considered as a line of the pencil

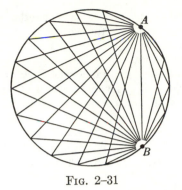

Fig. 2–31

of lines on A, the corresponding line of the pencil on B is the tangent to the conic at B. When the line AB is considered as a line of the pencil of lines on B, the corresponding line of the pencil on A is the tangent to the conic at A. If P is a variable point of the conic, then for each position of the point P the lines AP and BP are corresponding lines of the two pencils of lines. Accordingly, we may indicate the pencil of lines with center at A as the set of all lines $A[P]$, the pencil of lines with center at B as the set of lines $B[P]$, and the projectivity between the two pencils of lines as $A[P] \barwedge B[P]$.

THEOREM 2–9. THEOREM OF STEINER. *If A and B are any two given points of a conic, and P is a variable point of the conic, there exists a projective nonperspective correspondence $A[P] \barwedge B[P]$.*

The above paragraph constitutes a proof of this theorem when the points A and B are taken as the two points of the conic that are centers of the pencils of lines used to define the conic. In this sense the Theorem of Steiner states that a conic may be defined by pencils of lines having any two of its points as centers. We shall prove this statement in Section 4–10 when conics are considered from an algebraic point of view.

EXERCISES

1. Construct a projective nonperspective correspondence between the points of two lines.

2. Prove that a line in the plane of a conic is on at most two points of the conic.

3. Given five coplanar points, no three on the same line, construct a projectivity that will give a conic through the five given points.

4. Construct a conic when there are given
 (a) five points of the conic,
 (b) four points of the conic and a tangent at one of them,
 (c) three points of the conic and tangents at two of them.

5. In euclidean geometry, draw at least 20 lines of the form $ax + y\sqrt{1 - a^2} = 1$ (i.e., draw the lines for at least 20 values of a) and identify the conic associated with this set of lines.

6. Repeat Exercise 5 using the set of lines $ax + y/a = 1$.

7. Repeat Exercise 5 using the set of lines $2ax - y = a^2$.

2–13 Theorem of Pascal. The theory of conics is one of the most elegant and important aspects of classical projective geometry. We shall need some properties of conics in projective geometry not only as a basis for the conics of euclidean geometry but also as a basis for the development of the noneuclidean geometries from projective geometry. Most of the properties that we shall need will be developed from an algebraic point of view in Sections 4–9 and 4–10. We shall conclude the present chapter with a brief consideration of two famous theorems regarding conics.

THEOREM 2–10. THEOREM OF PASCAL. *If a hexagon is inscribed in a conic, its diagonal points are collinear.*

THEOREM 2–11. THEOREM OF BRIANCHON. *If a hexagon is circumscribed about a conic, the lines joining opposite vertices are concurrent.*

We first note that since the set of tangents of a point conic forms a line conic (Exercise 11, Section 4–9), the two theorems may be considered as plane duals and we need to prove only one of the theorems. We shall prove the Theorem of Pascal.

Let $P_1P_2P_3P_4P_5P_6$ be a simple hexagon inscribed in a conic. The diagonal points of the hexagon, i.e., the intersections of pairs of opposite sides, are $S = P_1P_2 \cdot P_4P_5$, $T = P_2P_3 \cdot P_5P_6$, and $U = P_3P_4 \cdot P_6P_1$ (Fig. 2–32). We shall apply the Theorem of Steiner to the pencils of lines $P_1[P]$ and $P_3[P]$ with centers at P_1 and P_3. The lines P_1P_2, P_1P_4, P_1P_5, and P_1P_6 of $P_1[P]$ intersect the line P_4P_5 in the points S, P_4, P_5, and $A = P_1P_6 \cdot P_4P_5$ respectively. The corresponding lines of $P_3[P]$ under the projectivity $P_1[P] \barwedge P_3[P]$ are P_3P_2, P_3P_4, P_3P_5, and P_3P_6 which intersect the line P_5P_6 in the points T, $D = P_3P_4 \cdot P_5P_6$, P_5, and P_6. The projectivity be-

tween the two pencils of lines
implies a projectivity

$$SP_4P_5A \barwedge TDP_5P_6,$$

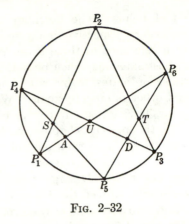

which is a perspectivity, since
P_5 is a self-corresponding point
(Exercise 1, Section 2–7). Thus
the lines ST, $P_4D = P_4P_3$, and
$AP_6 = P_1P_6$ are concurrent. The
point $U = P_3P_4 \cdot P_6P_1$, in other
words, is on the line ST and the diagonal points are collinear. This
proof of the Theorem of Pascal
is very similar to that of the

FIG. 2–32

Theorem of Pappus in Section 2–11. Indeed, the Theorem of Pappus
may be considered as a special case of the Theorem of Pascal when
the conic degenerates into two straight lines. We shall continue our
discussion of properties of conics in Sections 4–9 and 4–10.

EXERCISES

*1. Use the Theorem of Pascal to obtain a construction for a second point
on a conic on any secant line through one of five given points of the conic.
(*Note:* A secant line is a line that cuts the conic in two distinct points.)

2. Draw figures illustrating Theorems 2–10 and 2–11.

3. Prove that a conic may be determined by any five of its points.

2–14 Survey. We have developed many, but certainly not all, of
the properties of synthetic projective geometry. We have postulated
projective geometries having at least four points on each line and
proved many properties of these geometries. However, Postulates
P–1 through P–10 do not form a categorical set of postulates, since there
exist both finite projective geometries, such as the geometry with
four points on each line, and infinite projective geometries satisfying
these postulates. In Chapter 3 we shall adopt postulates of separation and continuity. The Postulates of Separation (Section 3–4)
imply that there are infinitely many points on any line and that
every line contains all points with rational coordinates relative to
any three points on the line. One categorical set of postulates may
then be obtained by restricting the projective space under consideration to the points with rational coordinates relative to a given

set of points. The Postulate of Continuity (Section 3–6) implies that every projective line contains all points having real coordinates with reference to three given points on the line. Categorical sets of postulates may then be obtained by considering only points with real coordinates, or by extending the sets of coordinates to complex and other number systems. We shall be primarily concerned with points having real numbers as coordinates.

In the present chapter we have defined projective figures and projective transformations, observed the principles of planar and space duality, considered complete and simple n-points, proved the Theorem of Desargues (a basic theorem of synthetic projective geometry), proved the Theorem of Pappus (another basic theorem), and considered some of the properties of conics. All these concepts are needed in our development of euclidean plane geometry from projective geometry; most of the concepts will be used in Chapter 3 as we continue our discussion of synthetic projective geometry and develop coordinate systems on projective spaces.

REVIEW EXERCISES

1. Draw two quadrilaterals that are perspective with respect to a point.

2. Draw a plane figure illustrating a point section of a line section of a complete quadrangle.

3. Draw two plane figures that are projective but not perspective.

4. Give the plane and space duals of each of the statements:

(a) Two figures F and G are perspective from a point P if the lines on pairs of corresponding points of F and G are all on P.

(b) Any three distinct points on a line m are projective with any three distinct points on a line m'.

5. Prove that two projective pencils of lines are perspective if their common line is invariant.

6. Prove that if two complete quadrangles are perspective from a line, they are perspective from a point.

7. Describe the figure obtained on a plane π by a plane section (by π) of a complete space four-point

(a) when π does not contain any of the vertices of the four-point;

(b) when π contains exactly two of the vertices of the four-point.

8. Construct a line conic.

9. Prove that in projective geometry any two conics are equivalent.

10. (a) Given any line AB, prove that there exists a line CD that does not intersect AB.

(b) Given any plane α, prove that every line CD intersects α.

CHAPTER 3

COORDINATE SYSTEMS

We now continue our development of properties of synthetic projective geometry and obtain coordinate systems for the points on projective lines, planes, and three-space. These coordinate systems can be based entirely upon geometric properties of projective spaces. We shall indicate how line sections of complete quadrangles may be used to obtain the points with rational coordinates. Then we shall use the reader's knowledge of the set of real numbers and postulate an isomorphism (Section 1–6) between the set of numbers in the extended real number system and a subset of the points on a projective line. In Chapter 4 we shall find that this isomorphism implies an isomorphism between the points of real synthetic projective geometry and the points of real analytic projective geometry.

3–1 Quadrangular sets. A complete quadrangle has been defined as the figure composed of four coplanar points (vertices), no three of which are on the same line, and the six lines (sides) determined by them. We now define a *quadrangular set of points* as any line section of a complete quadrangle by a line m that is not on any vertex of the quadrangle (Fig. 3–1). Since, by Postulate P–10, the diagonal points of a complete quadrangle are not collinear, every quadrangular set of points contains at least four distinct points. Also, since a quadrangular set of points cannot contain a vertex of the quadrangle, a quadrangular set contains exactly four distinct points if

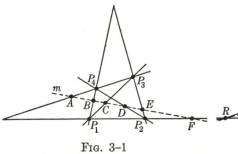

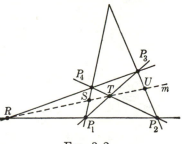

Fig. 3–1   Fig. 3–2

and only if the line m contains two diagonal points of the quadrangle, i.e., if and only if m is a side of the diagonal triangle of the quadrangle (Fig. 3–2). The section of a complete quadrangle by a side of its diagonal triangle is a very important special case of a quadrangular set of points and is called a *harmonic set of points* (Section 3–3). In the finite projective geometry with exactly four points on a line, all quadrangular sets are harmonic sets; i.e., the sides of the diagonal triangle are the only lines on the plane of the quadrangle that do not contain a vertex of the quadrangle. In Section 3–4, we shall use harmonic sets to postulate the existence of infinitely many points on a line.

The following definitions of quadrangular and harmonic sets of lines are plane duals of the above definitions of quadrangular and harmonic sets of points. A complete quadrilateral has been defined as the figure composed of four coplanar lines (sides), no three of which are on the same point, and the six points (vertices) determined by them. We now define a *quadrangular set of lines* as any point section of a complete quadrilateral by a point P that is not on any side of the quadrilateral (Fig. 3–3). Since, by the plane dual of Postulate P–10, the diagonal lines of a complete quadrilateral are not concurrent, every quadrangular set of lines contains at least four distinct lines. Also, since a quadrangular set of lines cannot contain a side of the quadrilateral, a quadrangular set contains exactly four distinct lines if and only if the point P is on two diagonal lines of the quadrilateral, i.e., if and only if P is a vertex of the diagonal triangle of the quadrilateral (Fig. 3–4). The section of a complete quadrilateral by a vertex of its diagonal triangle is called a *harmonic set of lines*.

Quadrangular sets will play a very important and precise role in the development of coordinate systems. Accordingly, we have a

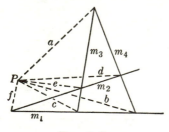

Fig. 3–3

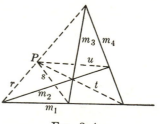

Fig. 3–4

very precise notation for quadrangular sets. We first consider quadrangular sets of points. We designate the complete quadrangle by its vertices $P_1P_2P_3P_4$ and attach significance to the order in which the vertices are given. In other words, even though the same figure (Fig. 3–1) is involved, there will be one notation for the quadrangular set of quadrangle $P_1P_2P_3P_4$ and another notation for the quadrangular set of quadrangle $P_2P_4P_1P_3$. The same set of points is involved in both cases, but it will be very convenient to use different notations. The pairing of the points corresponding to opposite sides of the quadrangle is the same in both cases. Indeed, although its importance may not be evident, this pairing of points is a fundamental relationship underlying many of the applications that we shall find for quadrangular sets.

Given any complete quadrangle $P_1P_2P_3P_4$, we shall consider in order the three lines P_1P_2, P_1P_3, P_1P_4 obtained by taking the first vertex P_1 with each of the others in turn. In Fig. 3–1, these three lines determine on m the ordered set of three points F, C, B which is called the *point triple* of the quadrangular set determined on m by the quadrangle $P_1P_2P_3P_4$. The remaining three points of the quadrangular set on m are on the sides of triangle $P_3P_4P_2$ and constitute the *triangle triple* of the quadrangular set. The six points are ordered in the symbol $Q(FCB, ADE)$ to show that F and A, C and D, B and E are on pairs of opposite sides of the given quadrangle. The quadrangular set of any complete quadrangle $P_1P_2P_3P_4$ consists of its point triple and its triangle triple. The point triple is always given first. The above description of the notation for a quadrangular set may be summarized as follows: The quadrangular set determined by a complete quadrangle $P_1P_2P_3P_4$ on a line m is designated by the symbol $Q(P_{12}P_{13}P_{14}, P_{34}P_{24}P_{23})$ where P_{jk} indicates the point of intersection of m with the line P_jP_k.

We now consider a few examples illustrating the above definition. It is highly recommended that the reader also work out several applications of the definition (Exercises 1 to 3). In Fig. 3–1, quadrangle $P_1P_4P_2P_3$ has quadrangular set $Q(BFC, EAD)$ on m, $P_3P_2P_1P_4$ has quadrangular set $Q(ECA, BDF)$, and $P_4P_1P_3P_2$ has quadrangular set $Q(BAD, EFC)$. In Fig. 3–2, quadrangle $P_1P_2P_3P_4$ has quadrangular set $Q(RTS, RTU)$ on m and $P_2P_4P_3P_1$ has quadrangular set $Q(TUR, TSR)$. Note that in each figure the pairing of the points on opposite sides is independent of the particular order in which the vertices of the quadrangle are considered. In each symbol for a quadrangular set of points the first and fourth, second and fifth, third and sixth points are on opposite sides of the quadrangle.

The notation for a quadrangular set of lines is very similar to that for a quadrangular set of points. We designate the complete quadrilateral by its sides $m_1m_2m_3m_4$ and attach significance to the order in which the sides are given. There will be a pairing of lines on opposite vertices of the quadrilateral. Given any complete quadrilateral $m_1m_2m_3m_4$, we shall consider in order the three points m_1m_2, m_1m_3, m_1m_4 obtained by taking the intersections of the first side m_1 with each of the others in turn. In Fig. 3–3, these three points determine with P the ordered set of three lines f, c, b which is called the *line triple* of the quadrangular set determined on P by the quadrilateral $m_1m_2m_3m_4$. The remaining three lines of the quadrangular set on P are on the vertices of the triangle $m_3m_4m_2$ and constitute the *triangle triple* of the quadrangular set. The six lines are ordered in the symbol $Q(fcb, ade)$ to show that f and a, c and d, b and e are on pairs of opposite vertices of the quadrilateral. The line triple is always given first. In each symbol for a quadrangular set of lines, the first and fourth, second and fifth, third and sixth lines are on opposite vertices of the quadrilateral.

A quadrangular set of points (or lines) may contain four (Figs. 3–2 and 3–4), five (Exercises 3 and 6), or six (Figs. 3–1 and 3–3) distinct points (or lines). In the next section we shall prove (Theorem 3–1) that, in our precise notation for a quadrangular set, whenever two quadrangular sets coincide in five of their corresponding points, they must coincide in their sixth points. This result is independent of the number of distinct points in the quadrangular set.

Exercises

1. Use Fig. 3–1 and indicate the quadrangular sets of each of the following quadrangles on m: $P_1P_3P_2P_4$, $P_2P_4P_1P_3$, and $P_4P_3P_2P_1$.

2. Repeat Exercise 1 for Fig. 3–2.

3. Draw a complete quadrangle $EFGH$ and give the quadrangular set of points on a line m that contains exactly one diagonal point of the quadrangle.

4. Use Fig. 3–3 and indicate the quadrangular sets of each of the following quadrilaterals on P: $m_1m_3m_2m_4$, $m_2m_4m_1m_3$, and $m_4m_3m_2m_1$.

5. Repeat Exercise 4 for Fig. 3–4.

6. State the plane dual of Exercise 3 and draw the figure.

3-2 Properties of Quadrangular Sets.   We now consider two important properties of quadrangular sets: the complete dependence of the sixth point on the other five (Theorem 3–1) and the invariance of quadrangular sets under projective transformations (Theorem 3–2). We first prove

THEOREM 3–1. FUNDAMENTAL THEOREM ON QUADRANGULAR SETS. *If two quadrangular sets coincide in five of their corresponding points, they are identical.*

Consider a quadrangle $P_1P_2P_3P_4$ with quadrangular set $Q(ABC, DEF)$ and another quadrangle $R_1R_2R_3R_4$ with quadrangular set $Q(ABC, DEG)$ as in Fig. 3–5. We are to prove that the points F and G coincide, that is, $F = G$. We note that under our precise definition of quadrangular sets the quadrangles must correspond point triple to point triple and triangle triple to triangle triple. The proof of the theorem consists of three applications of the Theorem of Desargues (Theorem 2–7).

Let m be the line containing the quadrangular sets. By hypothesis, triangles $P_1P_3P_4$ and $R_1R_3R_4$ are perspective from m, and therefore by the Theorem of Desargues the lines P_1R_1, P_3R_3, and P_4R_4 all pass through a point O.   Similarly, triangles $P_1P_2P_4$ and $R_1R_2R_4$

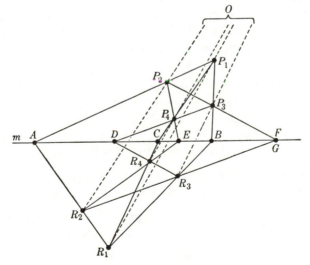

FIG. 3–5

are perspective from m, and P_2R_2 passes through $O = P_1R_1 \cdot P_4R_4$. Finally, triangles $P_1P_2P_3$ and $R_1R_2R_3$ are perspective from O and therefore from the line m through $A=P_1P_2 \cdot R_1R_2$ and $B=P_1P_3 \cdot R_1R_3$. Thus $P_2P_3 \cdot m = F$ and $R_2R_3 \cdot m = G$ must coincide, and $F = G$, as was to be proved. This proof is valid whether the quadrangular set has four, five, or six distinct points and whether the two quadrangles are on the same or different planes. When the quadrangles are assumed to be on the same plane, the statement and proof of the plane dual theorem are given as exercises (Exercises 1 and 2).

We may use Theorem 3–1 to construct the sixth point of any quadrangular set of points. For example, given the points A, B, C, D, and E of $Q(ABC, DEF)$ on a line m, we may construct the point F as follows:

(i) Select any point R_1 that is not on m.
(ii) Draw R_1A, R_1B, and R_1C.
(iii) Select any third point R_2 on R_1A.
(iv) Draw R_2E and let $R_2E \cdot R_1C = R_4$.
(v) Draw R_4D and let $R_4D \cdot R_1B = R_3$.
(vi) Draw R_2R_3 and let $R_2R_3 \cdot m = F$.

Then the line section of the complete quadrangle $R_1R_2R_3R_4$ by m is $Q(ABC, DEF)$. The plane dual of this construction may be used to construct the sixth line of a quadrangular set of lines. For example, given the lines a, b, c, d, and e of $Q(abc, def)$ on a plane π and a point P, we may construct the line f as follows:

(i) Select any line r_1 on π but not on P.
(ii) Label the points r_1a, r_1b, and r_1c.
(iii) On the plane π select any third line r_2 on r_1a.
(iv) Label the point r_2e and draw the line $r_4 = (r_2e) \cdot (r_1c)$.
(v) Label the point r_4d and draw the line $r_3 = (r_4d) \cdot (r_1b)$.
(vi) Label the point r_2r_3 and draw the line $f = (r_2r_3) \cdot P$.

Then the point section of the complete quadrilateral $r_1r_2r_3r_4$ by P is $Q(abc, def)$. These results will now be used to prove

THEOREM 3–2. *Any set of collinear points that is projective with a quadrangular set of points is a quadrangular set of points.*

Consider any quadrangular set of points $Q(ABC, DEF)$ on a line m and any point P that is not on m. We first construct a quadrangle

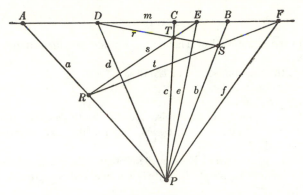

Fig. 3-6

having P as a first vertex and having the given quadrangular set of points. As above, let R be any third point on PA, $PB \cdot RF = S$ and $PC \cdot RE = T$ (Fig. 3-6). Then ST is on D by Theorem 3-1 and quadrangle $PRST$ has the given quadrangular set on m. We now consider the quadrilateral formed by the lines $RS = t$, $RT = s$, $ST = r$ and m. The quadrilateral $tsrm$ has quadrangular set $Q(abf, dec)$. Note that the positions of c and f have been interchanged relative to the given quadrangular set. This corresponds to an interchange of point and triangle triples, as one may verify by considering quadrangle $TSRP$ with quadrangular set $Q(DEC, ABF)$. The same interchange arises when we consider a line section of a quadrangular set of lines, i.e., the plane dual figure. Let any line m' that is not on P intersect the lines a, b, c, d, e, f in the points A', B', C', D', E', F' respectively. Then by the plane dual of the above procedure for obtaining $Q(abf, dec)$ from $Q(ABC, DEF)$, we may obtain a quadrangular set $Q(A'B'C', D'E'F')$ from $Q(abf, dec)$. Thus any six collinear points that are perspective with a quadrangular set of points form a quadrangular set of points. Theorem 3-2 then follows from the fact that any projectivity is equivalent to a finite sequence of perspectivities.

We have now proved that any quadrangular set of points is completely determined by five of the points (not necessarily distinct) and that quadrangular sets are invariant under projective transformations. In the next section we shall consider the special properties of quadrangular sets with exactly four distinct elements, i.e., harmonic sets.

1. State the plane dual of Theorem 3–1 when the quadrangles are in the same plane.

2. Prove the theorem stated in Exercise 1.

3. Select distinct points R, S, T, U, and V on a line and construct W satisfying $Q(RST, UVW)$.

4. State the plane dual of Exercise 3 and make the construction.

5. Select distinct points A, B, C, and D on a line and construct E satisfying $Q(ABC, ADE)$.

6. State the plane dual of Exercise 5 and make the construction.

7. Select distinct points A, B, and C on a line and construct D satisfying $Q(ABC, ABD)$.

8. State the plane dual of Exercise 7 and make the construction.

3–3 Harmonic sets. We have already defined (Section 3–1) a section of a complete quadrangle by a side of its diagonal triangle to be a harmonic set of points (Fig. 3–2), and a section of a complete quadrilateral by a vertex of its diagonal triangle to be a harmonic set of lines (Fig. 3–4). At present we are primarily concerned with properties of harmonic sets of points. The corresponding properties of harmonic sets of lines may be obtained using the principle of planar duality.

Let R, S, T, and U be any four coplanar points such that no three are on the same straight line. Let the diagonal points of the complete quadrangle $RSTU$ be A, B, and E as in Fig. 3–7. The section of the complete quadrangle $RSTU$ by the line AB gives a quadrangular set $Q(ABC, ABD)$, where $AB \cdot RU = C$ and $AB \cdot TS = D$. In general, any line section of a complete quadrangle by one of its diagonal lines (i.e., any harmonic set of points) corresponds to a quadrangular set of points of the form $Q(ABC, ABD)$. Conversely, if a quadrangular set of points $Q(ABC, ABD)$ is given on a line m, we may take any point that is not on the line m as R, take any third point on RA as S, let $RB \cdot SD = T$, let $RC \cdot SB = U$ (Fig. 3–7), and obtain a quadrangle $RSTU$ having the points of the given quadrangular set as a harmonic set of points. Thus given any quadrangular set of the form $Q(ABC, ABD)$, we may construct a quadrangle having the points of the quadrangular set as a harmonic set. We use the notation $H(AB, CD)$ to designate the harmonic set given by $Q(ABC, ABD)$. In the harmonic set $H(AB, CD)$, the point D is called *the harmonic conjugate of C with respect to A and B.*

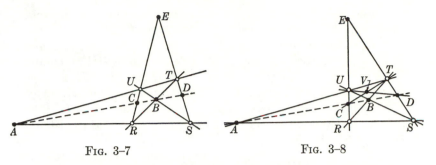

Fig. 3–7   Fig. 3–8

We now recall that our precise notation for the quadrangular set of a complete quadrangle depends upon the order in which the vertices are considered. In Fig. 3–7, quadrangle $RSTU$ has quadrangular set $Q(ABC, ABD)$ and quadrangle $SRUT$ has quadrangular set $Q(ABD, ABC)$. Thus the existence of $Q(ABC, ABD)$ implies the existence of $Q(ABD, ABC)$; that is, the existence of $H(AB, CD)$ implies the existence of $H(AB, DC)$. In other words, if D is the harmonic conjugate of C with respect to A and B, then C is the harmonic conjugate of D with respect to A and B. We indicate this relationship by writing

$$H(AB, CD) = H(AB, DC).$$

The two harmonic sets are equivalent in the sense that the existence of either one implies the existence of the other. The pair of points A,B and the pair C,D are called *conjugate pairs* of the harmonic set $H(AB, CD)$. The above relationship and the relationship

$$H(AB, CD) = H(BA, CD)$$

(Exercise 1) imply that the elements may be interchanged in either or both of the pairs of $H(AB, CD)$. We next prove that the conjugate pairs of a harmonic set may be interchanged; that is,

$$H(AB, CD) = H(CD, AB).$$

Given any harmonic set $H(AB, CD)$, as in Fig. 3–7, we wish to prove that there exists a harmonic set $H(CD, AB)$. In other words, given $Q(ABC, ABD)$, we wish to prove that we have $Q(CDA, CDB)$. If we draw TC and UD in Fig. 3–7 and let $TC \cdot UD = V$ (Fig. 3–8), then quadrangle $TVEU$ has quadrangular set $Q(CDA, CDB')$, where $EV \cdot CD = B'$. It now remains for us to prove that $B = B'$, that is, that RT, SU, and EV are concurrent. In order to do this, we look for two triangles inscribed in these three lines, such that their corresponding sides intersect on a line. Triangles RSE and TUV satisfy this condition, since $RS \cdot TU = A$, $SE \cdot UV = D$, and $RE \cdot TV = C$ on the line CD. Then, by the Theorem of Desargues, we know that RT, SU, and EV are concurrent and $B = B'$. This implies that we have $Q(CDA, CDB)$ from quadrangle $TVEU$. We have proved that the existence of $H(AB, CD)$ implies the existence of $H(CD, AB)$; that is,

$$H(AB, CD) = H(CD, AB).$$

This relationship and the two others considered above may be used to prove that the existence of one harmonic set implies the existence of seven other harmonic sets (Exercise 2).

The above properties of harmonic sets are based directly upon the special properties of the section of a complete quadrangle by a side of its diagonal triangle. We now consider properties of harmonic sets that are based upon the properties of quadrangular sets. When the Fundamental Theorem on Quadrangular Sets (Theorem 3–1) is applied to the set $Q(ABC, ABD)$, we have

THEOREM 3–3. *Given any three distinct collinear points A, B, C, the harmonic conjugate of C with respect to A and B may be constructed and is unique.*

This construction is a special case of the construction of the sixth point of a quadrangular set (Section 3–2) and may be outlined as follows:

(i) Select any point R_1 that is not on AB.
(ii) Draw R_1A, R_1B, and R_1C.
(iii) Select any third point R_2 on R_1A.
(iv) Draw R_2B and let $R_2B \cdot R_1C = R_4$.
(v) Draw R_4A and let $R_4A \cdot R_1B = R_3$.
(vi) Draw R_2R_3 and let $R_2R_3 \cdot AB = D$.

Then the line section by AB of the complete quadrangle $R_1R_2R_3R_4$ is $Q(ABC, ABD)$, and we have $H(AB, CD)$. The point D is unique by Theorem 3–1. This construction will be used frequently in our development of rational points on a line.

Theorem 3-2 may be restated for harmonic sets as follows:

Any set of collinear points that is projective with a harmonic set of points is a harmonic set of points.

This result is an immediate consequence of the definition of a harmonic set, the one-to-one correspondence of projective sets of points, and the proof of Theorem 3-2 (Exercise 7). When considered with Postulate P-9 and Theorem 2-3, the above result implies

Theorem 3-4. *Any two harmonic sets of points are projective. Any two harmonic sets of points having a common corresponding element are perspective.*

The details of the proof of this theorem are left as an exercise (Exercise 8).

We have considered the basic properties of harmonic sets in any geometry based upon Postulates P-1 through P-10. In such a geometry a line may contain only four points. In this special case all quadrangular sets are harmonic sets. We next use harmonic sets and make additional postulates that require infinitely many points on any line. In other words, we shall make additional assumptions and thereby limit the geometries under consideration to those in which a line has infinitely many points.

Exercises

1. Use Fig. 3-7 and prove that $H(AB, CD)$ implies $H(AB, DC)$, $H(BA, CD)$, and $H(BA, DC)$.

*2. Prove that the existence of $H(AB, CD)$ implies the existence of eight (including the given) harmonic sets.

3. State the eight harmonic sets implied by $H(FG, MN)$.

4. Repeat Exercise 3 for a harmonic set of lines $H(ab, cd)$.

5. Given three concurrent lines a, b, c, define and construct the harmonic conjugate of c with respect to a and b.

*6. State and give a direct proof of the plane dual of Theorem 3-3.

*7. Prove that harmonic sets correspond to harmonic sets under projective transformations.

*8. Prove Theorem 3-4.

9. Consider a number line in euclidean geometry having points P_x with coordinates x. Construct points P_x satisfying each of the following:

(a) $H(P_0P_2, P_3P_x)$;   (b) $H(P_1P_{-1}, P_2P_x)$;
(c) $H(P_1P_{-1}, P_{-2}P_x)$;   (d) $H(P_0P_4, P_3P_x)$;
(e) $H(P_0P_2, P_1P_x)$;   (f) $H(P_2P_4, P_3P_x)$.

3–4 Postulates of Separation. The concept of separation is at least vaguely familiar to everyone. Two countries may be separated by a river or mountain range on their common boundary. In euclidean geometry a point separates or divides a line into two parts; a line separates a plane; and a plane separates three-dimensional space. In this section we shall accept *separation* as an undefined relation and introduce three postulates that will imply that every line contains infinitely many points.

Given any four distinct points A, B, C, D on a line, there exists by Theorem 2–3 a projectivity $ABC \barwedge ABD$. Since D may be any point on the line distinct from A and B, it is clear that the order relations on a line in euclidean geometry are not preserved in projective geometry, i.e., under projective transformations. Intuitively, order relations on a real projective line are very similar to order relations on a simple closed curve (Section 9–2) in euclidean geom-

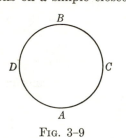

FIG. 3–9

etry (Fig. 3–9). Thus the phrase "A lies between B and C" has no meaning in projective geometry, and we must consider relations among four rather than three points on a projective line. Note that the existence of at least four points on every line is a consequence of Postulate P–10.

The order relations on a projective line will now be discussed using the intuitive representation of a projective line as a simple closed curve (Fig. 3–9). A single point A does not separate the line. Two points A, B separate the line into two segments. Thus the "segment AB" does not have a unique meaning on a projective line. Any point C that is on the line and is distinct from A and B must be on one of the segments determined by A and B. Thus we may consider either the segment AB that contains C or the segment AB that does not contain C. Given any four distinct points on a projective line, we may visualize the pair A, B as separated by the pair C, D, if C and D are on different segments determined by A and B. The order of the elements in the pair has no significance. These intuitive concepts are given to help the reader visualize the undefined relation, separation.

POSTULATES OF SEPARATION

P–11: *The pairs of a harmonic set $H(AB, CD)$ or $H(ab, cd)$ separate each other.*

P–12: *If the pairs A, B and C_1, D separate each other and the pairs A, C_1 and C_2, B separate each other, then the pairs A, B and D, C_2 separate each other.*

P–13: *If the pairs A, B and C, D separate each other, then A, B, C and D are distinct points.*

Separation may be used to define a line segment. Given three points A, B, D on a line, we adopt the notation of [4; 24] and define a *line segment AB/D* (read as "AB not containing D") to be the set of all points C on the line AB such that A, B and C, D separate each other.

We now prove that in any geometry satisfying Postulates P–1 through P–13 there are infinitely many points on every line. Let m be an arbitrary line (Postulate P–4) with distinct points A, B, D (Postulate P–5). We first construct (Theorem 3–3) a point C_1 satisfying $H(AB, DC_1)$, as in Fig. 3–10, where as usual only segments of the projective lines are shown. By Postulate P–11 and the definition of a line segment the point C_1 is on the segment AB/D. Next we construct C_2 on AC_1/B satisfying $H(AC_1, BC_2)$ and, in general, we construct C_j on AC_{j-1}/C_{j-2} satisfying $H(AC_{j-1}, C_{j-2}C_j)$. This gives a construction for a sequence of points C_1, C_2, $C_3, \ldots, C_j, \ldots$ on m. It remains for us to prove that the points of the sequence are distinct.

Consider the sequence of harmonic sets $H(AB, DC_1)$, $H(AC_1, BC_2)$, $H(AC_2, C_1C_3), \ldots, H(AC_{j-1}, C_{j-2}C_j), \ldots$. In each of these harmonic sets the conjugate pairs separate each other by P–11, and the four points are distinct by P–13 or P–10. We now wish to prove that under the above method of construction, the points of the set C_1, C_2, $C_3, \ldots, C_j, \ldots$ are all distinct. It will be sufficient to prove that for an arbitrary positive integer j the points C_j and C_b are distinct for all positive integers $b < j$. This result may be proved by using mathematical induction upon the positive integers $j - b$. If $j - b = 1$, then $b = j - 1$, and $H(AC_{j-1}, C_{j-2}C_j)$ may be expressed as $H(AC_b, C_{j-2}C_j)$, whence by P–11 and P–13 the points C_j and C_b are distinct. If $j - b = 2$, it is necessary to consider two harmonic sets $H(AC_j, C_{j-1}C_{j+1})$ and $H(AC_{j-1}, C_{j-2}C_j)$. We first use P–11 to prove that the pair A, C_j separates the pair C_{j-1}, C_{j+1} and the pair A, C_{j-1} separates the pair C_{j-2}, C_j. Then by P–12, the pair A, C_j separates the pair C_{j-2}, C_{j+1}, and by P–13 the points C_j and $C_{j-2} = C_b$ are distinct. If $j - b = 3$, we use three harmonic sets, whence the pairs A, C_j and C_{j-1}, C_{j+1} separate each other, the pairs A, C_{j-1} and C_{j-2}, C_j separate each other, and the pairs A, C_{j-2} and C_{j-3}, C_{j-1} separate each other. Then we apply P–12 to the last two sets of pairs to prove that the pairs A, C_{j-1} and C_{j-3}, C_j separate each other. Again we apply P–12 to this result and to the pairs from the first of the three harmonic sets to prove that the pairs A, C_j and C_{j-3}, C_{j+1} separate each other, whence by P–13 the points C_j and $C_{j-3} = C_b$ are distinct. If $j - b = 4$, we use four harmonic sets and apply P–12

    (i) to the pairs of the last two sets,

    (ii) to the result obtained in (i) and the pairs in the second harmonic set, and

    (iii) to the result obtained in (ii) and the pairs in the first harmonic set.

For any positive integer $j > 2$ and any positive integer $b < j$, $j - b$ is a positive integer, say k, and k harmonic sets are used. The method of proof of the desired result has now been established; the details are left as an exercise (Exercise 1).

The significance of Postulates P–10 through P–13 may be summarized briefly as follows: Postulate P–10 is needed to prove that there exists a harmonic set of four points. Postulate P–11 states that the conjugate pairs of a harmonic set separate each other. Postulate P–12 makes possible the construction of an infinite sequence of points C_j such that the pairs A, B and D, C_j separate each other for all positive integers j. Postulate P–13 then enables us to prove that under the construction used, the C_j are distinct for all j. Thus we may now associate distinct points on the line with each positive integer j and may say that there are infinitely many points on the line. If one line has infinitely many points, then every line has infinitely many points (Exercise 2). The points C_1, C_2, ... as constructed for Fig. 3–10 using harmonic sets are said to be harmonically related to the points A, B, D. This concept will be used in the definition of nets of rationality (Section 3–5).

Exercises

*1. Prove that C_j and C_b are distinct in the sequence $C_1, C_2, C_3, \ldots$, constructed as in Fig. 3–10, whenever

(a) $j - b = 1$;   (b) $j - b = 3$;   (c) $j - b = 4$;

(d) b is any positive integer less than j and j is any positive integer greater than 1.

*2. Prove that if a single line m contains infinitely many points, then every line r contains infinitely many points

(a) when m and r are coplanar;

(b) when m and r are not coplanar.

3–5 Nets of rationality. Nets of rationality are closely related to sets of points with rational coordinates, i.e., sets of rational points. After coordinate systems have been introduced, we shall find that a net of rationality on a line corresponds to a set of rational points on a line, a net of rationality on a plane corresponds to a set of rational points on a plane, and a net of rationality in space corresponds to a set of rational points in space. The definitions of nets of rationality on a line, on a plane, and in space depend upon the concept of harmonically related points.

We have observed that the points $C_1, C_2, C_3, \ldots$ are harmonically related to the points A, B, and D in Fig. 3–10. In general, a point P is said to be *harmonically related* to three given distinct points A, B, C on a line containing P if there exists a sequence $A, B, C, E_1, E_2, E_3, \ldots$ such that $P = E_k$ for some finite positive integer k; the relation $H(AB, CE_1)$ holds; and for all E_j we have a relation of the form $H(RS, TE_j)$ where R, S, and T are elements of the set A, B, $C, E_1, E_2, \ldots, E_{j-1}$. The points E_j need not all be distinct, since

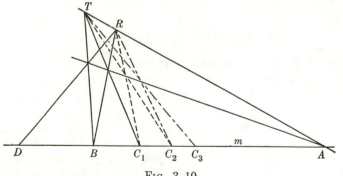

Fig. 3–10

these harmonic sets need not arise by a systematic procedure. For example, in Fig. 3–10 we could take $C = D$, $E_1 = C_1$, $E_2 = C_2$, and $E_3 = C_1 = E_1$. However, since the points C_j of Fig. 3–10 are all distinct and are all harmonically related to the points A, B, D, the set of all points harmonically related to any three distinct collinear given points is an infinite set of points. In general, on any line m the set of all points harmonically related to three distinct collinear points A, B, and C is called the *linear net of rationality* determined by A, B, and C and is denoted by $R(ABC)$.

The order of the points A, B, C has no significance in the symbol $R(ABC)$, since the points may be selected in any order for the harmonic sets used to determine the points of $R(ABC)$. Furthermore, we have

THEOREM 3–5. *A net of rationality on a line is completely determined by any three distinct points of the net.*

In other words, if S, T, and U are any three distinct points of $R(ABC)$, then $R(STU) = R(ABC)$. We shall prove Theorem 3–5 in this form. By the definition of $R(ABC)$, any point T in $R(ABC)$ may be obtained from A, B, and C, using a finite number of harmonic sets. At the first step of this process we obtain a point such as D, where $H(AB, CD) = H(AB, DC)$. This property of the harmonic sets implies that the points harmonically related to A, B, and C are precisely the points that are harmonically related to A, B, and D; that is, $R(ABC) = R(ABD)$. Since the order of the points in the symbol $R(ABC)$ has no significance, this result may be applied a finite number of times to obtain first

$$R(ABC) = R(ABT)$$

and then

$$R(ABC) = R(ABT) = R(ATS) = R(STU).$$

We next consider nets of rationality on a plane and in space. The construction of these nets is very similar to the determination of rational points on a plane and in space. Let $R(ABC)$ be a linear net of rationality on a line m. Let D be any point that is not on m, and let E be a third point on the line AD. The set of all points of intersection of lines joining points of $R(ABC)$ to points of $R(ADE)$ forms a *planar net of rationality*. From a slightly different point of view, a planar net of rationality is completely determined by any four points A, C, G, D on a plane such that no three of the points are on the same line. The linear nets of rationality considered above are obtained by taking $B = DG \cdot AC$ and $E = AD \cdot CG$. This determination of the planar net enables us to rephrase our definitions and thereby clarify the extension of nets of rationality from a line to a plane and to space. The new definitions will give more emphasis to some of the basic properties of nets of rationality.

A linear net of rationality is completely determined by any three points on a line such that no two are on the same point. It consists of the totality of points that are harmonically related to the given points. A planar net of rationality is completely determined by any four points A, C, G, D on a plane such that no three are on the same line. The points A, C, and $B = DG \cdot AC$ determine a linear net of rationality $R(ABC)$; the points A, D, and $E = AD \cdot CG$ determine a linear net of rationality $R(ADE)$. The planar net of rationality consists of the totality of points of intersection of lines determined by points of\ these two linear nets of rationality. A *spatial net of rationality* is completely determined by any five points A, C, D, F, H in space such that no four are on the same plane. Let the intersection of the line AF and the plane CDH be J, the intersection of the line FH and the plane ACD be G. Then the points A, C, G, and D determine a planar net as before, with linear nets $R(ABC)$ and $R(ADE)$. Also we may determine $R(AJF)$. The spatial net of rationality consists of the totality of points of intersection of planes determined by points of these three linear nets of rationality.

The definitions of linear, planar, and spatial nets of rationality provide a basis for the determination of rational points and the postulation of real points (Section 3-6). These nets of rationality and their properties depend upon all of the postulates and most of the definitions and theorems that we have considered in projective geometry. In one sense, the nets of rationality represent a "peak" in our development of synthetic projective geometry, and we now prepare for a change in the emphasis of our treatment of projective geometry. We are ready to associate points with numbers (coordinates). In the next section we shall consider the relationship between points and their coordinates and use this relationship to postulate the existence of real projective geometry. Then in the last two sections of this chapter we shall consider two types of coordinates — nonhomogeneous coordinates and homogeneous coordinates.

EXERCISES

*1. Prove that if $ABCD \barwedge STUV$ and D is a point of $R(ABC)$, then V is a point of $R(STU)$.

2. Prove that any set of collinear points that is projective with a linear net of rationality is itself a net of rationality.

*3. Prove that any two linear nets of rationality are projective.

*4. Prove that any two planar nets of rationality on the same plane are projective.

5. Prove that a planar net of rationality is completely determined by any four points of the net such that no three of the points are on the same line.

6. Select three collinear points A, B, C and construct six distinct points of $R(ABC)$.

3–6 Real projective geometry. We have seen that there exists a geometry satisfying Postulates P–1 through P–8 in which each line contains exactly three points and that if there are two such geometries in which each line contains exactly three points, then the geometries are isomorphic (Section 1–6). Similarly, there exist projective geometries satisfying Postulates P–1 through P–10 in which each line contains exactly four points and any two such geometries are isomorphic (Section 1–6).* Also there exist projective geometries satisfying Postulates P–1 through P–13 in which each line contains an infinite set of points; i.e., each line contains at least a set of points C_j that may be associated with the positive integers j such that $C_j \neq C_k$ when $j \neq k$ (Section 3–4). However, there exist such infinite projective geometries that are not isomorphic; i.e., Postulates P–1 through P–13 do not form a categorical set of postulates. We shall distinguish between infinite projective geometries that are not isomorphic by distinguishing between the lines of the geometries.

In euclidean geometry we have the following theorem:

CANTOR-DEDEKIND THEOREM: *There exists an isomorphism between the set of real numbers and the set of points on a line in euclidean geometry.*

---

* It should be noted that given any positive integer $n \geq 3$, it is not always possible to assert that there exists a projective geometry in which each line contains exactly n points and that any two such projective geometries are isomorphic. For some integers n (such as $n = 7$) there does not exist a projective geometry, for other integers (such as $n = 9$) there exist several different projective geometries, and for still other integers (such as $n = 11$) it is not yet known whether or not projective geometries exist.

This isomorphism enables us to use the set of real numbers as coordinates of the points on a euclidean line. In any geometry, an isomorphism between a set of points and a set of numbers gives rise to a coordinate system for the set of points. Any coordinate system for the points on a line may be used

   (i) to identify the points on the line;
   (ii) to establish coordinate systems for the points of the corresponding plane and space geometries;
   (iii) to obtain and to describe properties of the geometry.

Two reference points are needed to develop a coordinate system on a line in euclidean geometry. It is customary to select the origin with coordinate 0 and the unit point with coordinate 1 as reference points. However, formally any two distinct points with coordinates r and s $(r \neq s)$ may be selected. Different selections of the origin and unit point give rise to isomorphic coordinate systems on the line.

In projective geometry we have seen that any three distinct points A, B, C on a line determine a linear net of rationality $R(ABC)$. We shall consider the establishment of a coordinate system on a projective line in terms of the association of coordinates with the points of $R(ABC)$. The given points A, B, C will be called the *reference points* of the coordinate system. In euclidean geometry the reference points are the origin and the unit point. In projective geometry the reference points are called the "origin," the "unit point," and the "*ideal point*" on the line. The origin has coordinate 0 and the unit point has coordinate 1, as in euclidean geometry. The ideal point has as its coordinate the symbol ∞, which is a number but not a real number. The set of numbers composed of the symbol ∞ and the set of real numbers is called the *extended real number system*. In this number system the symbol ∞ has the following properties:

$$\infty + a = \infty, \quad \infty - a = \infty, \quad a - \infty = \infty, \quad \infty/a = \infty, \quad a/\infty = 0$$

whenever $a \neq \infty$, and the properties

$$\infty \cdot a = \infty, \quad a/0 = \infty$$

whenever $a \neq 0$. The operations indicated by

$$\infty + \infty, \quad \infty - \infty, \quad \infty/\infty, \quad \infty \cdot 0, \quad 0/0$$

are undefined (have no meaning) in the extended real number

system.   Note that we have defined a single number ∞ with the property $a + \infty = a - \infty$ (intuitively $+\infty = -\infty$), whereas in some branches of mathematics $+\infty$ and $-\infty$ are distinct numbers.

The above process of accepting a new number symbol and stating its properties should be familiar to you.   For example, after working with positive integers you probably accepted the number symbol 0 as a new number with the properties

$$a + 0 = a, \quad a - 0 = a, \quad 0 - a = -a, \quad a \cdot 0 = 0,$$

where a is any positive integer or zero, and the property

$$0/a = 0,$$

where $a \neq 0$ is any positive integer.   The operation indicated by $a/0$ was undefined in each subset of the real numbers.   Modifications of the above process may also be used to obtain the negative numbers, rational numbers, real numbers, and complex numbers [10; 20–43].

On any line in euclidean geometry there exists a coordinate system based upon any two given reference points, and we may designate the points on the line as P_x where x is any real number.   On any line in a geometry satisfying Postulates P–1 through P–13 there exist three distinct points A, B, C, and these points determine a linear net of rationality $R(ABC)$.   As in euclidean geometry, we may use subscripts to indicate the coordinates of points, designate A as the ideal point P_∞, designate B as the origin P_0, and designate C as the unit point P_1.   Then the relation $H(AC_1, BC_2)$ used in Fig. 3–10 corresponds to the relation $H(P_\infty P_1, P_0 P_2)$ and may be used to define the point P_2.   Similarly, P_3 may be determined by the relation $H(P_\infty P_2, P_1 P_3)$, and, in general, P_j may be determined by the relation $H(P_\infty P_{j-1}, P_{j-2} P_j)$ for any positive integer $j > 1$.   The relationship among a, b, and c in the harmonic sets $H(P_\infty P_a, P_b P_c)$ should be clear above, since there is a common pattern $H(P_\infty P_{j-1}, P_{j-2} P_j)$.   In general, we shall find in Section 5–3 that P_a is the mid-point of the line segment $P_b P_c$ whenever we have $H(P_\infty P_a, P_b P_c)$. This property underlies our selection of harmonic sets to determine points with integral coordinates.   For the positive integral points, P_1 is the mid-point of $P_0 P_2$, P_2 is the mid-point of $P_1 P_3$, and, in general, P_{j-1} is the mid-point of $P_{j-2} P_j$.   In the next paragraph we shall use the fact that P_0 is the mid-point of $P_{-j} P_j$ to determine negative integral points.

The above determination of the positive integral points P_j relative to the given reference points is precisely equivalent to the determination of the points C_j in Section 3–4. Thus the positive integral points are harmonically related to the reference points and are in the linear net of rationality $R(P_\infty P_0 P_1)$. However, there are also many other points in this linear net of rationality. The negative integral point P_{-k} may be determined by harmonic sets of the form $H(P_\infty P_0, P_k P_{-k})$. The points with fractional coordinates may be obtained using harmonic sets of the form $H(P_1 P_{-1}, P_b P_{1/b})$, $H(P_\infty P_{(a-1)/b}, P_{(a-2)/b} P_{a/b})$, and $H(P_\infty P_0, P_{a/b} P_{-a/b})$. If we interpret these harmonic sets as defining the points P_j, P_{-k}, $P_{1/b}$, $P_{a/b}$, and $P_{-a/b}$, we then find that for any given rational number a/b there exists a point $P_{a/b} \neq P_\infty$ in $R(P_\infty P_0 P_1)$. Later (Exercises 9 and 10, Section 4–11) we shall be able to use algebraic methods to prove this statement and its converse: If $P \neq P_\infty$ is a point of $R(P_\infty P_0 P_1)$, then $P = P_{a/b}$ for some rational number a/b. Both of these statements could be proved using methods now at our disposal, but such proofs would be long and tedious. We do not emphasize the proofs here because our forthcoming postulation of the existence of points with real numbers as coordinates will also imply the existence of points with rational numbers as coordinates.

We have now observed that under suitable definitions of $P_{a/b}$, Postulates P–1 through P–13 imply that on any line m there exists a linear net of rationality $R(P_\infty P_0 P_1)$, consisting of the point P_∞ and a set of points $P_{a/b}$ defined with reference to the points P_∞, P_0, and P_1. Since, by Theorem 3–5, $R(P_\infty P_0 P_1) = R(P_a P_b P_c)$ for any three distinct points P_a, P_b, P_c of $R(P_\infty P_0 P_1)$, any line m contains at least the totality of points harmonically related to any three of its points. In particular, any line m contains at least three distinct points, say P_∞, P_0, P_1, and the points of $R(P_\infty P_0 P_1)$. There may or may not be other points on m. If all points of m are points of $R(P_\infty P_0 P_1)$, that is, if each point $P \neq P_\infty$ on m is a point $P_{a/b}$ of $R(P_\infty P_0 P_1)$, then the line m is called a *line of rational points* and the corresponding geometry is called a *rational projective geometry*.

This definition of a rational projective geometry as a projective geometry having at least one line of rational points is justified, since, as we now prove,

THEOREM 3–6. *If a projective geometry has at least one line of rational points, then every line of the geometry is a line of rational points.*

Note that this theorem does not imply the existence of a line of rational points (for the present based upon the harmonic sets asserted above). Rather, the theorem states that if in a projective geometry (Postulates P-1 through P-13) there exists at least one line m consisting of exactly the points $P_{a/b}$ and P_∞ with reference to three of its points, then all lines of the geometry have this property.

Let m be a line of rational points with distinct points A, B, C. Then m contains the points of $R(ABC)$ and no other points. If s is any line distinct from m and coplanar with m, then s and m intersect in a point, say D. Let $M \neq D$ be a point of m, $S \neq D$ be a point of s, and O be a third point on the line MS. We may define points A', B', C' on s by the perspectivity

$$ABC \stackrel{O}{\wedge} A'B'C'.$$

Then every point E on m corresponds to a point E' on s such that E' is in $R(A'B'C')$ by Exercise 1, Section 3–5. Furthermore, if F' is any point on s, then by Postulate P–3 the line OF' intersects m in a point, say F; F is in $R(ABC)$, since all points of m are in $R(ABC)$; and, again by Exercise 1, Section 3–5, F' is in $R(A'B'C')$. This completes the proof of Theorem 3–6 for lines that are coplanar with m. We next use this result to prove the theorem for any line t of the geometry. Let M be any point of m and let $T \neq M$ be a point of t. The line TM is coplanar with m (any two intersecting lines are coplanar) and therefore is a line of rational points. Similarly, the line t is coplanar with the line of rational points TM and therefore is itself a line of rational points. This completes the proof of Theorem 3–6.

We now consider methods for obtaining lines of real projective geometry. As in algebra, the extension from rational elements to real elements requires a new approach. In algebra the rational numbers may be obtained from the integer 1 by using the rational operations $(+, -, \cdot, \div)$. In geometry the rational points may be obtained from the reference points by using either harmonic sets (as above) or projectivities [16; 141–157]. In algebra the irrational numbers, and therefore the set of real numbers containing them, cannot be obtained from the integer 1 by using rational operations. An additional assumption such as the existence of all infinite decimals (nonrepeating as well as repeating) as numbers is needed. Still another assumption is needed to obtain the set of complex numbers. Similarly, in geometry we need additional postulates to assure us of the existence of the set of real points and the set of complex points on a line. These postulates may be stated either in terms of geometric properties of the points or in terms of properties of the coordinates of the points. Since the geometric properties correspond precisely to the properties of numbers used to obtain the sets of real and complex numbers from the set of rational numbers [10; 24–34, 40–42], we shall take advantage of the reader's knowledge of the real number system and postulate the existence of a line of real points as follows:

POSTULATE OF CONTINUITY

P–14: *There exists a projective line m containing a set of points P_r isomorphic with the set of numbers of the extended real number system.*

Postulate P–14 implies that there exists at least one line m containing points P_r corresponding to the numbers r (in the extended real number system) with reference to any three points P_∞, P_0, and P_1 on the line. If the line m contains exactly the points P_r (i.e., these points and no others), then m is called a *line of real points* and the corresponding geometry is called a *real projective geometry*. Theorem 3–6 may be restated for real projective geometries. The existence of lines of complex points and a complex projective geometry may be similarly postulated and defined (Exercise 5). We are primarily interested in real projective geometry as a basis for euclidean geometry.

The word "continuity" is used in Postulate P–14 to indicate that the line contains points with irrational coordinates as well as points with rational coordinates. In terms of geometric properties of the euclidean plane, this implies intuitively that any segment of a continuous curve joining points on opposite sides of a straight line must contain a point of the line. For example, given a circle of radius $\sqrt{2}$ and with center at the origin, the minor arc joining the points $(1,1)$ and $(1,-1)$ intersects the x-axis at the point $(\sqrt{2},0)$ when (as in euclidean geometry) the x-axis is a line of real points. The arc does not intersect the x-axis when the axis is considered as a line of rational points.

Postulate P–14 asserts the existence of a line of real points and therefore of real projective geometry. A categorical set of postulates for real projective geometry may be obtained by postulating that there exists a line consisting exactly of the set of points P_r and taking this postulate with P–1 through P–13 (Exercise 4). Postulates P–1 through P–14 are satisfied by both real projective lines and complex projective lines. The remaining sections of this chapter are concerned with coordinate systems for lines, planes, and space in real projective geometry.

EXERCISES

1. Given a projective line with reference points P_∞, P_0, and P_1, use
   (a) $H(P_\infty P_1, P_0 P_2)$ and determine P_2,
   (b) $H(P_\infty P_2, P_1 P_3)$ and determine P_3,
   (c) $H(P_\infty P_3, P_2 P_4)$ and determine P_4,
   (d) $H(P_\infty P_4, P_3 P_5)$ and determine P_5.

2. Determine P_2 and P_3 as in Exercise 1 and then determine P_{-1}, P_{-2}, and P_{-3} using the harmonic sets $H(P_\infty P_0, P_k P_{-k})$ for $k = 1, 2$, and 3.

3. Determine P_2, P_3, and P_{-1} as in Exercise 2 and then determine $P_{1/2}$, $P_{1/3}$, $P_{2/3}$ and $P_{-2/3}$ using harmonic sets $H(P_1 P_{-1}, P_2 P_{1/2})$, $H(P_1 P_{-1}, P_3 P_{1/3})$, $H(P_\infty P_{1/3}, P_0 P_{2/3})$, and $H(P_\infty P_0, P_{2/3} P_{-2/3})$ respectively.

4. Give a set of postulates for (a) rational projective geometry, and (b) real projective geometry.

5. Give a set of postulates for complex projective geometry.

3–7 Nonhomogeneous coordinates. The isomorphism between the points P_r on a real projective line and the numbers r of the extended real number system associates each point P_r with a unique number r, called the coordinate or the *nonhomogeneous coordinate* of the point P_r. The second terminology is preferable because it emphasizes the distinction between these coordinates and those that we shall introduce in the next section. Nonhomogeneous coordinates on a real projective line have the advantage of associating each point on the line with a unique number and the disadvantage of requiring the use of the symbol or number ∞ with special properties under addition, subtraction, multiplication, and division. In the next section we shall remove this disadvantage by associating each point with a class of number pairs called the homogeneous coordinates of the point. In the present section we shall endeavor to extend our nonhomogeneous coordinate system on a line to obtain coordinate systems for projective planes and projective three-space. We shall find another disadvantage of nonhomogeneous coordinates that can also be removed by using homogeneous coordinates.

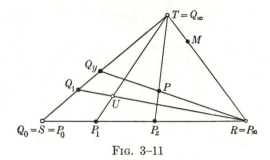

Fig. 3–11

The following procedure stated for real projective geometry may be readily adapted for rational and complex projective geometries. The set of numbers in the extended real number system consists of the set of real numbers and the number ∞. After the reference points have been selected, any point on a real projective line has a unique number of the extended real number system as its nonhomogeneous coordinate. One point has the number ∞ as its coordinate; all other points have real numbers as their nonhomogeneous coordinates. In other words, all points but one on a real projective line have real numbers as their nonhomogeneous coordinates. After one point has been deleted, the remaining set of points on a real projective line is isomorphic with the set of real numbers and, as we shall find, is isomorphic with the set of points on a line in euclidean geometry.

Consider the real projective plane determined by three non-collinear points R, S, T. Let U be a point that is on the plane but is not on any of the lines RS, ST, RT. We may set up nonhomogeneous coordinate systems P_x and Q_y on the lines RS and ST respectively by taking $P_0 = S = Q_0$, $P_\infty = R$, $Q_\infty = T$, $P_1 = TU \cdot RS$, and $Q_1 = RU \cdot ST$ (Fig. 3–11). Then any point P which is on the plane RST but not on the line RT may be uniquely associated with an ordered pair of real numbers (x, y) where $TP \cdot RS = P_x$ and $RP \cdot ST = Q_y$. If we endeavor to extend this procedure to include points M that are on the line TR but distinct from T and R, we find that for all such points $TM \cdot RS = P_\infty$ and $RM \cdot ST = Q_\infty$; that is, all the points M distinct from T and R on the line TR correspond (as above) to the same pair of numbers (∞, ∞). In other words, we cannot extend our nonhomogeneous coordinates on a projective line to obtain unique pairs of real numbers as nonhomogeneous coordinates for all points on a real projective plane. We can obtain pairs of real numbers as coordinates for all points on a real projective plane that are not on a particular line. The exceptional or *ideal line* depends upon the choice of reference points for the coordinate system. After one line has been deleted, the remaining set of points on a real projective plane is isomorphic with the set of ordered pairs of real numbers and also with the set of points on a euclidean plane. The four points R, S, T, and U that were selected as reference points determine a planar net of rationality, and our coordinate system for the plane may be based upon the concepts that were used in developing a planar net of rationality.

Any five points in real projective three-space such that no four are on the same plane determine a spatial net of rationality. The five points may be taken as P_∞, Q_∞, R_∞, $O = P_0 = Q_0 = R_0$, and U such that the plane $UP_\infty Q_\infty$ intersects the line OR_∞ in R_1, the plane $UQ_\infty R_\infty$ intersects OP_∞ in P_1, and the plane $UP_\infty R_\infty$ intersects OQ_∞ in Q_1. Then any point P in real projective three-space that is not on the plane $P_\infty Q_\infty R_\infty$ may be associated with an ordered triple of real numbers (x,y,z) where the plane $PQ_\infty R_\infty$ intersects the line OP_∞ in P_x, the plane $PP_\infty R_\infty$ intersects OQ_∞ in Q_y, and the plane $PP_\infty Q_\infty$ intersects OR_∞ in R_z. The real numbers x, y, z in the number triple (x,y,z) are called the nonhomogeneous coordinates of the point P. With reference to the five given points, there exists a unique point P corresponding to any ordered set of three real numbers and there exists a unique ordered set of three real numbers corresponding to any point P that is not on the plane $P_\infty Q_\infty R_\infty$. After one plane has been deleted, the remaining set of points on real projective three-space is isomorphic with the set of ordered triples of real numbers and with the set of points on euclidean three-space. The difficulties that have arisen with respect to the nonhomogeneous coordinate of one point on a real projective line, the nonhomogeneous coordinates of the points on one line on a real projective plane, and the non-homogeneous coordinates of the points on one plane in real projective three-space may all be removed by using homogeneous coordinate systems.

EXERCISES

1. Select reference points on a real projective plane and determine as defined above and in the preceding chapter the points with each of the following nonhomogeneous coordinates.

    (a) $(1,0)$, $(0,1)$, $(2,0)$, and $(2,1)$.   (b) $(3,1)$, $(\frac{1}{2},0)$, and $(\frac{1}{2},1)$.

                (c) $(4,2)$ and $(\frac{3}{2},3)$.

2. Select reference points on a real projective three-space and determine the points with each of the following nonhomogeneous coordinates.

(a) (1,0,0),   (b) (2,0,0),   (c) (1,1,0),
(d) (0,0,1),   (e) (1,0,1),   (f) (2,1,1).

3–8 Homogeneous coordinates. We have seen that nonhomogeneous coordinate systems require that special consideration be given to one point on each projective line, one line on each projective plane, and one plane on projective three-space. We now introduce homogeneous coordinate systems and remove the necessity of giving any points special consideration. This advantage is gained at the expense of the uniqueness of the sets of numbers corresponding to each point. We shall associate sets or classes of number pairs with the points on a line, classes of number triples with the points on a plane, and classes of number quadruples with points on a three-space.

We could introduce homogeneous coordinates (x_1,x_2) for points P_x on a line by their definition as pairs of real numbers, not both zero, such that (x_1,x_2) and (kx_1,kx_2) represent the same point P_x for any real number $k \neq 0$ and such that (x_1,x_2) represents the point with nonhomogeneous coordinate x if $x_1/x_2 = x$ when $x \neq \infty$, $x_2 = 0$ when $x = \infty$. Homogeneous coordinates for points on a plane and for points in space could be similarly introduced. However, we shall first use a few geometric concepts to help us visualize the relationship between homogeneous and nonhomogeneous coordinate systems. Our introduction of homogeneous coordinates for points on a line will involve the assumption that on a plane the points $(0,0)$, $(b,1)$, $(2b,2)$, $\ldots$, (ab,a), $\ldots$ are collinear for any given real number b. This assumption and the corresponding assumption in space are not needed as a basis for homogeneous coordinates. The assumptions are used only as an aid to our understanding of homogeneous coordinates.

Consider a real line m with reference points P_0, P_1, and P_∞. Let O be any point that is not on m. Then, as in Fig. 3–12, we may introduce a nonhomogeneous coordinate system in the plane π determined by O and m. Nonhomogeneous coordinates R_y on the line OP_0 may be determined by taking $R_0 = O$, $R_1 = P_0$ and selecting a third point R_∞. Similarly, nonhomogeneous coordinates Q_x on OP_∞ may be determined by taking $Q_0 = O$, $Q_\infty = P_\infty$, and $Q_1 = Q_0Q_\infty \cdot P_1R_\infty$. Then the coordinates of the points Q_x and R_y may be used as in Section 3–7 to determine nonhomogeneous coordi-

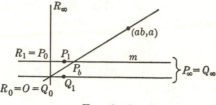

FIG. 3–12

nates for any point on the plane of O and m that is not on the line $Q_\infty R_\infty$. Since any point P_x on m determines with O a line m_x, there is a one-to-one correspondence between the pencil of points on the projective line m and the pencil of lines that is on the point O and on the plane π.

We next consider the real projective plane π determined by O and m, and associate the points of m with the reciprocals of the slopes of the corresponding lines through O. It is not formally necessary to introduce the concept of slope (undefined here), but this concept should add to the reader's intuitive understanding of homogeneous coordinates. The line $P_1 O$ has slope 1 and contains points $(1,1)$, $(2,2)$, $(-4,-4)$ and, in general, (a,a) for any real number a. The line $P_{-1} O$ has slope -1 and contains the set of points $(a,-a)$ where a is any real number. The line $P_\infty O$ has slope 0 and contains the set of points $(a,0)$; the line $P_0 O$ is said to have infinite slope and contains the set of points $(0,a)$. In general, the line $P_b O$ has slope $1/b$ and contains the set of points (ab,a) (Fig. 3–12). Thus each point P_b on the line m corresponds to the set of all points (ab,a) except $(0,0)$ on the line OP_b. In other words, there is a one-to-one correspondence between the points P_b on the line m and the sets of ordered pairs of real numbers (ab, a) corresponding to the points (ab,a) on the lines $P_b O$. Subject to certain conventions regarding the numbers 0 and ∞, we have a one-to-one correspondence between the points P_b on the line m and the pairs of numbers (x,y) where $x/y = b/1$.

Any real projective line m consists of the set of points P_r for all numbers r of the extended real number system. Under the above one-to-one correspondence between the set of points on the line m and the set of lines on O, each point P_s on m determines with O a line $P_s O$ which contains a point $T_s \neq O$. For example, suppose the point T_s has nonhomogeneous coordinates (u,v) obtained from Q_u and R_v. Then we may associate with each point P_s on the real projective line m at least one ordered pair of real numbers (u,v). Conversely, each ordered pair of real numbers (u,v) determines a unique point T_s on the plane. If u and v are not both zero, the point T_s determines with O a unique line on the plane, and this line intersects the line m in a unique point P_s. In this way every ordered pair of real numbers except $(0,0)$ determines exactly one point P_s on m, and every point on m may be determined in this way. However, several ordered pairs of real numbers may determine the same point P_s. Indeed, every point $T_s \neq O$ on the line P_sO is associated with the same point P_s. Thus we may associate each point P_s on m with the set of points on the line P_sO excluding the point O. This set of points is characterized by the fact that if (u,v) represents a point distinct from O on the line P_sO, then for any real number $k \neq 0$, the ordered pair (ku,kv) also represents a point distinct from O on the line P_sO. For any given real numbers u, v, the set of ordered real number pairs (ku,kv) where $k \neq 0$ is called a *class* of ordered real number pairs. We shall represent the class of number pairs by any one of its elements (u,v). The set of points P_s on m is in one-to-one correspondence with the set of classes of ordered number pairs (u,v) when the class $(0,0)$ is excluded.

Any point P_s on the real projective line m may be associated with a class of ordered pairs of real numbers (u,v), where u and v are not both zero. Any ordered pair of real numbers $(u,v) \neq (0,0)$ may be associated with a unique point P_s of m in such a way that for any real number $k \neq 0$ the pairs (u,v) and (ku,kv) are associated with the same point P_s. Thus the points on the real projective line m may be designated by classes of ordered pairs of real numbers (x_1,x_2), where

(i) $(x_1,x_2) \neq (0,0)$, and
(ii) $(x_1,x_2) = (kx_1,kx_2)$ for any real number $k \neq 0$.

Formally, we have an isomorphism between the points of a real projective line and the classes of ordered pairs of real numbers where the class $(0,0)$ is excluded and two ordered pairs are in the same class if and only if they are related, as in (ii) above. The numbers x_1 and x_2 are called *homogeneous coordinates* of the point associated with the class (x_1,x_2). The numbers kx_1 and kx_2 for any $k \neq 0$ are also homogeneous coordinates of the point.

Each point P_r on a real projective line has a unique nonhomogeneous coordinate r, where r is in the extended real number system.

Also each point P_r corresponds to a class (x_1,x_2) of ordered pairs of real numbers. When $r \neq \infty$, the nonhomogeneous coordinate r and each pair of the class of homogeneous coordinates (x_1,x_2) satisfy a relation of the form

$$r = x_1/x_2.$$

When $r = \infty$, the class of homogeneous coordinates of the point $P_r = P_\infty$ is the class $(1,0)$. We shall follow the convention of using letters $(x_1, x_2, \dots)$ with subscripts for homogeneous coordinates and letters $(x, y, \dots)$ without subscripts for nonhomogeneous coordinates.

We next consider homogeneous coordinates for points on a real projective plane. Given any real projective plane, there exists a nonhomogeneous coordinate system in projective three-space such that the set of points $(x,y,1)$ is on the given plane and we may again obtain an intuitive basis for the homogeneous coordinates of the points. Let the origin of this coordinate system be O. Then each point P on the plane may be associated with the class of ordered triples of real numbers that are the nonhomogeneous coordinates of points $T \neq O$ on the line OP (Exercise 2). In general, the points on any real projective plane may be designated by classes of ordered triples of real numbers (x_1,x_2,x_3), where

  (i) $(x_1,x_2,x_3) \neq (0,0,0)$ and
  (ii) $(x_1,x_2,x_3) = (kx_1,kx_2,kx_3)$ for any real number $k \neq 0$.

The numbers x_1, x_2, and x_3 are called *homogeneous coordinates* of the point associated with the class (x_1,x_2,x_3). The numbers kx_1, kx_2, and kx_3 for any $k \neq 0$ are also homogeneous coordinates of the point. These coordinates are similar to those used in Section 1–7 except that at that time only finite sets of numbers were allowed as possible values of the coordinates. Finally, as suggested in Exercise 5 of Section 2–1, the points of real projective three-space may be designated by classes of ordered quadruples of real numbers (x_1,x_2,x_3,x_4) where

  (i) $(x_1,x_2,x_3,x_4) \neq (0,0,0,0)$ and
  (ii) $(x_1,x_2,x_3,x_4) = (kx_1,kx_2,kx_3,kx_4)$ for any real number $k \neq 0$.

If a point on a real projective line has nonhomogeneous coordinate x and homogeneous coordinates (x_1,x_2), then, assuming that the reference points have been properly selected, $x = x_1/x_2$ when $x \neq \infty$

and $x_2 = 0$ when $x = \infty$. If a point on a real projective plane has nonhomogeneous coordinates (x,y) and homogeneous coordinates (x_1,x_2,x_3), then, assuming that the reference points have been properly selected,

$$x = x_1/x_3, \quad y = x_2/x_3, \quad \text{when } x_3 \neq 0,$$

and the points on the special line with reference to the system of nonhomogeneous coordinates have homogeneous coordinates of the form $(x_1,x_2,0)$. Similarly, if a point on real projective three-space has nonhomogeneous coordinates (x,y,z) and homogeneous coordinates (x_1,x_2,x_3,x_4), then, assuming that the reference points have been properly selected,

$$x = x_1/x_4, \quad y = x_2/x_4, \quad z = x_3/x_4, \quad \text{when } x_4 \neq 0,$$

and the points on the special plane have homogeneous coordinates of the form $(x_1,x_2,x_3,0)$.

We may now select reference points and associate homogeneous coordinates with all points of any real projective geometry. For each selection of the set of reference points, there is a corresponding system of homogeneous coordinates and a corresponding system of nonhomogeneous coordinates subject to the limitations that we have found necessary for them. Any two homogeneous coordinate systems on the same projective space are equivalent, in the sense of being isomorphic. A primary advantage of homogeneous coordinate systems for projective spaces lies in the fact that all points may be treated alike when homogeneous coordinates are used. Unless otherwise specified, we shall hereafter assume that one set of reference points has been selected and that all coordinates are with reference to these points; i.e., we shall assume that a coordinate system has been selected and fixed.

EXERCISES

1. Give five pairs of real numbers, homogeneous coordinates, that may be used to designate each of the following points on a projective line: P_1, P_2, P_5, P_0, P_{-1}, P_{-2}, P_{-5}, $P\sqrt{2}$, P_∞, P_π.

2. Rephrase the above development of homogeneous coordinates of points on a line and give a development of homogeneous coordinates of points on a plane.

3. Consider the sequence of points on a plane $(a,b,1)$, $(a,b,0.1)$, $(a,b,0.01)$, $(a,b,0.001)$ . . . with nonhomogeneous coordinates (a,b), $(10a,10b)$, $(100a,100b)$, $(1000a,1000b)$, . . . and discuss the interpretation of the triple $(x_1,x_2,0)$.

4. Give five sets of homogeneous coordinates for each of the following points on a projective plane:

(a) points with nonhomogeneous coordinates $(1,1)$, $(1,2)$, $(-1,-2)$, $(3,4)$, $(0,1)$, $(1,0)$, $(0,0)$;

(b) the ideal points on the lines obtained by joining to $(0,0)$ each of the points that is distinct from $(0,0)$ in part (a).

5. Give five sets of homogeneous coordinates for each of the following points in projective space:

(a) $(1,1,1)$, $(1,2,1)$, $(2,-1,3)$, $(0,1,2)$;

(b) $(1,0,2)$, $(\pi,e,0)$, $(0,0,1)$, $(1,0,0)$, $(0,0,0)$.

*6. Prove that any two sets of reference points on a projective line are projectively equivalent.

*7. Prove that any two sets of reference points on a projective plane are projectively equivalent.

*8. Prove that any two homogeneous coordinate systems are projectively equivalent (a) on a line and (b) on a plane.

3–9 Survey. We have now completed our development of coordinate systems for projective spaces and are ready for the algebraic representations of points by their coordinates, of lines and planes by linear equations in the coordinates, and of projective transformations by linear transformations of the coordinates. Since these algebraic representations of the elements of synthetic projective geometry are the elements of analytic projective geometry (Chapter 4), we are now ready to consider analytic projective geometry and to prove that analytic and synthetic projective geometries represent two points of view of the same geometry — projective geometry.

In the present chapter we have defined quadrangular sets, identified harmonic sets as a special case of quadrangular sets, postulated the existence of infinitely many points on each line in terms of separation and harmonic sets, used harmonic sets to develop nets of rationality, postulated the existence of real points on a line in terms of an isomorphism between a subset of the points on the line and the set of numbers of the extended real number system, and, finally, used this isomorphism as a basis for nonhomogeneous coordinates on a line and for homogeneous coordinates on a line, on a plane, and in space. In this development the properties of isomorphic correspondences are very important. If we defined addition, subtraction, multiplication, and division of points on a line in terms of perspectivities or quadrangular sets [16; 141–150], we would find that all sums, products, differences, and quotients in the geometry correspond to the corresponding results obtained in algebra for the coordinates. For example, the isomorphism between the set of points P_r and the set of numbers r in the extended real number system implies that on any line $P_a + P_b = P_{a+b}$, $P_a \cdot P_b = P_{ab} = P_b \cdot P_a$, and many other such relations. It is possible to prove such relations in geometry without reference to the coordinates, but such proofs are neither short nor obvious to the beginner. Accordingly, we use an isomorphism to assure us that in geometry, as in algebra, such relations as

$$P_2 + P_4 = P_1 + P_5 = P_5 + P_1 = P_2 \cdot P_3 = P_6$$

must hold. In this sense the isomorphism between points and numbers provides an important application of algebraic properties to our development of projective geometry. In particular, this isomorphism provides the basis for the coordinate systems which, in turn, provide the basis for the equivalence of analytic and synthetic projective geometry.

REVIEW EXERCISES

1. Give the quadrangular sets on m of the quadrangles $ABCD$ and $AFDE$ in Fig. 3–13.

2. Give the quadrangular sets on the line ET of the quadrangles $BACD$ and $WZAB$ in Fig. 3–13.

3. Given three points A, B, C on a line m (choose B about two-thirds of the way from A to C), construct D satisfying $H(AC, BD)$.

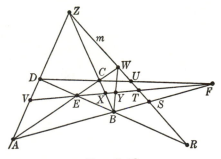

Fig. 3–13

4. State the plane dual of Exercise 3 and make the construction. (*Hint:* If necessary, list in order each step used in Exercise 3 and consider plane duals.)

5. Prove that a point section of a harmonic set of points on a line m by a point that is not on m is a harmonic set of lines.

6. Prove that a quadrangular set of lines is obtained whenever the two triples of lines are interchanged in a point section of a quadrangular set of points.

7. Use the statement in Exercise 6 and repeat Exercise 5.

8. Select four reference points on a plane and construct the points with nonhomogeneous coordinates $(2,+1)$ and $(2,-1)$.

9. Select three reference points on a line and construct the points with homogeneous coordinates $(1,2)$, $(0,5)$, and $(12,36)$.

10. Select four reference points on a plane and construct the points with homogeneous coordinates $(4,2,1)$ and $(9,12,3)$.

CHAPTER 4

ANALYTIC PROJECTIVE GEOMETRY

In the synthetic projective geometry of points (undefined), lines (undefined), and planes that we have discussed in Chapters 2 and 3, the points of real projective spaces may be associated with classes of real numbers — the homogeneous coordinates of the points. In real analytic projective geometry, we start with classes of real numbers without reference to coordinate systems or any other aspects of synthetic projective geometry. Thus *real analytic projective geometry* in space is the geometry of classes of quadruples of real numbers (points), sets of points satisfying homogeneous linear equations (planes), sets of points that are common points of two distinct planes (lines), and nonsingular linear homogeneous transformations of classes of quadruples of real numbers (projective transformations). In synthetic projective geometry the real numbers and their properties may be developed from the properties of the geometry; in analytic projective geometry the properties of synthetic projective geometry may be developed from the properties of the real numbers. In this chapter we shall find that the classes of numbers (points) of analytic projective geometry may be considered as homogeneous coordinates of points of synthetic projective geometry, and that analytic projective geometry and synthetic projective geometry are equivalent representations of projective geometry. We shall emphasize this equivalence throughout our development of analytic projective geometry.

4-1 Representations in space. The set of all quadruples of real numbers may be divided into subsets or classes of quadruples such that two quadruples

$$a_1, a_2, a_3, a_4 \quad \text{and} \quad b_1, b_2, b_3, b_4$$

are in the same class if and only if there exists a real number $k \neq 0$ such that $a_1 = kb_1$, $a_2 = kb_2$, $a_3 = kb_3$, and $a_4 = kb_4$, that is, such that $a_j = kb_j$ $(j = 1,2,3,4)$. After the class $0,0,0,0$ has been excluded, the remaining classes of quadruples of real numbers are the *points* of real analytic projective geometry in space. We shall represent a point by a symbol of the form

(4–1) $$(x_1, x_2, x_3, x_4),$$

where x_1, x_2, x_3, x_4 is any member of the class of quadruples that determines the point. Since these classes of quadruples of real numbers may also be interpreted as sets of homogeneous coordinates of points in space in synthetic projective geometry, we shall emphasize the relationship between synthetic and analytic projective geometry by calling the real numbers x_j the homogeneous coordinates of the points.

Planes, lines, and projective transformations in analytic projective geometry are defined in terms of linear equations in the coordinates of the points (4–1), i.e., in terms of equations of the form

(4–2) $$a_1 x_1 + a_2 x_2 + a_3 x_3 + a_4 x_4 = 0,$$

where the *coefficients* a_j $(j = 1,2,3,4)$ are real numbers, not all zero. The equation (4–2) is called a *homogeneous linear equation* in the x_j, since each term of the left member of the equation is of degree one in the numbers x_j $(j = 1,2,3,4)$ and the right member of the equation is 0. The set of coefficients a_j $(j = 1,2,3,4)$ of the equation (4–2) forms a quadruple of real numbers not all zero. Any two equations of the form (4–2) are equivalent in the sense of being satisfied by precisely the same points (x_1, x_2, x_3, x_4) if and only if the two quadruples of coefficients are from the same class of quadruples in the sense considered above (Exercise 1). Thus the quadruple of coefficients of an equation (4–2) may be considered as the representative of a class of quadruples. Two equations of the form (4–2) with coefficients from the same class of quadruples are equivalent and are called *dependent equations*. Two equations of the form (4–2) with coefficients from different classes of quadruples are not equivalent and are called *independent equations*.

A *plane* is composed of points (4–1) that satisfy a linear homogeneous equation of the form (4–2). Dependent equations represent the same plane; independent equations represent different planes (Exercise 2). Thus any plane may be represented by a class of quadruples of real numbers

(4–3) $$\{a_1, a_2, a_3, a_4\},$$

the set of coefficients of the corresponding equation (4–2). The principle of space duality in analytic projective geometry is based upon

(i) the representation of points by classes of quadruples of real numbers (4–1);

(ii) the representation of planes by classes of quadruples of real numbers (4–3);

(iii) the incidence of a point and a plane if and only if an equation of the form (4–2) is satisfied.

Note that all incidence relations remain unchanged when the notations for point and plane are interchanged.

The set of points that satisfy two independent equations of the form (4–2) is called a *line*. Thus any two independent linear homogeneous equations in the coordinates x_1, x_2, x_3, x_4 determine a unique line, and, as in synthetic projective geometry, any two distinct planes determine a unique line. We shall represent the line determined by the planes (4–2) and $b_1x_1 + b_2x_2 + b_3x_3 + b_4x_4 = 0$ by the rectangular array or matrix

$$(4\text{–}4) \qquad \begin{bmatrix} a_1 & a_2 & a_3 & a_4 \\ b_1 & b_2 & b_3 & b_4 \end{bmatrix}.$$

The real numbers a_j and b_j ($j = 1,2,3,4$) are called the *elements* of the matrix (4–4). The matrix has two rows and four columns.

The system of linear homogeneous equations

$$(4\text{–}5) \qquad \begin{aligned} x_1' &= a_{11}x_1 + a_{12}x_2 + a_{13}x_3 + a_{14}x_4, \\ x_2' &= a_{21}x_1 + a_{22}x_2 + a_{23}x_3 + a_{24}x_4, \\ x_3' &= a_{31}x_1 + a_{32}x_2 + a_{33}x_3 + a_{34}x_4, \\ x_4' &= a_{41}x_1 + a_{42}x_2 + a_{43}x_3 + a_{44}x_4 \end{aligned}$$

is uniquely solvable for x_1, x_2, x_3, x_4 in terms of the coefficients a_{ij} and the x_j' if and only if the determinant of the coefficients

$$(4\text{–}6) \qquad\qquad |a_{ij}| \neq 0 \qquad\qquad (i, j = 1,2,3,4);$$

that is, if and only if

$$\begin{vmatrix} a_{11} & a_{12} & a_{13} & a_{14} \\ a_{21} & a_{22} & a_{23} & a_{24} \\ a_{31} & a_{32} & a_{33} & a_{34} \\ a_{41} & a_{42} & a_{43} & a_{44} \end{vmatrix} \neq 0.$$

This familiar property of systems of equations may be found in texts, such as [10; 172–227], that include a detailed study of matrices and determinants. Most readers will recognize the above property from their study of Cramer's Rule in college algebra. When the condition (4–6) is satisfied, the system (4–5) is called a *nonsingular linear homogeneous transformation* or, in analytic projective geometry, a *projective transformation* of the points (4–1). We shall use the matrix of the coefficients of the equations (4–5), indicate that the corresponding determinant must be different from zero, and represent the projective transformation as in (4–7). The real numbers a_{ij} are called the elements of the matrix.

$$(4\text{--}7) \qquad \begin{bmatrix} a_{11} & a_{12} & a_{13} & a_{14} \\ a_{21} & a_{22} & a_{23} & a_{24} \\ a_{31} & a_{32} & a_{33} & a_{34} \\ a_{41} & a_{42} & a_{43} & a_{44} \end{bmatrix} \qquad \begin{array}{l} |\,a_{ij}\,| \neq 0, \\ (i,\, j = 1,2,3,4). \end{array}$$

We have now seen that in real analytic projective geometry of space, a point is represented by a class of quadruples of real numbers (4–1), a plane is represented by a linear equation (4–2) or a class of quadruples of real numbers (4–3), a line is represented by a matrix (4–4) of two rows and four columns where the quadruples of elements on different rows are from different classes, and a projective transformation is represented by a matrix (4–7) of four rows and four columns with determinant different from zero. We next consider the representations of points and lines on a plane in analytic projective geometry.

EXERCISES

1. Prove that any two equations of the form (4–2) are equivalent if and only if their quadruples of coefficients are from the same class of quadruples of real numbers.

2. Prove that two planes are distinct in the sense that each plane contains at least one point that is not on the other plane if and only if the sets of coefficients of their algebraic representations are from different classes of quadruples.

3. Write down a matrix of two rows and four columns with real numbers as elements such that the matrix
   (a) may be used to represent a line,
   (b) cannot be used to represent a line.

4. Write down a matrix of four rows and four columns with real numbers as elements such that the matrix

(a) may be used to represent a projective transformation,

(b) cannot be used to represent a projective transformation.

4–2 Representations on a plane.   The definitions of the points and lines of real analytic projective geometry on a plane in terms of classes of triples of real numbers are very similar to the definitions of points and planes in space.   The set of all triples of real numbers may be divided into subsets or classes of triples such that two triples

$$a_1, a_2, a_3 \quad \text{and} \quad b_1, b_2, b_3$$

are in the same class if and only if there exists a real number $k \neq 0$ such that $a_j = kb_j$ ($j = 1, 2, 3$).   After the class $0,0,0$ has been excluded, the remaining classes of triples of real numbers are the *points* of real analytic projective geometry on a plane.   We shall represent a point of plane geometry by a symbol of the form

$$(4–8) \qquad\qquad (x_1, x_2, x_3),$$

where x_1, x_2, x_3 is any member of the class of triples that determines the point.   Since these classes of triples of real numbers may also be interpreted as sets of homogeneous coordinates of points on a plane in synthetic projective geometry, we shall emphasize the relationship between synthetic and analytic projective geometry by calling the real numbers x_j the homogeneous coordinates of the points.

A *line* is composed of points (4–8) that satisfy a linear homogeneous equation of the form

$$(4–9) \qquad\qquad a_1 x_1 + a_2 x_2 + a_3 x_3 = 0,$$

where the coefficients a_j are real numbers not all zero.   Dependent equations represent the same line; independent equations represent different lines.   Thus any line may be represented by a class of triples of real numbers

$$(4–10) \qquad\qquad [a_1, a_2, a_3],$$

the set of coefficients of the corresponding equation (4–9).   The principle of planar duality in analytic projective geometry is based upon

   (i) the representation of points on a plane by classes of triples of real numbers (4–8);
   (ii) the representation of lines on a plane by classes of triples of real numbers (4–10);
   (iii) the incidence of a point and a line on a plane if and only if an equation of the form (4–9) is satisfied;
   (iv) the equivalence (Exercise 5) of the geometry of classes of triples and the geometry of points and lines on any plane in the geometry of quadruples (4–1).

The system of linear homogeneous equations

(4–11)
$$x'_1 = a_{11}x_1 + a_{12}x_2 + a_{13}x_3,$$
$$x'_2 = a_{21}x_1 + a_{22}x_2 + a_{23}x_3,$$
$$x'_3 = a_{31}x_1 + a_{32}x_2 + a_{33}x_3$$

is uniquely solvable for x_1, x_2, x_3 in terms of the coefficients a_{ij} and the x'_j if and only if the determinant of the coefficients

(4–12)
$$|a_{ij}| \neq 0 \qquad (i,j = 1,2,3),$$

that is, if and only if

(4–13)
$$\begin{vmatrix} a_{11} & a_{12} & a_{13} \\ a_{21} & a_{22} & a_{23} \\ a_{31} & a_{32} & a_{33} \end{vmatrix} = \begin{aligned} & a_{11}(a_{22}a_{33} - a_{23}a_{32}) \\ & - a_{12}(a_{21}a_{33} - a_{23}a_{31}) \\ & + a_{13}(a_{21}a_{32} - a_{22}a_{31}) \neq 0. \end{aligned}$$

When the condition (4–12) is satisfied, the system (4–11) is called a nonsingular linear homogeneous transformation or a *projective transformation* of the points (4–8). We shall represent the transformation by the matrix (4–14) of the coefficients of the equations (4–11) and indicate that the determinant must be different from zero.

(4–14)
$$\begin{bmatrix} a_{11} & a_{12} & a_{13} \\ a_{21} & a_{22} & a_{23} \\ a_{31} & a_{32} & a_{33} \end{bmatrix} \qquad \begin{aligned} & |a_{ij}| \neq 0, \\ & (i,j = 1,2,3). \end{aligned}$$

We have now seen that in real analytic projective geometry on a plane, a point is represented by a class of triples of real numbers (4–8), a line by a class of triples of real numbers (4–10), and a projective transformation by a matrix (4–14) of three rows and three columns with determinant different from zero. We shall use these representations and prove that the properties of synthetic projective geometry on the plane are also properties of analytic projective geometry. Several of these properties will be considered in the present section; others will be considered after additional concepts of analytic projective geometry have been discussed.

If R: (r_1,r_2,r_3) and S: (s_1,s_2,s_3) are points, there exists a unique line RS if and only if $R \neq S$. The form of the expansion of the determinant (4–13) may be used (Exercise 9) to prove that, when $R \neq S$, the equation of the line RS may be expressed in the form

$$(4\text{–}15) \qquad \begin{vmatrix} x_1 & x_2 & x_3 \\ r_1 & r_2 & r_3 \\ s_1 & s_2 & s_3 \end{vmatrix} = 0.$$

Thus three points R, S, T are collinear if and only if the determinant of their homogeneous coordinates

$$(4\text{–}16) \qquad \begin{vmatrix} r_1 & r_2 & r_3 \\ s_1 & s_2 & s_3 \\ t_1 & t_2 & t_3 \end{vmatrix}$$

is zero (Exercise 10). This condition for the collinearity of points has many applications and several forms (Exercises 11 and 12).

The above discussion indicates that Postulates P–1 and P–2 of synthetic projective geometry are properties of analytic projective geometry. Several other properties of synthetic projective geometry that are also properties of analytic projective geometry are considered in the following set of exercises.

EXERCISES

*1. Prove that any plane in the geometry of quadruples is equivalent to the plane $x_4 = 0$ under a projective transformation (4–7).

2. Prove that the set of points on the plane $x_4 = 0$ is equivalent to the set of classes of triples (4–8).

3. Prove that the set of lines on the plane $x_4 = 0$ is equivalent to the set of lines (4–10).

4. Prove that the geometry of points and lines on the plane $x_4 = 0$ is equivalent to the geometry of points and lines in the geometry of classes of triples of real numbers.

*5. Prove that the geometry of classes of triples is equivalent to the geometry of points and lines on any plane of the geometry of quadruples.

6. Prove that lines correspond to lines under any projective transformation (4–14).

*7. Prove that any line in the geometry of triples is equivalent to the line $x_3 = 0$ under a projective transformation (4–14).

8. Prove that any line in analytic projective geometry of space is equivalent to the line $x_3 = 0$ in analytic projective plane geometry.

*9. Prove that there exists a unique line RS with equation (4–15) if and only if the points R and S are distinct.

*10. Prove that three points of analytic projective plane geometry are collinear if and only if the determinant of their homogeneous coordinates (4–16) is zero.

11. Prove that three points R, S, and T are collinear if and only if there exist real numbers j and k such that the coordinates of the points satisfy the relations

$$t_i = jr_i + ks_i \qquad\qquad (i = 1,2,3).$$

12. Prove that any point on the line RS (4–15) has the coordinates $x_i = jr_i + ks_i$ ($i = 1,2,3$) for some pair of real numbers j and k not both zero.

13. State the plane dual of the statement of Exercise 12.

14. State the space duals of the statement of Exercise 12 and the statement obtained in Exercise 13.

*15. Prove Postulates P–1, P–2, P–4 through P–8, and P–14 of synthetic projective geometry as theorems of analytic projective geometry.

16. Prove that a line is determined by any two of its points.

17. State the plane and space duals of the statement in Exercise 16.

†18. Prove that planes correspond to planes and lines correspond to lines under any projective transformation (4–7).

4–3 Representations on a line. The points of real analytic projective three-space are classes of quadruples of real numbers; the points of real analytic projective plane geometry are classes of triples of real numbers. We now consider classes of pairs of real numbers. The set of all pairs of real numbers may be divided into classes of pairs such that two pairs

$$a_1, a_2 \quad \text{and} \quad b_1, b_2$$

are in the same class if and only if there exists a real number $k \neq 0$

---

† The exercises that are concerned with transformations in space and that are *not* used in our development of euclidean plane geometry are designated by daggers.

such that $a_1 = kb_1$ and $a_2 = kb_2$. After the class $0,0$ has been excluded, the remaining classes of pairs of real numbers are the *points* of real analytic projective geometry on a line. We shall represent a point of the geometry on a line by a symbol of the form

$$(x_1, x_2),$$

where the pair x_1, x_2 is any member of the class of pairs that determines the point. Since these classes of pairs of real numbers may also be interpreted as sets of homogeneous coordinates of points on a line in synthetic projective geometry, we shall emphasize the relationship between synthetic and analytic projective geometry by calling the real numbers x_j the homogeneous coordinates of the points.

A linear homogeneous equation of the form

$$a_1 x_1 + a_2 x_2 = 0,$$

where $(a_1, a_2) \neq (0, 0)$, is satisfied by and indeed determines a unique point $(-a_2, a_1)$. The relation

(4–17) $$x_1 / x_2 = x$$

represents an isomorphism between the classes of pairs of real numbers and the numbers of the extended real number system. Thus, on a line, the points of real analytic projective geometry may be considered either in terms of their homogeneous coordinates (x_1, x_2) or in terms of single numbers from the extended real number system, $x = x_1 / x_2$, their nonhomogeneous coordinates. Furthermore, the points on any given line in real analytic projective geometry on a plane or in three-space have this same property (Exercise 3).

The system of linear homogeneous equations

$$x_1' = a_{11} x_1 + a_{12} x_2,$$
$$x_2' = a_{21} x_1 + a_{22} x_2$$

is uniquely solvable for x_1 and x_2 in terms of the coefficients a_{ij} and the x_j' if and only if the determinant of the coefficients

$$|a_{ij}| \neq 0 \qquad (i, j = 1, 2),$$

that is, if and only if

$$\begin{vmatrix} a_{11} & a_{12} \\ a_{21} & a_{22} \end{vmatrix} = a_{11} a_{22} - a_{12} a_{21} \neq 0.$$

This condition is satisfied if and only if the two sets of coefficients are from different classes of pairs of real numbers. When the determinant is different from zero, the above system of equations is called a nonsingular linear homogeneous transformation or a *projective transformation* of the points (x_1, x_2). Any projective transformation on a line may be expressed in terms of the homogeneous coordinates of the points

$$(4\text{--}18) \qquad \begin{aligned} x_1' &= ax_1 + bx_2, \\ x_2' &= cx_1 + dx_2 \end{aligned} \qquad (ad - bc \neq 0),$$

where we have for convenience changed the notation for the coefficients to one that does not involve subscripts. For points with real nonhomogeneous coordinates the transformation (4–18) may also be expressed in the form

$$(4\text{--}19) \qquad x' = \frac{ax + b}{cx + d} \qquad (ad - bc \neq 0).$$

Both (4–18) and (4–19) may be represented by the matrix

$$\begin{bmatrix} a & b \\ c & d \end{bmatrix},$$

where $ad - bc \neq 0$.

We now consider two important properties of projective transformations on a line. Given any projective transformation (4–18), there exists a unique *inverse transformation* of the same form such that if P corresponds to P' under the given transformation, then P' corresponds to P under the inverse transformation for every point P on the line (Exercise 5). The coefficients of the system of equations of the inverse transformation may be obtained by solving the equations of the given transformation for x_1 and x_2 in terms of x_1' and x_2', and interchanging the primed and unprimed letters. For example, the transformation

$$(4\text{--}20) \qquad \begin{aligned} x_1' &= x_1 - x_2, \\ x_2' &= x_1 + x_2 \end{aligned}$$

has as its inverse transformation

$$(4\text{--}21) \qquad \begin{aligned} x_1' &= x_1 + x_2, \\ x_2' &= -x_1 + x_2, \end{aligned}$$

since any point $R:(r_1,r_2)$ corresponds to $R':(r_1 - r_2, r_1 + r_2)$ under the transformation (4–20) and R' corresponds to the point

$$(r_1 - r_2 + r_1 + r_2, -r_1 + r_2 + r_1 + r_2) = (2r_1,2r_2) = (r_1,r_2)$$

under the transformation (4–21). Note that the complete set of coefficients of any transformation may be multiplied or divided by any real number different from zero without changing the transformation. In particular, the coefficients of the inverse transformation may be divided by the determinant of the transformation.

We have proved that when the transformation (4–20) is followed by the transformation (4–21), the transformation resulting from these two transformations taken in order leaves every point R on the line invariant, i.e., is the *identity transformation*

$$x'_1 = x_1, \quad x'_2 = x_2$$

on the line. In general, a transformation obtained by taking any two given transformations of the form (4–18) in order is a unique transformation of the form (4–18) and is called the *ordered product* of the two transformations (Exercise 7). When an ordered product of two projective transformations on a line is the identity transformation I, the two transformations are called *inverse transformations*. The inverse of a transformation T is denoted by T^{-1} and satisfies the relation

$$TT^{-1} = I = T^{-1}T$$

(Exercise 8). Note that the order is unimportant when one finds the product of a transformation and its inverse transformation. However, in general, we shall find that the order must be considered (Exercise 9).

We have now defined the elements of analytic projective geometry in space, on a plane, and on a line. We have found that analytic projective geometry may be considered as the geometry of the homogeneous coordinates of the points of synthetic projective geometry. Under this interpretation we shall find that the two geometries are equivalent.

EXERCISES

1. Discuss the properties of the isomorphism (4–17).

2. Prove that the set of points on the line $x_3 = 0$ in analytic plane geometry is equivalent to the set of classes of pairs of real numbers.

*3. Prove that the geometry of pairs of real numbers is equivalent to the geometry of points on (a) any line of the geometry of triples of real numbers, and (b) any line of the geometry of quadruples of real numbers.

*4. Prove that the transformations (4–18) and (4–19) are equivalent under the isomorphism (4–17).

*5. Find a general representation for the inverse transformation of any transformation (4–18) and prove that the inverse transformation of any given projective transformation on a line is uniquely defined.

6. Find the inverse transformations of each of the following transformations:

(a) $x_1' = x_1 + x_2$
$x_2' = x_1 - x_2,$

(b) $x_1' = 2x_1 - 3x_2$
$x_2' = 3x_1 + x_2,$

(c) $x_1' = x_2$
$x_2' = x_1,$

(d) $x_1' = -x_1 + x_2$
$x_2' = x_2,$

(e) $x_1' = x_1$
$x_2' = x_2.$

*7. Prove that the ordered product of any two given projective transformations on a line is a unique projective transformation on the line.

*8. When T has the form (4–18), prove that if T^{-1} is the inverse of T, then T is the inverse of T^{-1}.

*9. Use the transformations in Exercise 6(c) and (d) to prove that, in general, the order in which transformations are taken in a product must be considered.

10. Find the nonhomogeneous coordinate of each of the following points on a line. Each point is given in terms of its homogeneous coordinates.

(a) $(1,1)$,
(b) $(3,2)$,
(c) $(5,-1)$,
(d) $(-4,-6)$,
(e) $(17,0)$,
(f) $(0,75)$.

11. Find homogeneous coordinates for each of the following points on a line. Each point is given in terms of its nonhomogeneous coordinate.

(a) 3,
(b) -2,
(c) $+5$,
(d) 0,
(e) 1.7,
(f) ∞,
(g) $-\infty$,
(h) $-\frac{1}{12}$.

4–4 Matrices. We have used matrices to represent the projective transformations of analytic projective geometry. We now consider matrices in general, define products of matrices, and define the equality of two matrices. These definitions will enable us to represent ordered products of projective transformations as ordered products of matrices and to represent the systems of linear homogeneous equations of projective transformations as matrix equations.

In general, a matrix is a rectangular array of m rows and n columns:

$$\begin{bmatrix} a_{11} & a_{12} & \cdots & a_{1n} \\ a_{21} & a_{22} & \cdots & a_{2n} \\ \cdot & \cdot & \cdots & \cdot \\ a_{m1} & a_{m2} & \cdots & a_{mn} \end{bmatrix}$$

The numbers a_{ij} are called the *elements* of the matrix. The subscripts of the elements have been chosen so that the first subscript indicates the row on which the element occurs and the second subscript indicates the column on which the element occurs. Thus a_{21} is on the second row and the first column, and, in general, a_{ij} is on the ith row and the jth column.

A matrix with m rows and m columns is called a square matrix of *order m*. Any square matrix with real numbers as its elements has a unique associated real number called its *determinant*. For example, the matrix

$$\begin{bmatrix} a & b \\ c & d \end{bmatrix}$$

has determinant $ad - bc$, which may also be written in the form

$$\begin{vmatrix} a & b \\ c & d \end{vmatrix}.$$

The matrix (4–14) has determinant (4–13) and the matrix (4–7) has determinant (4–6). Methods for expanding determinants of the fourth order [i.e., for finding the real numbers associated with determinants expressed in the form (4–6)] may be found in [10; 183–204]. We shall be primarily concerned with determinants of matrices of the second and third orders. Expansions of such determinants have been considered above and in (4–13).

Two matrices are said to be *equal* if and only if they have the same number of rows, they have the same number of columns, and their corresponding elements are equal. For example, the matrix equation

$$\begin{bmatrix} x_1' \\ x_2' \end{bmatrix} = \begin{bmatrix} ax_1 + bx_2 \\ cx_1 + dx_2 \end{bmatrix}$$

is equivalent to the system of equations (4–18). We shall make extensive use of such matrix equations.

The product of two matrices may be defined in terms of the inner products of pairs of ordered sets of elements. In general, the *inner product* of two ordered n-tuples, such as

$$a_1, a_2, a_3, \ldots, a_n,$$
$$b_1, b_2, b_3, \ldots, b_n,$$

is defined to be the sum of the products of corresponding elements,

$$a_1b_1 + a_2b_2 + a_3b_3 + \cdots + a_nb_n.$$

For example, the product of two matrices

$$\begin{bmatrix} a & b \\ c & d \end{bmatrix} \begin{bmatrix} x_1 \\ x_2 \end{bmatrix}$$

is defined to be the matrix

$$\begin{bmatrix} ax_1 + bx_2 \\ cx_1 + dx_2 \end{bmatrix},$$

where the element on the first row and first column of the product matrix is obtained as the inner product of the set of elements on the first row of the first matrix and the set of elements on the first column of the second matrix; the element on the second row and first column is obtained as the inner product of the set of elements on the second row of the first matrix and the set of elements on the first column of the second matrix.

In general, the ordered product AB of a matrix $A = [a_{ij}]$ of m rows and n columns and a matrix $B = [b_{ij}]$ of n rows and s columns is a matrix $C = [c_{ij}]$ of m rows and s columns where

$$c_{ij} = a_{i1}b_{1j} + a_{i2}b_{2j} + a_{i3}b_{3j} + \cdots + a_{in}b_{nj}.$$

The number of columns in the first matrix must be equal to the number of rows in the second matrix. The phrase "ordered product" is used, since for matrices we may have $AB \neq BA$ (Exercise 9, Section 4–3). We shall be primarily concerned with products of matrices such as

$$\begin{bmatrix} a_{11} & a_{12} & a_{13} \\ a_{21} & a_{22} & a_{23} \\ a_{31} & a_{32} & a_{33} \end{bmatrix} \begin{bmatrix} x_1 \\ x_2 \\ x_3 \end{bmatrix} = \begin{bmatrix} a_{11}x_1 + a_{12}x_2 + a_{13}x_3 \\ a_{21}x_1 + a_{22}x_2 + a_{23}x_3 \\ a_{31}x_1 + a_{32}x_2 + a_{33}x_3 \end{bmatrix}$$

and

$$\begin{bmatrix} a & 0 & b \\ 0 & a & c \\ 0 & 0 & 1 \end{bmatrix} \begin{bmatrix} a' & 0 & b' \\ 0 & a' & c' \\ 0 & 0 & 1 \end{bmatrix} = \begin{bmatrix} aa' & 0 & ab' + b \\ 0 & aa' & ac' + c \\ 0 & 0 & 1 \end{bmatrix}.$$

We have proved (Exercise 7, Section 4–3) that the ordered product of any two projective transformations on a line is a unique projective transformation on the line. Matrix notations facilitate our proof of the corresponding results on a plane (Exercise 14) and in space (Exercise 15). The condition that the determinant of the product matrix be different from zero is always satisfied, since the determinant of the product of two matrices is equal to the product of their determinants [10; 201–202].

In Section 4–3 we calculated the product of the transformations (4–20) and (4–21) in order to prove that they were inverse transformations (i.e., to prove that their product was the identity transformation). The product of their matrices is

$$\begin{bmatrix} 1 & 1 \\ -1 & 1 \end{bmatrix} \begin{bmatrix} 1 & -1 \\ 1 & 1 \end{bmatrix} = \begin{bmatrix} 2 & 0 \\ 0 & 2 \end{bmatrix}.$$

The position of the matrix of (4–20) on the right indicates that the corresponding transformation was used first. In other words, the order in which the transformations are applied to the points is indicated by taking the matrices from right to left.

The above product matrix

$$\begin{bmatrix} 2 & 0 \\ 0 & 2 \end{bmatrix}$$

represents the identity transformation on the line. In general, a square matrix of order m with elements a_{ij} where

$$a_{ij} = k \quad \text{when} \quad i = j,$$
$$= 0 \quad \text{when} \quad i \neq j$$

for some fixed number $k \neq 0$, represents the *identity transformation* on a line when $m = 2$, on a plane when $m = 3$, and in space when $m = 4$. Since homogeneous coordinates represent classes of numbers and satisfy relations of the form

$$(x_1, x_2) = (2x_1, 2x_2),$$

these matrices for identity transformations in terms of homogeneous coordinates are often distinct from the identity matrices. The *identity matrix* of order m has elements a_{ij} where

$$a_{ij} = 1 \quad \text{when} \quad i = j,$$
$$= 0 \quad \text{when} \quad i \neq j.$$

For example,

$$\begin{bmatrix} 1 & 0 & 0 \\ 0 & 1 & 0 \\ 0 & 0 & 1 \end{bmatrix}$$

is the identity matrix of order 3.

The inverse transformation of any projective transformation is defined by the fact that the product of any projective transformation and its inverse transformation is the corresponding identity transformation. For example, the product of any two inverse transformations on the line is the identity transformation on the line. We have proved (Exercise 5, Section 4–3) that any projective transformation on a line has a unique projective transformation as its inverse. Similar results hold for projective transformations on a plane (Exercise 16) and in space (Exercise 17). We now use these properties of transformations to prove two basic properties of analytic projective geometry.

Many of the properties of synthetic projective geometry were based upon Postulate P–9 (Section 2–7) stating the uniqueness of any projective transformation of three distinct points on a given line onto three distinct points on the line. The corresponding statement in analytic projective geometry on a line is

THEOREM 4–1.   *There exists a unique projective transformation of any three distinct points on a line m onto any three distinct points on the line m.*

We first prove that given any three distinct points $R:(r_1,r_2)$, $S:(s_1,s_2)$, and $T:(t_1,t_2)$, there exists a unique transformation (4–18) of the points $(1,0)$, $(0,1)$, and $(1,1)$ onto R, S, and T, respectively. In other words, we shall prove that there exists essentially only one set of coefficients a, b, c, d for the transformation (4–18) such that

(4–22) $(r_1,r_2) = (a,c)$ is the image of $(1,0)$,
          $(s_1,s_2) = (b,d)$ is the image of $(0,1)$, and
          $(t_1,t_2) = (a + b, c + d)$ is the image of $(1,1)$

under the transformation. Since the number pairs represent classes of number pairs, we wish to prove that there exist numbers k, m, and n different from zero such that

$$kr_1 = a, \quad ms_1 = b, \quad nt_1 = a + b,$$
$$kr_2 = c, \quad ms_2 = d, \quad nt_2 = c + d.$$

In other words, given r_1, r_2, s_1, s_2, t_1, and t_2, we seek numbers k, m, n, different from zero such that

(4–23) $a = kr_1, \qquad b = ms_1,$
          $c = kr_2, \qquad d = ms_2,$

where

$$(4\text{–}24) \qquad \begin{aligned} kr_1 + ms_1 &= nt_1, \\ kr_2 + ms_2 &= nt_2. \end{aligned}$$

We may find numbers satisfying these conditions by taking $n = 1$, and solving the corresponding equations (4–24) for k and m (Exercise 18). The values obtained for k and m may then be used in the equations (4–23) to obtain values for a, b, c, and d such that the relations (4–22) hold. Since the values for a, b, c, and d are unique except for a constant factor $n \neq 0$ and since $(x'_1, x'_2) = (nx'_1, nx'_2)$, there is essentially only one set of coefficients for (4–18); i.e., the transformation is uniquely determined.

We have proved that given any three distinct points on a line, there exists a unique projective transformation of that line onto itself such that the points $(1,0)$, $(0,1)$, and $(1,1)$ correspond respectively to the three given points. In terms of the coordinate systems of synthetic projective geometry, we have proved that the reference points on the line may be transformed onto any three distinct points on the line. Then, since any projective transformation on a line has a unique inverse transformation on the line (Exercise 5, Section 4–3), any three distinct points on the line may be transformed onto the reference points. Finally, since the ordered product of any two projective transformations on the line is a projective transformation on the line (Exercise 7, Section 4–3), there exists a projective transformation of any three distinct points on the line onto any three distinct points on the line. For example, if A, B, and C are three distinct points and R, S, and T are three distinct points, there exists a projective transformation of the points A, B, C onto the reference points and a projective transformation of the reference points onto R, S, T. The product of these two transformations is a projective transformation of the points A, B, C onto the points R, S, T. This completes the proof of Theorem 4–1. The corresponding theorem on a plane is

THEOREM 4–2. *On a plane there exists a unique projective transformation of any four points, no three on the same line, onto any four points, no three on the same line.*

The proof of this theorem is similar to that of Theorem 4–1 and is left as an exercise (Exercise 19). The corresponding statement in space is given as Exercise 20.

Theorem 4–2 provides a basis for the following proof of Postulate P–3 as a theorem of analytic projective geometry. Given any three noncollinear points A, B, and C, a point D on the line BC, and a point $E \neq D$ on the line AC, we must establish the existence of a point $F = AB \cdot DE$ (Fig. 2–1). By Theorem 4–2 there exists a projective transformation of the plane onto itself (a change of coordinate systems) such that we have $A:(1,0,0)$, $B:(0,1,0)$, $C:(0,0,1)$, and $D:(0,1,1)$. Then the point E on the line $AC:(x_2 = 0)$ must have coordinates of the form $(e_1,0,e_2)$, the line DE has equation

$$e_2 x_1 + e_1 x_2 - e_1 x_3 = 0$$

as in (4–15), and the lines DE and $AB:(x_3 = 0)$ intersect in a point $F:(e_1,-e_2,0)$. The existence of this point F completes our proof of Postulate P–3. The proofs of Postulates P–9 and P–10 in analytic projective geometry are given as an exercise (Exercise 22).

The only postulates of synthetic projective geometry that we have not considered as theorems of analytic projective geometry are the Postulates of Separation. These postulates will be considered in the next section where separation is defined in terms of cross ratio. Since Postulates P–1 through P–10 and P–14 of synthetic projective geometry are theorems of real analytic projective geometry, we may use these postulates and all theorems implied by them as we consider additional properties of analytic projective geometry.

EXERCISES

1. Rewrite the following systems of equations as matrix equations:

(a) $x_1' = x_1 + x_2$ (b) $x_1' = x_1 + x_2 + x_3$ (c) $x_1' = 2x_1 + 3x_2$
    $x_2' = x_1 - x_2,$ $x_2' = x_1 - x_2$ $x_2' = 3x_1 - x_2.$
                      $x_3' = \qquad\ x_3,$

2. Express each of the following products as a single matrix:

(a) $\begin{bmatrix} 1 & 2 & 3 \\ a & b & c \end{bmatrix} \begin{bmatrix} x_1 \\ x_2 \\ x_3 \end{bmatrix},$ (b) $\begin{bmatrix} 1 & 0 & 0 \\ 1 & 0 & 1 \\ 0 & 1 & 1 \end{bmatrix} \begin{bmatrix} a & b & c \\ d & 0 & e \\ 0 & 0 & 1 \end{bmatrix},$

(c) $\begin{bmatrix} 1 & 0 \\ 0 & 1 \end{bmatrix} \begin{bmatrix} x_1 \\ x_2 \end{bmatrix},$ (d) $\begin{bmatrix} a_1 & a_2 & a_3 \end{bmatrix} \begin{bmatrix} x_1 \\ x_2 \\ x_3 \end{bmatrix}.$

3. Write the condition that a point (4–8) be on a line (4–10) as a matrix equation.

4. Write the condition that a point (4–1) be on a plane (4–3) as a matrix equation.

5. Write the systems of equations (4–5) and (4–11) as matrix equations.

6. Write the matrix for each of the transformations given in Exercise 6, Section 4–3.

7. Multiply the matrix of each transformation considered in Exercise 6 by the matrix of its inverse transformation and verify that each of the product matrices is of the form

$$\begin{bmatrix} k & 0 \\ 0 & k \end{bmatrix}$$

for some real number $k \neq 0$.

8. Find a rule for writing down the matrix of an inverse transformation for any given transformation of the form (4–18).

9. Prove that in general the multiplication of matrices of the third order is noncommutative $(AB \neq BA)$ by considering the two ordered products of the matrices

$$\begin{bmatrix} 2 & 0 & 0 \\ 0 & 2 & 0 \\ 0 & 0 & 1 \end{bmatrix} \quad \text{and} \quad \begin{bmatrix} 1 & 0 & a \\ 0 & 1 & b \\ 0 & 0 & 1 \end{bmatrix}.$$

10. Express the determinant of each of the following matrices as an integer.

(a) $\begin{bmatrix} 1 & 0 & 0 \\ 1 & 0 & 1 \\ 0 & 1 & 1 \end{bmatrix},$ (b) $\begin{bmatrix} 1 & 2 & 3 \\ 1 & 1 & 1 \\ 2 & 1 & 2 \end{bmatrix},$ (c) $\begin{bmatrix} 1 & 0 & 0 \\ 0 & 1 & 0 \\ 0 & 0 & 1 \end{bmatrix}.$

*11. Write down the identity matrices of orders 1, 2, 3, and 4.

12. Write down three matrices of order two such that no two matrices are equal as matrices but each of the matrices represents the identity transformation on a line.

13. Repeat Exercise 12 for matrices of order 3 and transformations on a plane.

*14. Prove that the ordered product of any two given projective transformations on a plane is a unique projective transformation on the plane.

†15. Repeat Exercise 14 for transformations in space.

*16. Prove that any given projective transformation on a plane has a unique projective transformation on the plane as its inverse transformation.

†17. Repeat Exercise 16 for transformations in space.

*18. Prove that the system of equations (4–24) with $n = 1$ may be solved for k and m whenever R, S, and T are distinct points on the line.

*19. Prove Theorem 4–2.

†20. Restate Theorem 4–2 for projective transformations in space.

†21. Prove the theorem stated in Exercise 20.

*22. Prove Postulates P–9 and P–10 as theorems of analytic projective geometry.

4–5 Cross ratio. The term cross ratio signifies a ratio of ratios such as the cross ratio in euclidean geometry

$$\Re(AB,CD) = AC/BC : AD/BD$$

of four collinear points in terms of the specified line segments. The cross ratio has significance in projective geometry because it is an invariant (is unchanged) under projective transformations. The following definition holds in analytic projective geometry and also, after a coordinate system has been selected, in synthetic projective geometry. Given four points A, B, C, D on a line such that at least the first three points are distinct, there exists a unique projective transformation

$$ABCD \mathbin{\overset{\scriptscriptstyle\wedge}{\scriptstyle -}} P_\infty P_0 P_1 P_k$$

for some point P_k (Theorems 4–1 and 2–5). The point P_k is uniquely determined by the four given points, and its nonhomogeneous coordinate k is called the *cross ratio* of the four given points,

$$\Re(AB,CD) = k.$$

The cross ratio has the property that if two ordered sets of four collinear points are projective, they have the same cross ratio (Exercise 1). Furthermore, if two ordered sets of four collinear points have the same cross ratio, there exists a projective transformation of one set onto the other (Exercise 2). This relationship between cross ratio and projective transformations on a line implies that Postulate P–9 of synthetic projective geometry may be replaced by a postulation of the invariance of cross ratios (Exercise 3). It also implies that any one-to-one transformation of a line onto a line such that cross ratios are preserved is a projective transformation (Exercise 4).

Let the given points A, B, C, D have real nonhomogeneous coordinates a, b, c, d respectively. Also, as in euclidean geometry, let us associate numbers with symbols such as $AB = b - a$ and, in general, $P_r P_s = s - r$. Then subject to conventions, such as $1/0 = \infty$, that are required whenever the extended real number system is used, we have (Exercise 5)

$$(4\text{–}25) \qquad \Re(AB,CD) = AC/BC : AD/BD = \frac{c-a}{c-b} : \frac{d-a}{d-b}.$$

This representation of the cross ratio may be used to prove that for any four distinct points we have (Exercise 6)

$$(4\text{–}26) \quad \Re(AB,CD) = \Re(CD,AB) = \Re(BA,DC) = \Re(DC,BA).$$

These relationships may be used to define the cross ratio of any ordered set of four collinear points whenever three of the points (not necessarily the first three) are distinct (Exercise 7).

The twenty-four cross ratios (Exercise 8) of four given collinear points fall into six groups of four each under the following rules.

(i) The cross ratio is unaltered if the two pairs are interchanged; that is, $\Re(AB,CD) = \Re(CD,AB)$.

(ii) The cross ratio is unaltered if the elements of each pair are interchanged within the pair; that is, $\Re(AB,CD) = \Re(BA,DC)$.

If $\Re(AB,CD) = k$, then the cross ratios of the six groups are k, $1/k$, $1 - k$, $1/(1 - k)$, $k/(k - 1)$, $(k - 1)/k$ and may be obtained according to the following rules (Exercise 9):

(i) The interchange of the elements of one pair of points changes the cross ratio from k to $1/k$; that is, $\Re(AB,DC) = 1/k = \Re(BA,CD)$.

(ii) The interchange of either the inner or the outer points in the symbol $\Re(AB,CD)$ changes the cross ratio from k to $1 - k$; that is, $\Re(AC,BD) = 1 - k = \Re(DB,CA)$.

On any line in real analytic projective geometry a pair of points A, B may be defined to *separate* a pair of points C, D if and only if $\Re(AB,CD)$ is negative. Postulates P–11 (Exercise 16) and P–13 (Exercise 17) are immediate consequences of this definition. Thus P–12 is the only postulate of synthetic projective geometry which remains to be proved as a theorem in analytic projective geometry.

Postulate P–12 may be verified by proving that if $\Re(AB,C_1D) < 0$ and $\Re(AC_1,C_2D) < 0$, then $\Re(AB,DC_2) < 0$. We shall use the method of indirect proof (Section 1–2); i.e., we shall assume that the above statement is false and prove that this assumption leads to a contradiction. Then, since the statement cannot be false, it must be valid.   Suppose $\Re(AB,DC_2) \geq 0$.   Then from

$$\Re(AB,C_1D) \cdot \Re(AB,DC_2) = \Re(AB,C_1C_2)$$

(Exercise 19), we have $\Re(AB,C_1C_2) \leq 0$.   Also, since

$$\Re(AC_1,BC_2) = 1 - \Re(AB,C_1C_2),$$

we have $\Re(AC_1,BC_2) > 0$; and, from

$$\Re(AC_1,BC_2) \cdot \Re(AC_1,C_2D) = \Re(AC_1,BD),$$

we have $\Re(AC_1,BD) < 0$.   But $\Re(AC_1,BD) = 1 - \Re(AB,C_1D) > 0$. This contradiction completes our proof of Postulate P–12.

We have now seen that Postulates P–1 through P–14 of synthetic projective geometry are valid as theorems of analytic projective geometry.   Accordingly, we are ready to consider the relationship between the two geometries.

EXERCISES

1. Prove that if $ABCD \barwedge A'B'C'D'$ and $\Re(AB,CD) = k$, then $\Re(A'B', C'D') = k$.

2. Prove that if $\Re(AB,CD) = \Re(A'B',C'D')$, then $ABCD \barwedge A'B'C'D'$.

3. Prove that Postulate P–9 may be replaced by a postulation of the invariance of cross ratios.

*4. Prove that any one-to-one transformation of a line onto a line such that cross ratios are preserved is a projective transformation.

5. Prove that under the relationship (4–25) we have $\Re(AB,CA) = \infty$, $\Re(AB,CB) = 0$, and $\Re(AB,CC) = 1$.

6. Use the relationship (4–25) and prove that

(a) the cross ratio is unaltered if the two pairs are interchanged;

(b) the cross ratio is unaltered if the elements of each pair are interchanged within the pair;

(c) the relationships in (4–26) hold.

7. Use the relationships in (4–26) and define the cross ratio of four collinear points whenever three of the points (not necessarily the first three) are distinct.

8. Write down the twenty-four permutations of the points A, B, C, D.

9. Prove that

(a) the interchange of the elements of one pair of points changes the cross ratio from k to $1/k$;

(b) the interchange of either the inner or the outer points in the symbol $\Re(AB,CD)$ changes the cross ratio from k to $1 - k$.

10. Group the twenty-four permutations of A, B, C, D into six groups of four each, according to their cross ratios.

11. Prove that every harmonic set of points or lines has cross ratio -1. (*Note:* The cross ratio of four coplanar lines on a point O is defined to be the cross ratio of their traces on any line that is on the plane but not on O.)

12. Define the harmonic conjugate of C with respect to A and B in terms of a cross ratio.

13. Prove that we have $H(AB, CD)$ if and only if $\Re(AB,CD) = -1$.

14. Discuss the cross ratios of the special sets of four points for which the cross ratios are not all distinct.

*15. Prove that separation is a projective invariant.

*16. Verify Postulate P–11 in analytic projective geometry.

*17. Verify Postulate P–13 in analytic projective geometry.

18. Define the segment AB/D (Section 3–4) in terms of cross ratios.

*19. Prove that for any five distinct collinear points

$$\Re(AB,CD) \cdot \Re(AB,DE) = \Re(AB,CE).$$

*20. Prove that if the pairs A, B and C, D separate each other, then the pairs A, C and B, D do not separate each other.

21. Prove that the converse of the statement in Exercise 20 is false by giving a particular set of four points for which it fails.

4–6 Analytic and synthetic geometries. Let us now endeavor to increase our understanding of projective geometry by comparing analytic and synthetic projective geometries. We developed synthetic projective geometry in Chapters 2 and 3 as a deductive system based upon Postulates P–1 through P–14. We also observed that real projective geometry could be postulated by rephrasing Postulate P–14. Analytic projective geometry may also be developed as a postulational system. However, we took advantage of the reader's knowledge of the real numbers and algebra and simply defined the elements of real analytic projective geometry in terms of classes of real numbers and equations. The elements point and line, as well as the relations of incidence and separation that were undefined in synthetic projective geometry, were defined in analytic projective geometry in terms of numbers and number relations. Furthermore, all the postulates of synthetic projective geometry may be proved as theorems in analytic projective geometry. Thus all properties of the geometry based upon these postulates must be properties of analytic projective geometry. In other words, all properties of synthetic projective geometry are properties of analytic projective geometry.

We have also seen that the points, lines, planes, projective transformations, and indeed all elements of analytic projective geometry, may be considered as algebraic representations of corresponding elements of synthetic projective geometry. Therefore all properties of analytic projective geometry are properties of synthetic projective geometry. Thus synthetic projective geometry and analytic projective geometry are equivalent in the sense that they have the same properties. The two geometries simply represent different approaches to the study of projective geometry. They correspond to isomorphic representations of projective geometry.

In a more comprehensive treatment of synthetic projective geometry, the real numbers and their properties may be developed from properties of projective geometry; in analytic projective geometry, the properties of projective geometry may be developed from the real numbers and their properties. Synthetic and analytic projective geometries may each be based upon undefined elements and undefined relations (Exercise 1); i.e., both geometries may be developed from sets of postulates. Thus projective geometry may be developed as a postulational system based upon either a set of postulates in terms of geometric relations (synthetic projective geometry) or a set of postulates in terms of number relations (analytic projective geometry).

Consider the statement: *Two distinct points determine a unique line.* In synthetic projective geometry, this statement is a logical consequence of Postulates P–1 and P–2. In analytic projective geometry, any two distinct points (r_1, r_2, r_3) and (s_1, s_2, s_3) determine a unique line (Exercise 9, Section 4–2)

$$\begin{vmatrix} x_1 & x_2 & x_3 \\ r_1 & r_2 & r_3 \\ s_1 & s_2 & s_3 \end{vmatrix} = 0.$$

In other words, the triple of coefficients of the equation

$$(r_2 s_3 - r_3 s_2) x_1 + (r_3 s_1 - r_1 s_3) x_2 + (r_1 s_2 - r_2 s_1) x_3 = 0$$

is different from the triple 0,0,0 and the equation represents a unique line whenever the triples of numbers (r_1, r_2, r_3) and (s_1, s_2, s_3) are from different classes of number triples representing points.

Several of the comparisons between the elements and methods of synthetic and analytic projective geometry are considered in the exercises at the end of this section. We shall conclude our present discussion with an algebraic proof of the Theorem of Desargues on a plane. Since the principle of planar duality holds in analytic as well as synthetic projective geometry (Exercise 3), we shall consider the theorem in the following form:

If two coplanar triangles ABC *and* $A'B'C'$ *are perspective from a point, they are perspective from a line.*

Suppose $ABC \overset{P}{\barwedge} A'B'C'$. Then by Theorem 4–2 there exists a projective transformation on the plane such that we have

$$A':(1,0,0), \quad B':(0,1,0), \quad C':(0,0,1), \quad P:(1,1,1).$$

The conditions A on $A'P:(x_2 = x_3)$, B on $B'P:(x_1 = x_3)$, C on $C'P:(x_1 = x_2)$ imply that

A has coordinates of the form (a_1,a_2,a_2),
B has coordinates of the form (b_1,b_2,b_1), and
C has coordinates of the form (c_1,c_1,c_3),

where

$$\begin{vmatrix} a_1 & a_2 & a_2 \\ b_1 & b_2 & b_1 \\ c_1 & c_1 & c_3 \end{vmatrix} \neq 0,$$

since, by hypothesis, A, B, C are not collinear. We then have the lines

$A'B': x_3 = 0,$　　　　$B'C': x_1 = 0,$　　　　$C'A': x_2 = 0,$
$AB: a_2(b_1 - b_2)x_1 + b_1(a_2 - a_1)x_2 + (a_1b_2 - a_2b_1)x_3 = 0,$
$BC: (b_2c_3 - b_1c_1)x_1 + b_1(c_1 - c_3)x_2 + c_1(b_1 - b_2)x_3 = 0,$
$CA: a_2(c_3 - c_1)x_1 + (a_2c_1 - a_1c_3)x_2 + c_1(a_1 - a_2)x_3 = 0,$

and the points

$$AB \cdot A'B' = C'': \big(b_1(a_1 - a_2), \quad a_2(b_1 - b_2), \quad 0 \quad \big),$$
$$BC \cdot B'C' = A'': \big(\quad 0, \quad -c_1(b_1 - b_2), \quad b_1(c_1 - c_3)\big),$$
$$AC \cdot A'C' = B'': \big(c_1(a_1 - a_2), \quad 0, \quad a_2(c_1 - c_3)\big).$$

Finally, the points A'', B'', C'' are collinear, since the determinant of their coordinates is equal to

$$(a_1 - a_2)(b_1 - b_2)(c_1 - c_3)(a_2b_1c_1 - a_2b_1c_1) = 0.$$

The above algebraic proof of the Theorem of Desargues on a plane illustrates the effectiveness of algebraic methods. Throughout the remainder of our development of euclidean plane geometry, we shall be primarily concerned with algebraic methods. However, since analytic and synthetic projective geometries are equivalent (isomorphic), we may use either synthetic or algebraic methods at any time.

EXERCISES

1. Compare the undefined elements and undefined relations of synthetic projective geometry with those of analytic projective geometry.

2. Define or identify each of the following in analytic projective geometry on a plane:

(a) point,	(b) line,
(c) incidence of a point and a line,	(d) line AB,
(e) collinear points A, B, C,	(f) $H(AB, CD)$,
(g) segment AB/D,	(h) $R(ABC)$,
(i) projective transformation,	
(j) concurrent lines m, s, t.	

*3. Prove that the principle of planar duality holds in analytic projective geometry.

†4. Repeat Exercise 3 for the principle of space duality.

5. Prove the following theorems in analytic projective geometry:

(a) Theorem 2–1 (Section 2–2)	(b) Theorem 2–2
(c) Theorem 2–3 (Section 2–6)	(d) Theorem 2–4
(e) Theorem 2–5 (Section 2–7)	(f) Theorem 2–8 (Section 2–11)
(g) Theorem 3–3 (Section 3–3)	(h) Theorem 3–4
(i) Theorem 3–5 (Section 3–5).	

6. Repeat the following exercises in analytic projective geometry:

(a) Exercise 4, Section 2–1	(b) Exercise 1, Section 2–2
(c) Exercise 1, Section 3–3	(d) Exercise 1, Section 3–4
(e) Exercises 2 and 3, Section 3–5	(f) Exercises 6 and 7, Section 3–8

4–7 Groups.   We have seen that projective geometry may be considered as a postulational system from the point of view of either synthetic or analytic projective geometry. In this section, we shall define a group and consider projective geometry as a study of properties that are invariant under the group of projective transformations.

Consider the set of projective transformations on a line (4–18). The product of any two projective transformations on a line is a unique projective transformation on the line (Exercise 7, Section 4–3). In other words, the set of projective transformations on a line is *closed.* Given any three projective transformations on a line, say T_1, T_2, and T_3, with matrices

$$\begin{bmatrix} a_1 & b_1 \\ c_1 & d_1 \end{bmatrix}, \quad \begin{bmatrix} a_2 & b_2 \\ c_2 & d_2 \end{bmatrix}, \quad \begin{bmatrix} a_3 & b_3 \\ c_3 & d_3 \end{bmatrix},$$

respectively, we have

$$T_3(T_2 T_1) = (T_3 T_2) T_1.$$

In other words, the transformation obtained when the product $T_2 T_1$ is multiplied by T_3 is the same as the transformation obtained when T_1 is multiplied by the product $T_3 T_2$; that is, the set of projective transformations on a line is *associative*. There exists an *identity* transformation (Section 4–4) in the set of transformations (4–18). Finally, every transformation in the set has a unique transformation of the set as its *inverse* transformation (Exercise 5, Section 4–3). Any set of transformations with these four properties (closed, associative, contains the identity, and contains the inverse of each of its elements) is called a group of transformations. Formally, we say that a set of transformations forms a *group* under multiplication if

　　(i) the product of any two transformations of the set is a uniquely determined transformation of the set;

　　(ii) the transformations of the set are associative under multiplication, i.e., if $(AB)C = A(BC)$;

　　(iii) the set contains an identity transformation I such that $BI = IB = B$ for every transformation B of the set; and

　　(iv) every transformation B of the set has an inverse B' under multiplication where B' is a transformation of the set. Then $BB' = I = B'B$.

We have already observed that the set of projective transformations on a line forms a group. We may also prove that the set of projective transformations on a plane (4–14) forms a group (Exercise 5) and the set of projective transformations in space (4–7) forms a group (Exercise 6).

When we select transformations (some or all) from a given set of transformations, as for example when we select the transformations of the form

(4–27)
$$\begin{aligned} x_1' &= ax_1 + bx_2, \\ x_2' &= x_2 \end{aligned} \qquad (a \neq 0)$$

from the set of projective transformations on a line, we obtain a *subset* of the given set of transformations. If a subset of a given group of transformations forms a group, we call it a *subgroup* of the given group. Throughout our development of euclidean plane geometry we shall be primarily concerned with subgroups of the group of projective transformations on a plane.

Let S represent the subset (4–27) of the group (4–18) of projective transformations on a line. Since the transformations of S are also transformations of the form (4–18), they are associative and the inverse of each transformation exists, although we must prove that the inverse transformations are in the set S. In general, we have (Exercise 7)

THEOREM 4–3. *A subset S of a group of transformations forms a group if the subset is closed and contains the inverse of each of its transformations.*

The properties of projective geometry that are invariant under all projective transformations are called *projective invariants*. As we mentioned in Section 1–8, Felix Klein suggested that any geometry could be considered as the study of the invariant properties under a group of transformations. Under any projective transformation (considered either in synthetic or analytic projective geometry), points correspond to points and lines correspond to lines. Thus the set of all points and the set of all lines of the space are invariant under projective transformations. A particular subset of the set of all points on the space such as the set of points on a particular line may or may not be invariant. If we wish to make the set of points on a particular line invariant, we must restrict the transformations and accordingly consider only a subset of the group of projective transformations. We shall do this in Chapter 5. Throughout our development of euclidean geometry, we shall make extensive use of invariant properties and invariant elements under sets of projective transformations. A few invariant properties and invariant sets of elements of projective geometry are considered in the following set of exercises.

EXERCISES

1. Write down matrices for the general projective transformations on a line, on a plane, and in space.

2. Write down matrices for the identity transformation on a line, on a plane, and in space.

3. Prove that the set of projective transformations is closed on a line, on a plane, and in space.

4. Prove that the set of projective transformations is associative (a) on a plane and †(b) in space.

*5. Prove that the set of projective transformations (4–14) on a plane forms a group.

†6. Prove that the set of projective transformations (4–7) in space forms a group.

*7. Prove Theorem 4–3.

8. Prove that the set of transformations (4–27) forms a group.

9. Prove that the point (1,0) is invariant under the group of transformations (4–27).

10. Prove that each of the following is a projective invariant:
  (a) cross ratio;   (b) separation;
  (c) incidence of a point on a line, a point on a plane, and a line on a plane.

*11. Prove that every invariant of a given group of transformations is an invariant of every subgroup of the given group.

12. Prove that the total set of each of the following figures is a projective invariant on a plane (i.e., prove that each of the following figures is transformed under any projective transformation on the plane into another figure of the same type):
  (a) points,   (b) lines,
  (c) triangles,   (d) complete quadrangles,
  (e) quadrangular sets,   (f) harmonic sets,
  (g) point conics,   (h) line conics.

13. Determine whether or not the subsets of the set of transformations (4–18) satisfying each of the following conditions form groups.
  (a) $a = 0$,   (b) $b = 0$,   (c) $c = 0$,
  (d) $d = 0$,   (e) $a = d = 0$,   (f) $a = c = 0$,
  (g) $a = d = 0$, and $b = c = 1$,   (h) $a = d = 1$, and $b = c = 0$.

14. Determine which of the following sets of transformations form groups:

(a) $x_1' = a_{11}x_1 + a_{13}x_3$
   $x_2' = a_{22}x_2 + a_{23}x_3$ $(a_{11}a_{22} \neq 0)$
   $x_3' = x_3$

(b) $x_1' = -x_1 + a_{13}x_3$
   $x_2' = -x_2 + a_{23}x_3$
   $x_3' = x_3$

(c) $x_1' = x_1 + a_{13}x_3$
   $x_2' = x_2 + a_{23}x_3$
   $x_3' = x_3$

(d) $x_1' = x_2$
   $x_2' = x_1$
   $x_3' = x_3$

(e) $x_1' = x_1$
   $x_2' = x_2$
   $x_3' = x_3$

15. Given any plane figure F, prove that the set of all projective transformations leaving F invariant forms a group.

4-8 Classification of projective transformations. We may identify groups of transformations in terms of the elements of the matrices representing the transformations. For example, each transformation in the group of projective transformations on a line may be represented by a matrix of the form

$$\begin{bmatrix} a & b \\ c & d \end{bmatrix},$$

where $ad - bc \neq 0$ (Section 4–3). Furthermore, the origin corresponds to itself (is an invariant point) under a projective transformation on a line if and only if $b = 0$ in the matrix of the transformation (Exercise 1). In this section we shall consider the elements of the matrices of the group of projective transformations on a line and classify the transformations as hyperbolic, parabolic, or elliptic, according as there are two, one, or no real invariant points under the transformations. We shall also consider a few projective transformations on a plane.

Any projective transformation on a line may be expressed in the form (4–18) in terms of homogeneous coordinates, or in the form (4–19) in terms of nonhomogeneous coordinates of the points. The invariant points of any transformation (4–19) may be identified by the relation $x' = x$. Thus the coordinates of the invariant points of any transformation (4–19) are the points with coordinates that satisfy the equation

$$cx^2 + (d - a)x - b = 0.$$

Under the isomorphism (4–17) the invariant points of any transformation (4–18) are the points with coordinates that satisfy the equation

(4–28) $$cx_1^2 + (d - a)x_1x_2 - bx_2^2 = 0.$$

If $c = 0$, then $x_2 = 0$ is a root of the equation (4–28) and the point $(1,0)$ is an invariant point. If $c = 0$ and $d - a = 0$, then $x_2 = 0$ is a double root of the equation. In general, when a, b, c, and d are real numbers, each of the above equations has real and distinct, real and equal, or two imaginary roots (invariant points) according as $(d - a)^2 + 4bc$ is positive, zero, or negative. The corresponding projective transformations of the line onto itself are defined to be *hyperbolic*, *parabolic*, or *elliptic*, according as there are two, one, or no real distinct invariant points.   Notice that when the points are represented by classes of pairs of complex numbers (i.e., in complex projective geometry), every projective transformation on a line has two (not necessarily distinct) invariant points on the complex line.   We shall be primarily concerned with projective transformations of real projective lines and planes, i.e., with real projective geometry.   The distinction between real and complex projective geometry may be easily made from the point of view of analytic projective geometry, in which case complex projective geometry is based upon classes of complex numbers and real projective geometry is based upon classes of real numbers.   Matrices of transformations have complex numbers as elements in complex projective geometry and real numbers as elements in real projective geometry.   By the above definition we have

> THEOREM 4–4.   *A projective transformation,* (4–18) *or* (4–19), *of a real projective line onto itself is hyperbolic, parabolic, or elliptic, according as* $(d - a)^2 + 4bc$ *is positive, zero, or negative.*

We now consider a special type of projective transformation.   Any transformation that is not an identity transformation but which when applied twice (i.e., its square) gives rise to the identity transformation, is called an *involution*.   For example, on a line the transformation

$$x_1' = x_2, \quad x_2' = x_1$$

is an involution and has the form $x' = 1/x$ in terms of nonhomogeneous coordinates.   In general, any involution of a projective line onto itself may be expressed in the form (Exercise 3)

$$(4\text{--}29) \qquad \begin{aligned} x_1' &= ax_1 + bx_2, \\ x_2' &= cx_1 - ax_2 \end{aligned} \qquad (a^2 + bc \neq 0).$$

We shall consider the construction of involutions on a line in Section 4–11 and shall use involutions in the definition of perpendicular lines in euclidean geometry (Section 6–1).   Any projective transformation on a line may be expressed as a product of two involutions (Exercise 5).

Projective transformations on a real line have been classified in terms of the number of invariant points on the line under the transformation.   Also this classification has been expressed in terms of the elements of the matrices of the transformations (Theorem 4–4). We next consider projective transformations on a plane and prove that under a transformation of a real projective plane onto itself there must be at least one invariant point and at least one invariant line. A point (x_1,x_2,x_3) is invariant under a projective transformation (4–11) if and only if $(x'_1,x'_2,x'_3) = (x_1,x_2,x_3)$, i.e., if and only if there exists a real number $k \neq 0$ such that $x'_j = kx_j$ $(j = 1,2,3)$. Thus the invariant points of any projective transformation (4–11) of a plane onto itself must satisfy the system of equations

$$0 = (a_{11} - k)x_1 + a_{12}x_2 + a_{13}x_3,$$
$$0 = a_{21}x_1 + (a_{22} - k)x_2 + a_{23}x_3,$$
$$0 = a_{31}x_1 + a_{32}x_2 + (a_{33} - k)x_3.$$

This system of linear homogeneous equations has a solution different from $x_j = 0$ $(j = 1,2,3)$ if and only if the determinant of the coefficients is equal to zero [10; 209], i.e., if and only if

$$(4\text{--}30) \qquad \begin{vmatrix} a_{11} - k & a_{12} & a_{13} \\ a_{21} & a_{22} - k & a_{23} \\ a_{31} & a_{32} & a_{33} - k \end{vmatrix} = 0.$$

Since the equation (4–30) is a cubic equation in k with real coefficients, there always exists at least one real value of k for which the equation is satisfied [10; 142]. Such a value of k gives rise to an invariant point of the transformation (4–11). Thus every projective transformation of a plane onto itself has at least one invariant point. By the principle of planar duality, every projective transformation on a plane also has at least one invariant line. Note that a line is invariant if the line corresponds to itself. A particular line on an invariant point is not necessarily invariant; a particular point on an invariant line is not necessarily invariant. The projectivity imposed on the points of an invariant line by the projective transformation on the plane may be hyperbolic, parabolic, or elliptic. When the transformation on the line is the identity transformation (i.e., when each point on the line corresponds to itself), the line is said to be *pointwise invariant*.

We conclude this chapter with a discussion of conics. We shall develop an algebraic representation for conics, use this representation to obtain properties of conics, and use some of these properties to construct involutions on a line. These concepts will provide a basis not only for the properties of conics in euclidean geometry but also for the development of euclidean (Chapter 6) and noneuclidean (Chapter 8) geometries from projective geometry.

EXERCISES

1. Prove that the origin on a projective line is an invariant point under the group of transformations (4–18) if and only if $b = 0$.

2. Classify the projective transformation associated with each of the following matrices:

(a) $\begin{bmatrix} 1 & 0 \\ 1 & -1 \end{bmatrix}$, (b) $\begin{bmatrix} 2 & 3 \\ 3 & 6 \end{bmatrix}$, (c) $\begin{bmatrix} 2 & 1 \\ 1 & 2 \end{bmatrix}$,

(d) $\begin{bmatrix} 3 & 7 \\ 5 & -3 \end{bmatrix}$, (e) $\begin{bmatrix} 1 & 0 \\ 0 & 1 \end{bmatrix}$, (f) $\begin{bmatrix} 0 & 1 \\ 1 & 0 \end{bmatrix}$.

*3. Prove that a transformation (4–18) represents an involution if and only if it has the form (4–29).

4. Consider the matrices given in Exercise 2 and list the ones that correspond to involutions.

*5. Prove that any transformation of the form (4–18) may be expressed as the product of two involutions [16; 223–224].

6. Prove that all involutions (4–29) are either hyperbolic or elliptic, i.e., that there are no parabolic involutions on a line.

7. Prove that any involution on a line having P_0 and P_∞ as invariant points may be expressed in the form

$$x' = sx, \qquad\qquad (s \neq 0)$$

in terms of the nonhomogeneous coordinates of the points on the line.

8. Prove that a projective transformation of a line onto itself may be used to express the involution $x_1' = x_2,\ x_2' = x_1$ in the form (4–29).

9. Prove that a projective transformation of a line onto itself may be used to express any involution (4–29) in the form (a) $x_1' = x_2,\ x_2' = x_1$, (b) $x_1' = x_2,\ x_2' = -x_1$.

*10. Prove that any two involutions on a line are equivalent under a projective transformation.

4–9 Polarities and conics.   Many classical theorems may be considered in a thorough study of polarities and conics. We shall consider a few of these theorems in preparation for the development of euclidean plane geometry and the noneuclidean geometries.

A class of triples of real numbers may be considered either as a point (x_1, x_2, x_3) or as a line $[u_1, u_2, u_3]$. The projective transformations of real analytic projective geometry on a plane were defined as nonsingular linear homogeneous transformations of triples considered as points into triples considered as points. Since points correspond to points and lines correspond to lines under these transformations, projective transformations are often called *collineations*. We now consider nonsingular linear homogeneous transformations of triples considered as points into triples considered as lines. These transformations are called *correlations* and may be expressed in the form

$$(4\text{-}31) \qquad \begin{aligned} u_1 &= a_{11}x_1 + a_{12}x_2 + a_{13}x_3, \\ u_2 &= a_{21}x_1 + a_{22}x_2 + a_{23}x_3, \\ u_3 &= a_{31}x_1 + a_{32}x_2 + a_{33}x_3, \end{aligned} \qquad \begin{aligned} |\, a_{ij}\,| &\neq 0 \\ (i,j &= 1,2,3), \end{aligned}$$

where $[u_1,u_2,u_3]$ represents a line, and (x_1,x_2,x_3) represents a point.

The correlation (4-31) transforms each point (x_1,x_2,x_3) into a line $[u_1,u_2,u_3]$ containing points (y_1,y_2,y_3) such that $u_1y_1 + u_2y_2 + u_3y_3 = 0$; that is,

$$(4\text{-}32) \quad (a_{11}x_1 + a_{12}x_2 + a_{13}x_3)y_1 + (a_{21}x_1 + a_{22}x_2 + a_{23}x_3)y_2 \\ + (a_{31}x_1 + a_{32}x_2 + a_{33}x_3)\,y_3 = 0.$$

Thus the equation (4-32) is the condition that the point (y_1,y_2,y_3) be on the line corresponding to the point (x_1,x_2,x_3) under the correlation (4-31). Similarly, the equation

$$(4\text{-}33) \quad (a_{11}y_1 + a_{12}y_2 + a_{13}y_3)x_1 + (a_{21}y_1 + a_{22}y_2 + a_{23}y_3)x_2 \\ + (a_{31}y_1 + a_{32}y_2 + a_{33}y_3)x_3 = 0$$

is the condition that the point (x_1,x_2,x_3) be on the line corresponding to the point (y_1,y_2,y_3) under the correlation (4-31). When (4-32) and (4-33) are equivalent for all points, i.e., when $a_{ij} = a_{ji}$ (Exercise 1), the correlation is called a *polarity*. In other words, a polarity is a correlation such that any point Y is on the line corresponding to a point X if and only if X is on the line corresponding to Y.

Any polarity may be expressed in the form

$$(4\text{-}34) \qquad \begin{aligned} u_1 &= a_{11}x_1 + a_{12}x_2 + a_{13}x_3, \\ u_2 &= a_{12}x_1 + a_{22}x_2 + a_{23}x_3, \\ u_3 &= a_{13}x_1 + a_{23}x_2 + a_{33}x_3, \end{aligned} \qquad \begin{aligned} |\, a_{ij}\,| &\neq 0 \\ (i,j &= 1,2,3). \end{aligned}$$

Under a polarity the line $[u_1,u_2,u_3]$ is called the *polar* of the point (x_1,x_2,x_3) and the point is called the *pole* of the line. Then, from the above definition, a point X is on the polar of a point Y if and only if Y is on the polar of X. A point that is on its own polar is called a *self-conjugate point* of the polarity. The condition that a point (x_1, x_2, x_3) be self-conjugate under a polarity (4–34) is (Exercise 6)

$$(4\text{–}35) \quad a_{11}x_1^2 + 2a_{12}x_1x_2 + 2a_{13}x_1x_3 + a_{22}x_2^2 + 2a_{23}x_2x_3 + a_{33}x_3^2 = 0.$$

The polarity is said to be *elliptic* when (4–35) is not satisfied by any triples of real numbers (real points) and *hyperbolic* when there exist real points satisfying the equation (4–35). Thus a polarity that has no self-conjugate points on a real projective plane is called an elliptic polarity, and a polarity that has self-conjugate points is called a hyperbolic polarity.

The importance of polarities in our development of euclidean geometry lies in the following definition in analytic projective geometry: The set of self-conjugate points of a hyperbolic polarity is a *point conic*. The equation (4–35) represents a conic in real analytic projective geometry if and only if there exists at least one real point satisfying the equation. Note that the classification of polarities as elliptic and hyperbolic does not refer to the conic as an ellipse or a hyperbola since the conic exists on a real plane only for hyperbolic polarities. The condition that the polarity have determinant different from zero is precisely the condition in euclidean geometry that the conic be nondegenerate. Thus, as in synthetic projective geometry, the degenerate conics of euclidean geometry are not considered as conics in analytic projective geometry.

The equation of the polar of any point (x_1', x_2', x_3') has the form (4–33) where $a_{ij} = a_{ji}$; that is,

$$a_{11}x_1x_1' + a_{12}(x_2x_1' + x_1x_2') + a_{13}(x_3x_1' + x_1x_3') + a_{22}x_2x_2' \\ + a_{23}(x_3x_2' + x_2x_3') + a_{33}x_3x_3' = 0.$$

This equation is sometimes called the *scratch equation* of (4–35), since it may be written down directly from (4–35) by "scratching" "half" of the x_j's. In euclidean geometry we may use the scratch equation of a conic to determine the equation of the tangent at a point on the conic or the equation of the line joining the points of tangency of the tangents from a point outside the conic. For example, the circle $x^2 + y^2 = 25$ has scratch equation $xx_1 + yy_1 = 25$. When (x_1, y_1) is a point on the circle such as $(3,4)$, we may use the scratch equation to obtain the equation $3x + 4y = 25$ of the tangent

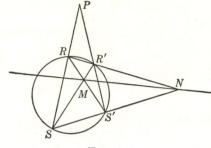

FIG. 4-1

line at that point. When (x_1,y_1) is a point outside the circle such as (7,5), we may use the scratch equation to obtain the equation $7x + 5y = 25$ of the line joining the points of tangency (0,5) and $(\frac{175}{37},-\frac{60}{37})$ of the two tangents from the point (7,5) to the conic. In general, the scratch equation represents the polar of the point with respect to the polarity defining the conic or, as we usually say, the polar with respect to the conic.

The above applications of the scratch equation (i.e., the equation of the polar of a point with respect to a conic) illustrate the following definitions. The polar p of a point P on a conic intersects the conic in exactly one point, the point P (Exercise 13). The line p is called the *tangent* to the conic at P. If, on a real projective plane, the polar of a point with respect to a conic does not intersect the conic, the point is said to be an *interior point* of the conic. If the polar of a point with respect to a conic intersects the conic in two distinct points, the point is said to be an *exterior point* of the conic.

A construction for the polar of any point on the plane of a given conic may be given in terms of the above concepts. Given any conic and any point P that is not on the conic, we may draw two lines through P such that one line intersects the conic in distinct points, say R and S, and the other line intersects the conic in distinct points, say R' and S'. The polar of the given point P is then completely determined by the two points $M = RS' \cdot R'S$ and $N = RR' \cdot SS'$ (Fig. 4-1). In other words, the polar of any point P on the plane of a conic, but not on the conic, may be obtained as the side MN opposite P in the diagonal triangle of a complete quadrangle $RR'SS'$ inscribed in the conic and having P as a diagonal point (Exercises 18 and 19). The pole of any line p on the plane of the conic may be constructed as the intersection of the polars of two of its points (Exercise 20). Finally, the polar of a point P on a conic (i.e., the tangent to the conic at a point P) may be constructed as the line joining P and the point of intersection of the polars of any two points collinear with P (Exercise 21).

EXERCISES

*1. Prove that (4–32) and (4–33) are equivalent for all points (x_1,x_2,x_3) if and only if $a_{ij} = a_{ji}$.

2. Prove that the product of any two correlations is a collineation.

3. Prove that the product of any even number of correlations is a collineation.

4. Prove that the product of any odd number of correlations is a correlation.

5. Prove that if P is the pole of p under a given polarity, then p is the polar of P under that polarity.

*6. Prove that the self-conjugate points under a polarity (4–34) satisfy (4–35).

7. Prove that if the polars of P and Q intersect at R, then the line PQ is the polar of R.

8. Prove that the polars of distinct points are distinct.

9. Prove that under any hyperbolic polarity there exists a triangle ABC on the real plane such that A is the pole of BC, B is the pole of AC, and C is the pole of AB. Such a triangle is called a *self-polar* triangle under the polarity.

*10. Define a line conic as the plane dual of a point conic.

*11. Prove that the set of tangents of a point conic form a line conic.

12. Prove that the matrix of a hyperbolic polarity cannot be the identity matrix.

*13. Given a conic and a point P on the conic, prove that the polar of P with respect to the conic intersects the conic in exactly one point, the point P.

14. Prove that any conic contains at least six points.

15. Consider the determinant of the corresponding hyperbolic polarity and prove that the equation

$$4x_1^2 + 4x_1x_2 + 2x_1x_3 + 6x_2^2 - 4x_2x_3 + 1.5x_3^2 = 0$$

does not represent a conic in projective geometry.

16. Given a conic and an exterior point P, let the points at which the polar of P intersects the conic be R and S. Prove that the lines PR and PS are tangent to the conic.

*17. Given a conic on a real projective plane, prove that every point on the plane is a point on the conic, an exterior point, or an interior point.

18. Given a conic on a real projective plane and any point P that is on the plane but not on the conic, prove that there exists a complete quadrangle inscribed in the conic and having P as a diagonal point.

19. Assume that the Theorem of Pascal (Section 2–13) may be applied to limiting cases of hexagons such as $RRS'SSR'$ in Fig. 4–1 where RR and SS are taken as tangent lines to the conic. Under this assumption that a quadrangle may be taken as a limiting case of a hexagon, prove that if P is a diagonal point of a complete quadrangle inscribed in a conic, then the polar of P with respect to the conic is the side opposite P in the diagonal triangle of the complete quadrangle.

20. Given any conic, prove that the pole of any line p on the plane of the conic is the point of intersection of the polars of any two of its points.

21. Given any conic and a point P on the conic, prove that the tangent to the conic at P may be constructed as the line determined by P and $m \cdot n$, where m and n are the polars with respect to the conic of any two points M, N that are collinear with P.

22. In euclidean geometry, write down the scratch equation of each of the following conics:

(a) $x^2 + y^2 - 36 = 0$,   (b) $3x^2 - y^2 + 5 = 0$,
(c) $x^2 - 4y + 5 = 0$,   (d) $y^2 + 6x = 0$,
(e) $x^2 + y^2 - 4x + 2y + 4 = 0$.

23. Find the polar line of the point $(5,3)$ with respect to each of the conics in Exercise 22 and identify the point $(5,3)$ as a point of the conic, an exterior point, or an interior point with respect to each conic.

4–10 Conics. We have defined a point conic or conic in synthetic projective geometry (Section 2–12) as the set of all points of intersection of two projective nonperspective pencils of lines that are on the same plane but not on the same point. In Section 4–9 we defined a conic in analytic projective geometry as the set of self-conjugate points under a hyperbolic polarity, i.e., the set of points satisfying an equation of the form (4–35)

$$a_{11}x_1^2 + 2a_{12}x_1x_2 + a_{22}x_2^2 + 2a_{13}x_1x_3 + 2a_{23}x_2x_3 + a_{33}x_3^2 = 0,$$

when such points exist on a real projective plane and

$$\begin{vmatrix} a_{11} & a_{12} & a_{13} \\ a_{12} & a_{22} & a_{23} \\ a_{13} & a_{23} & a_{33} \end{vmatrix} \neq 0.$$

The second condition, $|a_{ij}| \neq 0$, is needed to ensure the existence of a hyperbolic polarity having the conic as its set of self-conjugate points, since the determinant $|a_{ij}|$ is the determinant of the matrix of the polarity. In this section we shall prove the Theorem of Steiner (Theorem 2–9, Section 2–12) in analytic projective geometry and thereby prove that the above definitions of a conic are equivalent.

We first prove that in analytic projective geometry any conic contains at least three points. Then we shall use three points of a conic and a change of coordinate system to simplify the equation of the conic. By definition any conic contains at least one point, say A. Let $C \neq A$ be a point on the polar of A with respect to the conic. Since the line AC is the tangent to the conic at A, the point C is not on the conic (Exercise 13, Section 4–9) and the polar of C is a line through A but not through C. Then C is an exterior point of the conic and the polar of C intersects the conic in distinct points A and B. Furthermore, if E is a third point on the line AC, then the polar of E is a line AD where $D \neq B$ and D is a point of the conic. We have now proved that any conic in analytic projective geometry contains at least three distinct points.

Consider any conic and let A, B, D be any three distinct points on the conic. Let the intersection of the polars of A and B be C and consider the coordinate system determined by the four points

$$A:(0,0,1), \quad B:(1,0,0), \quad C:(0,1,0), \quad D:(1,1,1)$$

(Exercise 1). In the new coordinate system the polar of A is AC:[1,0,0] and the polar of B is BC:[0,0,1]. These results may be used as follows to determine the new equations of the polarity (4–34) and the conic (4–35).

The polar of the point $A:(0,0,1)$ under the polarity (4–34) has the line coordinates $[u_1,u_2,u_3] = [a_{13},a_{23},a_{33}]$. Then, since AC is the polar of A,

$$[a_{13},a_{23},a_{33}] = [1,0,0],$$

and therefore

$$a_{13} \neq 0, \quad a_{23} = 0, \quad a_{33} = 0.$$

Similarly, the polar of B is

$$[a_{11},a_{12},a_{13}] = [0,0,1],$$

and therefore

$$a_{11} = 0, \quad a_{12} = 0, \quad a_{13} \neq 0.$$

These results imply that the equation of the given conic may be expressed in the form

$$2a_{13}x_1x_3 + a_{22}x_2^2 = 0.$$

The fact that $D:(1,1,1)$ is on the conic then implies that the equation of the given conic may be expressed in the form

(4–36) $$x_2^2 - x_1x_3 = 0$$

in the new coordinate system. We have proved that any conic may be expressed in the form (4–36).

The Theorem of Steiner states that if A and B are any two distinct points on a conic and if P is a variable point on the conic, then the pencils of lines $A[P]$ and $B[P]$ are projective but not perspective. Given any conic and any two points A and B on the conic, we may select a coordinate system such that we have $A:(0,0,1)$, $B:(1,0,0)$, and the equation of the given conic is of the form (4–36). Then all lines on the point A have equations of the form $u_1x_1 + u_2x_2 = 0$; all lines on the point B have equations of the form $v_2x_2 + v_3x_3 = 0$. When the line coordinates $[u_1,u_2,0]$ and $[0,v_2,v_3]$ of these lines are considered, we may associate nonhomogeneous coordinates $u_1/u_2 = k$ with the lines a_k on A and nonhomogeneous coordinates $v_2/v_3 = j$ with the lines b_j on B. The lines

$$u_2x_2 = -u_1x_1 \quad \text{and} \quad v_2x_2 = -v_3x_3$$

of the two pencils intersect on the given conic $x_2^2 - x_1x_3 = 0$ if and only if $u_2v_2 = u_1v_3$, that is, if and only if the nonhomogeneous coordinates k and j are equal. In other words, when P is a variable point on the conic, the correspondence of AP to BP is precisely the correspondence of the lines a_k and b_k in the above coordinate systems on the pencils of lines. This correspondence is a one-to-one correspondence that preserves cross ratio

$$\Re(a_\infty a_0, a_1 a_k) = k = \Re(b_\infty b_0, b_1 b_k),$$

and, by the plane dual of Exercise 4, Section 4–5, the correspondence is a projective correspondence. Since the line AB with equation $x_2 = 0$ is the line $[0,u_2,0]$ on A and also the line $[0,v_2,0]$ on B, we have $a_0 = b_\infty$. Finally, we recall that two projective pencils of points (lines) are perspective if and only if their common point (line) corresponds to itself. Then, since the line $AB = a_0 = b_\infty$ does not correspond to itself under the correspondence of a_k to b_k, the correspondence $A[P] \mathrel{\underset{\wedge}{-}} B[P]$ is a projective nonperspective correspondence. This completes the proof of the Theorem of Steiner in analytic projective geometry.

We have proved that any conic may be represented (using a suitable coordinate system) by the equation (4–36) such that any two given points on the conic may be taken as $A:(0,0,1)$ and $B:(1,0,0)$. Furthermore, if P is a variable point on the conic, the lines $AP = a_k$ and $BP = b_k$ correspond in a projective nonperspective correspondence $A[P] \mathrel{\underset{\wedge}{-}} B[P]$. Conversely, we may now consider any four points A, B, C, D; establish a coordinate system for the pencil of lines on A such that

$$AB = a_0, \quad AC = a_\infty, \quad \text{and} \quad AD = a_1;$$

establish a coordinate system for the pencil of lines on B such that

$$BA = b_\infty, \quad BC = b_0, \quad \text{and} \quad BD = b_1;$$

and use the projective nonperspective correspondence of a_k to b_k to define the conic. In this case three points A, B, and D on the conic determine the lines $AB = a_0$, $AD = a_1$, $BA = b$, and $BD = b_1$; the tangent lines $AC = a_\infty$ and $BC = b_0$ determine the fourth point C and therefore complete the determination of the conic. In other words, a conic is completely determined by any three of its points and the tangents at two of them. Since A and B are any two points on the given conic, we have proved that in analytic projective geometry a conic may be defined by two projective nonperspective pencils of lines. Finally, since any two coordinate systems for a pencil of lines are equivalent under a projective transformation (Exercise 8, Section 3–8), the definitions of conics in synthetic and analytic projective geometries are equivalent.

The Theorem of Pascal (Theorem 2–10) may also be proved for the conics of analytic projective geometry (Exercise 7). The construction for points of a conic when five of its points are given (Exercise 1, Section 2–13) may be extended to obtain a construction for points of a conic having given (Exercise 9)

    (i) five of its points;
    (ii) four of its points and a tangent at one of them; or
    (iii) three of its points and the tangents at two of them.

Additional properties of conics may be found in advanced texts of either synthetic or analytic projective geometry. We shall conclude this chapter with a brief discussion of the use of conics in the construction of involutions on a line.

EXERCISES

1. Prove that if A, B, and D are distinct points on a conic and if the tangents to the conic at A and B intersect at C, then no three of the points A, B, C, D are collinear.

2. Prove that if $a_{13} \neq 0$ and $a_{23} = a_{33} = a_{11} = a_{12} = 0$, then the polar of $(0,1,0)$ is $[0,1,0]$.

3. Prove that a conic is completely determined by any triangle that is self-polar (Exercise 9, Section 4–9) with respect to the polarity that defines the conic.

4. Prove that any five distinct points on a conic completely determine the conic.

5. Prove that a conic is completely determined by any four points of the conic and the tangent at one of them.

6. Prove that a conic is completely determined by any three points of the conic and the tangents at two of them.

7. Prove the Theorem of Pascal in analytic projective geometry.

8. Prove the Theorem of Pappus in analytic projective geometry.

9. Give a construction for additional points of a conic having given
    (a) five points of the conic,
    (b) four points and the tangent at one of them,
    (c) three points and the tangents at two of them.

4–11 Involutions on a line. We have defined an involution as a transformation that is not the identity but whose square is the identity (Section 4–8). On a projective line any involution may be represented in the form (4–29)

$$\begin{aligned} x_1' &= ax_1 + bx_2, \\ x_2' &= cx_1 - ax_2 \end{aligned} \qquad (a^2 + bc \neq 0).$$

If $a^2 + bc$ is positive, the involution is hyperbolic; if $a^2 + bc$ is negative, the involution is elliptic (Theorem 4–4). In this section we shall give a construction for hyperbolic and elliptic involutions on a real projective line.

The notion of a conic cut by a line in two points is familiar to all. The lines of a pencil of lines with a point O on the conic as center

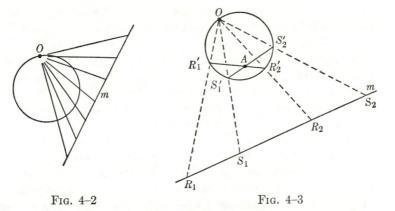

FIG. 4–2   FIG. 4–3

are in one-to-one correspondence (Fig. 4–2) with the points P of the conic, since each line on the pencil cuts the conic in two points, say O and P. Thus each point P on the conic is on exactly one line of the pencil of lines through O. As the point P approaches the point O on the conic, the line OP approaches the tangent line to the conic at O. Thus under the above correspondence the point O corresponds to the tangent line to the conic at O.

Let m be any line on the plane of the conic but not on O (Fig. 4–2). Then every line of the pencil of lines on O intersects m at a unique point P' giving a one-to-one correspondence between the points P on the conic and the points P' on the line m. We say that the conic is projected from O or *mapped* onto the line m and, conversely, the line is mapped onto the conic.

Consider a pencil of lines on the plane of a conic and such that the center A of the pencil is an interior point of the conic. Each line of the pencil intersects the conic in a pair of distinct points, say R_1' and R_2', which may be mapped from a point O on the conic onto points R_1 and R_2 on any given line m. Also given any point S_1 on the line m, we may find a corresponding point S_2 on m by taking the second intersection of the line OS_1 and the conic as S_1', the second intersection of $S_1'A$ and the conic as S_2', and the intersection of m and OS_2' as S_2 (Fig. 4–3). Under this correspondence S_1 corresponds to S_2 and S_2 corresponds to S_1. Such a correspondence between points S_1 and S_2 on the line m is a projective correspondence and an involution. The points S_1 and S_2 are called *conjugate points* of the involution.

Given any conic and a line m on the plane of the conic, any involution on m is uniquely determined by two pairs of points, say R_1, R_2 and S_1, S_2 on m. Let O be a point on the conic; let R_1', R_2', S_1', and S_2' be determined as intersections of the lines OR_1, OR_2, OS_1, and OS_2 with the conic; and let $R_1'R_2' \cdot S_1'S_2' = A$. Then the conjugate Q_2 of any point Q_1 on m may be constructed as above. We call the point A the *center* of the involution for the given conic, point O, and line m. If A is an exterior point, the two points of contact of the tangents from A to the conic project (or map) onto the fixed points of the involution. Thus the involution is hyperbolic if A is an exterior point and elliptic if A is an interior point. We shall use an elliptic involution when we define perpendicular lines. The above mappings are also useful in the sense that many properties of involutions are intuitively evident from the mappings (Exercise 4).

EXERCISES

1. Prove that an involution on a line is completely determined by any two pairs of corresponding points.

2. Derive algebraic representations for involutions on a line having
    (a) P_0 as an invariant point,
    (b) P_∞ as an invariant point,
    (c) P_0 and P_∞ as a pair of the involution.

3. Given a projective transformation on a line and a pair of distinct points A, B such that A corresponds to B and B corresponds to A under the transformation, prove that the transformation is an involution.

4. Sketch a conic and draw figures illustrating each of the following:
    (a) an elliptic involution on m;
    (b) a hyperbolic involution on m;
    (c) an involution in which the pairs of corresponding points separate each other;
    (d) an involution in which the pairs of corresponding points do not separate each other;
    (e) two distinct involutions having a common real pair;
    (f) two distinct involutions having a common imaginary pair;
    (g) two distinct involutions having a real double point as their common pair;
    (h) an involution uniquely determined by
        (i) two pairs of conjugate elements,
        (ii) two double points,
        (iii) one double point and one other pair of conjugate elements.

5. Given any two distinct points P_q and P_r where q and r are elements of the extended real number system, prove that the pairs of points P_x and P'_x satisfying the relation $H(P_qP_r, P_xP'_x)$ are pairs of a hyperbolic involution with fixed points P_q and P_r.

6. Prove that the involution in Exercise 5 may be expressed in the form

$$x'_1 = (q + r)x_1 - 2qrx_2,$$
$$x'_2 = 2x_1 - (q + r)x_2$$

in terms of the homogeneous coordinates of points on a line and in the form

$$x' = \frac{(q + r)x - 2qr}{2x - (q + r)}$$

in terms of nonhomogeneous coordinates.

7. Prove that in nonhomogeneous coordinates the relation $H(P_qP_r, P_sP_t)$ holds if and only if

$$2st - (q + r)(s + t) + 2qr = 0$$

and in homogeneous coordinates

$$H(P_{(q_1,q_2)}P_{(r_1,r_2)}, P_{(s_1,s_2)}P_{(t_1,t_2)})$$

holds if and only if

$$2s_1t_1r_2q_2 - (q_1r_2 + r_1q_2)(s_1t_2 + t_1s_2) + 2q_1r_1s_2t_2 = 0.$$

8. Prove that the following relationships hold:
   (a) $H(P_\infty P_0, P_h P_{-h})$,   (b) $H(P_1 P_{-1}, P_b P_{1/b})$,
   (c) $H(P_\infty P_{(a-1)/b}, P_{(a-2)/b}P_{a/b})$,   (d) $H(P_\infty P_0, P_{a/b}P_{-a/b})$.

9. Prove that if P_t is in $R(P_\infty P_0 P_1)$, then either $t = \infty$ or t is a rational number.

10. Prove that if P_t is a point on a line with reference points P_∞, P_0, and P_1 and t is a rational number, then P_t is in $R(P_\infty P_0 P_1)$.

11. Prove that the following sets of points on a given line are equivalent:
   (a) $R(P_\infty P_0 P_1)$,
   (b) the set of rational points and P_∞ on the line with reference points P_∞, P_0, P_1,
   (c) the set of points harmonically related to P_∞, P_0, P_1, and
   (d) the set of points projectively related to P_∞, P_0, P_1.

4–12 Survey. We have now considered the properties of projective geometry that we shall need as we develop euclidean and non-euclidean geometries. In the following chapters we shall think of a geometry as a study of properties that are invariant under a group of transformations and shall consider several subgroups of the group of projective transformations.

Our development of projective geometry has been based upon a postulational development of synthetic projective geometry (Chapters 2 and 3), the definition of analytic projective geometry, and a proof of the equivalence (isomorphism) of analytic and synthetic projective geometries. We have used both synthetic and algebraic representations in our study of properties of projective geometry. In particular, we have found (among other results) that

(i) the set of projective transformations on a line forms a group;

(ii) the set of projective transformations on a plane forms a group;

(iii) projective transformations on a line may be classified in terms of their invariant points;

(iv) conics may be defined in terms of invariant points under correlations;

(v) conics may be used to construct involutions on a line.

Throughout the last three chapters we have been primarily concerned with properties that are invariant under groups of projective transformations. In the next chapter, we shall study properties that are invariant under subgroups of the group of projective transformations which leave a particular plane invariant in projective space.

REVIEW EXERCISES

1. Write down the matrix representations for general projective transformations on a line, on a plane, and in space.

2. Repeat Exercise 1 for identity transformations.

3. Find $\Re(AB,CD)$ when
   (a) $\Re(AB,CD) = \Re(AC,BD)$,
   (b) $\Re(AB,CD) = \Re(AB,DC)$.

4. Indicate which of the following matrices represents groups of transformations:

(a) $\begin{bmatrix} 0 & b \\ c & d \end{bmatrix}$,   (b) $\begin{bmatrix} a & 0 \\ 0 & -a \end{bmatrix}$.

5. State equations for an elliptic polarity.

6. Prove that if A, B, and C are distinct points on a conic, then the points A, B, C are not collinear.

7. Prove that any conic contains infinitely many points.

8. Given a conic and a line m on the plane of the conic, construct the pole of m with respect to the conic.

9. Construct an elliptic involution on a line. Give equations for an elliptic involution on a line.

10. Compare analytic and synthetic projective geometries.

11. Find the scratch equation of each of the following conics:

    (a) $a^2x^2 + b^2y^2 = a^2b^2$,   (b) $a^2x^2 - b^2y^2 = a^2b^2$,

    (c) $2y = ax^2 + 2bx + c$,   (d) $xy = 2a$.

[*Note:* The answers obtained in Exercise 11 may be checked either by considering the polar lines of points (x_1, y_1) or by using the methods of differential calculus and determining the equations of tangent lines at points (x_1, y_1) on the conics].

12. If A and B are distinct self-conjugate points under a polarity, prove that no other point on the line AB is self-conjugate under the polarity.

13. Given a polarity and a line m that is not self-conjugate under the polarity, prove that for all points P on m with polars p the correspondence of the points P with the points $P' = p \cdot m$ is an involution.

14. Prove that any polarity induces an involution on any point that is not a self-conjugate point.

15. If A and B, C and D, E and F are pairs of an involution on a line and if $ABCDEF \overset{\wedge}{=} A'B'C'D'E'F'$, prove that A' and B', C' and D', E' and F' are pairs of an involution.

CHAPTER 5

AFFINE GEOMETRY

We now proceed to place further restrictions on the projective space and projective transformations discussed in Chapters 2, 3, and 4. The points of an arbitrary plane $x_4 = 0$ will be taken as ideal points or points at infinity. Then every projective line that is not on the plane $x_4 = 0$ will have exactly one ideal point. These ideal points will be used in the definitions of parallelism, mid-point, and several other familiar concepts. Throughout this chapter we shall consider a geometry as a study of properties that are invariant under a group of transformations. The group of projective transformations on a plane will be specialized to obtain the subset of transformations leaving a fixed ideal line invariant, the group of affine transformations. Then the sets of homothetic transformations, translations, dilations, point reflections, line reflections, equiaffine transformations, and equiareal transformations will be considered as subsets of the set of affine transformations. Some of these subsets form groups, others do not. We shall use synthetic as well as algebraic methods as we study properties that are invariant under these sets of transformations. In this way we shall compare and illustrate the advantages of these two methods.

5–1 Ideal points. We now designate an arbitrary plane in projective space as the *ideal plane* or *plane at infinity*. Since any two planes are equivalent in projective geometry, we may use a change of coordinates (projective transformation of space onto itself) so that the ideal plane has the equation $x_4 = 0$. Then the points $(x_1, x_2, x_3, 0)$ are on the ideal plane and are called *ideal points* or points at infinity. All other points have homogeneous coordinates of the form $(x_1, x_2, x_3, 1)$, may be represented by nonhomogeneous coordinates (x, y, z), and are called *ordinary points*. Similarly, the lines on the ideal plane are called *ideal lines* or lines at infinity. All other lines are called *ordinary lines*. All planes that are distinct from the ideal plane are called *ordinary planes*. Throughout the remainder of our development of euclidean geometry from projective geometry, all points, lines, and planes are assumed to be ordinary unless otherwise specified.

The ideal plane $x_4 = 0$ has now been selected and will be assumed to be fixed throughout our development of euclidean geometry. Any plane could have been selected as the ideal plane, but once the choice has been made, it is not changed. Every ordinary plane contains exactly one ideal line since two distinct planes in projective three-space have exactly one line in common (Exercise 6, Section 2-2). Similarly, every ordinary line contains exactly one ideal point (Exercise 5, Section 2-2).

If we delete the ideal plane from projective three-space, the remaining space is called *affine three-space*. Similarly, a projective plane with the ideal line deleted is called an *affine plane;* a projective line with its ideal point deleted is called an *affine line*. We shall frequently refer to the ideal point of an affine line and mean thereby the ideal point on the projective line whose ordinary points are isomorphic to the affine line. Similarly, the ideal line of an affine plane is understood to be the ideal line on the corresponding projective plane. In this sense we shall assume throughout our discussion of affine geometry that we are concerned with an affine space that is embedded in a projective space.

Any projective transformation under which ordinary points correspond to ordinary points and ideal points correspond to ideal points is said to be an *affine transformation*. In this chapter we are primarily concerned with properties left invariant under affine transformations, i.e., with *affine geometry*. We shall find that affine geometry has many of the properties of projective geometry. However, since we have selected a particular point on each line as an ideal point and have not selected a particular plane on each line in space or a particular line on each point on a plane, the principles of duality do not apply to the ordinary points, lines, and planes of affine geometry. They apply only to the projective space in which the affine space is embedded. It is also clear that the point F in Postulate P-3 may be on the ideal plane. In this sense the selection of an ideal plane detracts from the inherent simplicity of the geometry as it makes possible the definition of new concepts based upon parallelism.

From the point of view of the coordinate systems, the selection of an ideal plane makes it possible to use either homogeneous or non-homogeneous coordinates for all ordinary points. In other words, the exceptional point P_∞ on each projective line (Section 3-7), the exceptional line $P_\infty Q_\infty$ on each plane (Fig. 3-11), and the exceptional plane in three-space may be taken as ideal points, lines, and plane respectively. Accordingly, we shall hereafter assume that the coordinate system has been chosen such that on any line of points (x_1,x_2) the ideal point is taken as $(1,0)$, on any plane of points (x_1,x_2,x_3) the ideal line is taken as the line $x_3 = 0$ with points $(x_1,x_2,0)$, and on the three-space of points (x_1,x_2,x_3,x_4) the ideal plane is taken as the plane $x_4 = 0$ with points $(x_1,x_2,x_3,0)$.

EXERCISES

1. Prove that any two distinct coplanar lines intersect in an ordinary point or have an ideal point in common.

2. Prove that any two planes have at least one ideal point in common.

3. Given a line m and a point P that is not on m, prove that on the plane of P and m there exists exactly one line through P that does not intersect m in an ordinary point.

4. On an affine line, assume that $x_2 = 0$ is the ideal point and prove that the ordinary points are precisely the points $(x,1)$ with nonhomogeneous coordinates x.

5. On an affine plane, assume that $x_3 = 0$ is the ideal line and prove that the ordinary points are precisely the points $(x,y,1)$ with nonhomogeneous coordinates (x,y).

6. Find the nonhomogeneous coordinates of each of the following points. Each point is given in terms of its homogeneous coordinates.

    (a) $(2,3,1)$,   (b) $(6,-4,2)$,   (c) $(5,0,5)$,

    (d) $(0,b,1)$,   (e) $(u,0,7)$,   (f) $(5x,3y,2)$.

7. Find homogeneous coordinates for each of the following points. Each point is given in terms of its nonhomogeneous coordinates

    (a) $(6,5)$,   (b) $(7,0)$,   (c) $(0,0)$.

8. Under the assumption that all lines are ordinary unless otherwise specified, prove that on a plane

    (a) any line $a_1x_1 + a_2x_2 + a_3x_3 = 0$ has ideal point $(a_2,-a_1,0)$,

    (b) any line $ax + by + c = 0$ has ideal point $(b,-a,0)$.

9. Prove that on a plane all lines on the ideal point $(b,-a,0)$ may be expressed in the form $ax + by + c = 0$.

10. Prove that two planes are on the same ideal line if and only if they may be expressed in the form

$$ax + by + cz + d = 0 \quad \text{and} \quad ax + by + cz + e = 0.$$

5–2 Parallels. The ideal plane $x_4 = 0$ has been chosen, and it is now possible to define parallel lines. Two (ordinary) lines that intersect in an ideal point are said to be *parallel lines*. Similarly, two (ordinary) planes that intersect in an ideal line are said to be *parallel planes.* A line is parallel to a plane if and only if it intersects the plane in an ideal point. These definitions provide the basis for many of the figures and properties that are studied in high-school (euclidean) geometry. Several of these figures and properties will now be introduced. Others may be found in the exercises at the end of this section and in our later discussions.

We first observed that by definition the ideal plane $x_4 = 0$ is invariant (remains fixed) under any affine transformation. The ideal plane remains invariant as a set of points; i.e., every ideal point $(x_1,x_2,x_3,0)$ must correspond to an ideal point $(x_1',x_2',x_3',0)$, but a particular ideal point does not necessarily correspond to itself. Thus under all affine transformations, parallel lines must correspond to parallel lines and parallel planes must correspond to parallel planes. We say that parallelism is an invariant of affine geometry.

The presence in affine geometry of many figures and properties of euclidean plane geometry is in part based upon the fact that the real projective plane, with a particular line designated as the ideal line and that line deleted, is isomorphic (Section 1–6) to the euclidean plane. In other words, a real plane in affine geometry corresponds to a plane in euclidean geometry; a real line in affine geometry corresponds to a line in euclidean geometry. However, if two euclidean lines are used to represent two parallel affine lines, they will appear as parallel lines if and only if the ideal line of the affine plane is represented by that of the euclidean plane.

Theorem 2–2 (Section 2–2) states that any two distinct lines on a projective plane intersect in a unique point. Therefore, two lines on an affine plane must intersect (i.e., have a common ordinary point) or be parallel (i.e., have a common ideal point). In other words, any two coplanar lines in affine geometry intersect or are parallel. We consider any line parallel to itself and have

THEOREM 5–1. *There is one and only one line parallel to a given line and through a given point.*

The proof of Theorem 5–1 is given as Exercise 1. We may also prove that two lines parallel to the same line are parallel to each other (Exercises 4 and 5) and that on a plane a line intersecting one of two parallel lines must intersect the other also (Exercise 6). These are common properties of euclidean geometry. The choice of a plane as

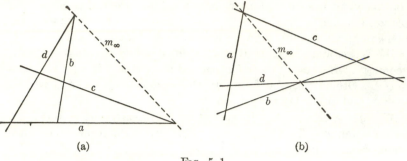

(a)   (b)

Fig. 5-1

the locus of ideal points gives rise to one of the principal properties
(Theorem 5-1) that distinguishes euclidean geometry from the noneuclidean geometries (Section 7-10 and Chapter 8). Thus when we
discuss the classical noneuclidean geometries, we shall base our work
upon projective geometry rather than upon affine geometry.

The most common closed figure based upon parallel lines is probably the parallelogram. In euclidean plane geometry, a parallelogram is often defined as a quadrilateral with its opposite sides parallel.
We have defined complete and simple plane quadrilaterals in Chapter 2 and have now defined parallel lines. A complete quadrilateral
was defined as the figure consisting of four coplanar lines, no three of
which are concurrent, and their six points of intersection (vertices).
We may make the opposite sides parallel by requiring that two of
the opposite vertices be ideal points. Thus we define a *parallelogram*
to be a complete quadrilateral with two of its opposite vertices on the
ideal line of the plane of the quadrilateral. When the ideal line m_∞
of affine geometry is represented by a line on a euclidean plane, a
parallelogram *abcd* may appear as in either of the two figures in
Fig. 5-1.

A parallelogram may also be defined as a complete quadrilateral
having the ideal line as one of its diagonal lines. The *diagonals* of a
parallelogram are then the two ordinary lines that are sides of the
diagonal triangle of the quadrilateral. We may also define a trapezoid (Exercise 9), a parallelepiped (i.e., a six-sided "box" with
opposite sides parallel) (Exercise 10) and other figures based upon
parallelism and the incidence relations of points, lines, and planes.
In the next section, we shall define mid-point and find several more
familiar properties of euclidean geometry that depend only upon
parallelism and incidence relations.

*1. Prove Theorem 5–1.

2. Prove that on a plane all lines parallel to a given line $ax + by + c = 0$ may be expressed in the form $ax + by + d = 0$, and conversely.

3. Prove that all planes parallel to a given plane $ax + by + cz + d = 0$ may be expressed in the form $ax + by + cz + e = 0$, and conversely.

4. Use a synthetic method of proof and prove that if two lines are parallel to the same line, they are parallel to each other.

5. Repeat Exercise 4 using an algebraic method of proof.

*6. Prove that on a plane if a line intersects one of two parallel lines, it intersects the other also.

7. Prove that there exists a unique plane parallel to any given plane and through any given point.

8. Draw a figure indicating two parallel lines cut by a transversal.

*9. Define a trapezoid.

10. Define a parallelepiped in space. (Note that you cannot yet define a cube or a rectangular parallelepiped.)

11. Draw a figure illustrating the fact that a complete quadrangle with the ideal line as a side of its diagonal triangle forms a parallelogram and its diagonals. (Note that the diagonals of this parallelogram based upon a quadrangle instead of a quadrilateral are *not* the ordinary lines that are sides of the diagonal triangle of the quadrangle.)

5–3 Mid-point. We have seen that two distinct points A, B determine a unique line in projective space (Postulates P–1 and P–2) and divide it into two segments (Section 3–4). In order to identify a particular segment, we considered three distinct collinear points A, B, D and the segment AB/D. Such a notation is not necessary in affine geometry. Any two (ordinary) points A, B determine a unique line AB with an ideal point P_∞. On the projective line, one of the segments determined by the points A, B contains P_∞ and the other does not. In affine geometry we define the segment AB/P_∞ to be the *segment AB*. The points A and B are called *end points* of the segment AB; all other points of the segment are called *interior points*. From an algebraic point of view, any two points A and B, with non-homogeneous coordinates a and b, respectively, on the line AB, are end points of a line segment AB having the points X with coordinates x as interior points, where $a < x < b$ if $a < b$ and $b < x < a$ if $a > b$.

The mid-point of a segment AB may be defined from either a synthetic or an algebraic point of view. In synthetic affine geometry we define the *mid-point* of a segment AB to be the harmonic conjugate with respect to A and B of the ideal point P_∞ on the line AB; that is, the mid-point of the segment AB is the point D, where we have $H(AB, P_\infty D)$. The mid-point is then unique by Theorem 3–3. In analytic affine geometry we define the mid-point of a segment AB where $A:(x,y,z)$ and $B:(x',y',z')$ to be the point

$$\left(\frac{x + x'}{2}, \frac{y + y'}{2}, \frac{z + z'}{2}\right).$$

This definition in space implies the usual mid-point formulas on a line and on a plane (Exercise 1). On a line the synthetic and algebraic definitions may be proved to be equivalent by using cross ratio (Exercise 2). Then, since any two lines are equivalent under an affine transformation, and since affine transformations preserve cross ratios and harmonic sets, the definitions are equivalent in general.

We next use the above definitions and the invariance of harmonic sets (Exercise 7, Section 3–3) under projective and, in particular, affine transformations to prove three common theorems (Theorems 5–2, 5–3, and 5–4). These theorems are proved in most high-school plane geometry classes. However, they are valid in affine geometry as well as in all special cases of affine geometry, such as euclidean geometry.

THEOREM 5–2. *The diagonals of a parallelogram bisect each other.*

THEOREM 5–3. *A line bisecting two sides of a triangle is parallel to the third side.*

THEOREM 5–4. *The three medians of a triangle are concurrent.*

A *median* of a triangle is defined to be a line joining a vertex of the triangle to the mid-point of the segment determined by the other two vertices of the triangle, i.e., the mid-point of the opposite side of the triangle. We shall consider a synthetic proof of Theorem 5–2 and an algebraic proof for Theorem 5–4. The proof of Theorem 5–3 is given as Exercise 3.

Let $ABCD$ be a parallelogram on a plane with ideal line m_∞ (Fig. 5–2) where $AB \cdot CD = A_\infty$, $AD \cdot BC = B_\infty$, $AC \cdot m_\infty = C_\infty$, $BD \cdot m_\infty = D_\infty$, and $AC \cdot BD = E$. To prove Theorem 5–2, we must prove that E is the mid-point of the segments AC and BD. Consider the quadrangle $ABCD$ with $H(A_\infty B_\infty, C_\infty D_\infty)$ and the perspectivities

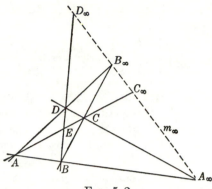

FIG. 5–2

$$DBED_\infty \overset{C}{\underset{\wedge}{=}} A_\infty B_\infty C_\infty D_\infty \overset{D}{\underset{\wedge}{=}} CAC_\infty E.$$

Then by Exercise 7, Section 3–3, we have $H(DB, D_\infty E)$ and $H(CA, C_\infty E)$, whence by definition E is the mid-point of BD and of AC. This completes the proof of Theorem 5–2.

Theorem 5–4 may be proved by synthetic methods, using the Theorem of Desargues and Theorem 5–3 (Exercise 4). The following proof is based upon the methods of analytic projective geometry. Consider any triangle ABC on a plane with ideal line m_∞. Let $AB \cdot m_\infty = C_\infty$, $BC \cdot m_\infty = A_\infty$, and $AC \cdot m_\infty = B_\infty$ (Fig. 5–3). Then $H(C_\infty B_\infty, A_\infty D_\infty)$ determines a unique point D_∞ on m_∞ (Theorem 3–3). Also there exists a homogeneous coordinate system on the plane such that we have $A:(0,0,1)$, $B:(2,0,1)$, $C:(0,2,1)$, and $D_\infty:(1,1,0)$. Then the mid-point of BC is $AD_\infty \cdot BC = A_1:(1,1,1)$; the mid-point

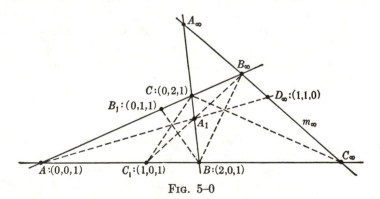

FIG. 5–0

of AB is $C_1:(1,0,1)$; and the mid-point of AC is $B_1:(0,1,1)$. The line AA_1 has equation

$$\begin{vmatrix} x & y & 1 \\ 1 & 1 & 1 \\ 0 & 0 & 1 \end{vmatrix} = x - y = 0$$

(Section 4–2), BB_1 has equation $x + 2y - 2 = 0$, and CC_1 has equation $2x + y - 2 = 0$. All three equations are satisfied by the point $(\frac{2}{3}, \frac{2}{3}, 1)$, and therefore the medians AA_1, BB_1, and CC_1 are concurrent. Since ABC was an arbitrary triangle, this completes the proof of Theorem 5–4.

The above properties of triangles and parallelograms depend only upon the incidence and existence relations of projective geometry and the parallelism of affine geometry. Since we have not yet introduced a distance relationship or metric on a line or on a plane, the above properties are independent of the size of the figures. In projective geometry any three distinct collinear points can be made to correspond to any three distinct collinear points; any three noncollinear points can be made to correspond to any three noncollinear points. Thus any two triangles are equivalent under projective transformations. Furthermore, in projective geometry any two complete quadrilaterals are equivalent (Theorem 2–6). Under affine transformations the invariance of the ideal plane restricts the quadrangles that may be made to correspond to a given quadrangle, but any triangle in the affine plane is equivalent to any other triangle in the affine plane (Exercise 6).

In affine geometry parallel lines must correspond to parallel lines and the mid-point of any segment must correspond to the mid-point of any corresponding segment (Exercise 7). If we define opposite sides of a parallelogram to be "equal," we may compare segments on parallel lines. We shall not be able to compare segments on intersecting lines, i.e., in different directions, until we essentially have rotations in euclidean geometry.

EXERCISES

1. Use the algebraic definition of mid-point and prove that the mid-point of any line segment AB where A and B, respectively, have
   (a) nonhomogeneous coordinates a and b is $(a + b)/2$,
   (b) homogeneous coordinates (x_1, x_2) and (x_1', x_2') is $(x_1 + x_1', x_2 + x_2')$,

(c) nonhomogeneous coordinates (x_1, y_1) and (x_2, y_2) is

$$\left(\frac{x_1 + x_2}{2}, \frac{y_1 + y_2}{2}\right),$$

(d) homogeneous coordinates (x_1, x_2, x_3) and (x_1', x_2', x_3') is

$$(x_1 + x_1', \ x_2 + x_2', \ x_3 + x_3').$$

2. Use (4–25) in the form

$$\mathcal{R}(AB, P_\infty D) = \frac{d - b}{d - a}$$

and prove that $H(AB, P_\infty D)$ implies $d = (a + b)/2$.

3. Prove Theorem 5–3.

4. Use Theorems 2–7 and 5–3 to prove Theorem 5–4 by a synthetic method.

5. Let m and n be any two given parallel lines. Prove that for all points M and N on m and n respectively, the mid-points of the segments MN lie on a line p parallel to m and n.

6. Use Theorem 2–6 and prove that any two ordinary triangles are equivalent in affine plane geometry.

7. Prove that if a segment AB corresponds to a segment $A'B'$ under an affine transformation, then the mid-point of AB corresponds to the mid-point of $A'B'$ under that transformation.

8. Assume that the opposite sides of any parallelogram are equal and prove that the line segment determined by the mid-points of two sides of any triangle is equal to half the third side.

9. As in Exercise 8, give a construction on a line AB for a segment three times as long as the segment AB.

10. Repeat Exercise 9 for a segment four times as long as AB on any given line CD that is parallel to AB.

5–4 Classification of conics. Conics have been defined in synthetic projective geometry and in analytic projective geometry. The two definitions were proved to be equivalent in Section 4–10. We now give a partial classification of conics in terms of their ideal points on the real projective plane, that is, in terms of the ideal points on the conic that have triples of real numbers as their homogeneous coordinates. We shall call such ideal points *real ideal points*. A conic is defined to be a *hyperbola*, a *parabola*, or an *ellipse* according as it contains two distinct, one distinct, or no real ideal points (Fig. 5–4). This easily visualized basis for the classification of conics is useful as an aid to remembering the classification in analytic geom-

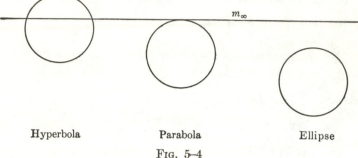

Hyperbola   Parabola   Ellipse

FIG. 5–4

etry.   The general equation (4–35) of a conic, where the determinant of the polarity (4–34) is different from zero, may be expressed in the form

$$Ax_1^2 + Bx_1x_2 + Cx_2^2 + Dx_1x_3 + Ex_2x_3 + Fx_3^2 = 0$$

in homogeneous coordinates and in the form

(5–1) $$Ax^2 + Bxy + Cy^2 + Dx + Ey + F = 0$$

in nonhomogeneous coordinates.   In each case we have the condition

$$\begin{vmatrix} 2A & B & D \\ B & 2C & E \\ D & E & 2F \end{vmatrix} \neq 0,$$

since this determinant is 8 times the determinant of the corresponding polarity (4–34).   Under this condition the equation (5–1) represents a conic in affine geometry if and only if there exist points that satisfy the equation on the real affine plane.   As mentioned in Section 4–9, the condition that the above determinant be different from zero implies that the conic is nondegenerate (does not reduce to two lines, a single line, or a single point).

The homogeneous coordinates of the ideal points $(x_1, x_2, 0)$ on the conic (5–1) must satisfy the equation

$$Ax_1^2 + Bx_1x_2 + Cx_2^2 = 0.$$

Accordingly (from the properties of quadratic equations), there are two distinct, one distinct, or no real ideal points on the conic (5–1) according as the discriminant $B^2 - 4AC$ is positive, zero, or negative. Thus in affine geometry (and also in euclidean geometry) a conic (5–1) is called a hyperbola, a parabola, or an ellipse according as $B^2 - 4AC$ is positive, zero, or negative.   The classification of conics provides an example of the insight into fundamental concepts of euclidean geometry that may be obtained by studying more general geometries.

Many common properties of conics may be proved with or without the algebraic representation of conics. For example, consider the fact that any line meets a conic in at most two points in a real projective plane. This property may be proved algebraically in terms of the number of common roots of a linear and a quadratic equation. It may also be proved without recourse to the algebraic representation. By Postulate P–9 a projectivity between two pencils of points is uniquely determined by three distinct pairs of points. Thus two pencils of points are perspective if and only if the lines joining three pairs of corresponding points are concurrent. By the principle of planar duality (Section 2–4), two pencils of lines are perspective if and only if the points of intersection of three pairs of corresponding lines are collinear. Therefore, since the pencils of lines determining a conic are nonperspective, a conic (assumed to be nondegenerate in Section 2–12) cannot contain three collinear points; that is, any line meets a conic in at most two points. Thus in affine geometry every conic is a hyperbola, a parabola, or an ellipse.

The hyperbola and the ellipse are often called central conics. The basis for this terminology of euclidean geometry may be easily understood in affine geometry. The *center* of any conic (hyperbola, parabola, or ellipse) is the pole of the ideal line with respect to the conic. The polar of any ideal point with respect to a conic is a *diameter* of the conic. The center of a conic may be determined as the intersection of any two distinct diameters of the conic. For example, the equation of the parabola $x^2 - 2y = 0$ may be expressed in homogeneous coordinates in the form

$$x_1^2 - 2x_2x_3 = 0.$$

The corresponding scratch equation (Section 4–9)

$$x_1x_1' - x_2x_3' - x_2'x_3 = 0$$

may be used to obtain the polar of any point. When $(x_1',x_2',x_3') = (1,0,0)$, we obtain the diameter $x_1 = 0$. When $(x_1',x_2',x_3') = (0,1,0)$, we obtain the diameter $x_3 = 0$. The pole of any line is the intersection of the polars of any two of its points (Section 4–9). The pole of the ideal line with respect to the given parabola is the intersection of the two diameters $x_1 = 0$, $x_3 = 0$. Thus the center of the given parabola is the ideal point $(0,1,0)$. In general, any parabola has an ideal point as its center, since it is tangent to the ideal line. Any hyperbola or ellipse has an ordinary point as center, since hyperbolas and ellipses are not tangent to the ideal line. A conic that has an ordinary point as its center is called a *central conic*. Thus we have another concept of euclidean geometry that may be easily explained in a more general geometry.

<h3 style="text-align:center">EXERCISES</h3>

1. Classify the following conics:
     (a) $4x^2 + 9y^2 - 36 = 0$,   (b) $xy - 4 = 0$,
     (c) $x^2 + y^2 - 2x - 4y + 4 = 0$.

2. Find two diameters of each conic in Exercise 1.

3. Find the center of each conic in Exercise 1.

4. Prove that the center of an ellipse is an interior point of the ellipse.

5. Prove that the center of a hyperbola is an exterior point of the hyperbola.

6. Prove that the center of any central conic (5–1) has nonhomogeneous coordinates

$$x = \frac{2CD - BE}{B^2 - 4AC}, \qquad y = \frac{2AE - BD}{B^2 - 4AC}.$$

7. Prove that any conic with two distinct parallel tangent lines is a central conic.

5–5 Affine transformations. We now start our formal process of restricting or specializing the transformations of projective geometry to obtain the transformations of less general geometries. In this section we shall use the invariance of an ideal plane to obtain the transformations of affine geometry. We assume that the coordinate systems of affine spaces are selected such that the ideal plane has equation $x_4 = 0$ in affine space, the ideal line has equation $x_3 = 0$ on any affine plane, and the ideal point has equation $x_2 = 0$ on any affine line. We shall restrict the projective transformations by placing conditions upon the elements of the matrices of the transformations.

A projective transformation on a plane is an affine transformation if and only if the ideal line is invariant as a set of points. Particular ideal points do not need to be invariant, but each ideal point must correspond to a point on the ideal line. In general, any projective transformation is an affine transformation if ideal points correspond to ideal points and ordinary points correspond to ordinary points. We now seek to express this condition in terms of the coefficients of the systems of equations, i.e., in terms of the elements of the matrices of the transformations.

The ideal point on a line of points with homogeneous coordinates (x_1, x_2) is characterized by the relation $x_2 = 0$. A projective transformation on a line (4–18)

$$\begin{aligned} x_1' &= ax_1 + bx_2, \\ x_2' &= cx_1 + dx_2 \end{aligned} \qquad (ad - bc \neq 0)$$

is an affine transformation if and only if $x_2 = 0$ implies $x_2' = 0$, that is, if and only if $c = 0$. Then from the relation $ad - bc \neq 0$ we know that $d \neq 0$. Finally, since $(x_1, x_2) = (kx_1, kx_2)$ for any real number $k \neq 0$, we shall usually represent the homogeneous coordinates of any ordinary point by the particular set of its homogeneous coordinates for which $d = 1$. In this particular case any affine transformation of a line onto itself may be expressed in the form

(5–2)
$$\begin{aligned} x_1' &= ax_1 + bx_2, \\ x_2' &= x_2 \end{aligned} \qquad (a \neq 0).$$

The corresponding matrix equation is

$$\begin{bmatrix} x_1' \\ x_2' \end{bmatrix} = \begin{bmatrix} a & b \\ 0 & 1 \end{bmatrix} \begin{bmatrix} x_1 \\ x_2 \end{bmatrix} \qquad (a \neq 0),$$

and the corresponding equation in terms of the nonhomogeneous coordinates of the points on the line is

$$x' = ax + b \qquad (a \neq 0).$$

Affine transformations on a plane and in space may be considered from the same point of view as the transformations on a line. The ideal points on a plane with homogeneous coordinates (x_1, x_2, x_3) are characterized by the relation $x_3 = 0$. These are the points without corresponding nonhomogeneous coordinates. The ideal line is invariant as a set of points if and only if every point $(x_1, x_2, 0)$ corresponds to a point $(x_1', x_2', 0)$, i.e., if and only if $x_3 = 0$ implies $x_3' = 0$. In order for $x_3 = 0$ to imply $x_3' = 0$ in the system of equations (4–11), it is necessary and sufficient that $a_{31} = a_{32} = 0$. Then, since the determinant of the matrix of the coefficients is different from zero, $a_{33} \neq 0$, and we may, for convenience, consider for each point the particular set of homogeneous coordinates for which $a_{33} = 1$. Thus any affine transformation of a plane onto itself may be expressed in the form

$$(5\text{-}3) \quad \begin{aligned} x_1' &= a_{11}x_1 + a_{12}x_2 + a_{13}x_3, \\ x_2' &= a_{21}x_1 + a_{22}x_2 + a_{23}x_3, \quad (a_{11}a_{22} - a_{12}a_{21} \neq 0). \\ x_3' &= \phantom{a_{21}x_1 + a_{22}x_2 +} x_3 \end{aligned}$$

The corresponding matrix equation is

$$\begin{bmatrix} x_1' \\ x_2' \\ x_3' \end{bmatrix} = \begin{bmatrix} a_{11} & a_{12} & a_{13} \\ a_{21} & a_{22} & a_{23} \\ 0 & 0 & 1 \end{bmatrix} \begin{bmatrix} x_1 \\ x_2 \\ x_3 \end{bmatrix}.$$

The corresponding system of equations in terms of the nonhomogeneous coordinates of the points on the affine plane is

$$(5\text{-}4) \quad \begin{aligned} x' &= a_1 x + b_1 y + c_1, \\ y' &= a_2 x + b_2 y + c_2 \end{aligned} \quad (a_1 b_2 - b_1 a_2 \neq 0),$$

which may be represented by the matrix equation

$$\begin{bmatrix} x' \\ y' \\ 1 \end{bmatrix} = \begin{bmatrix} a_1 & b_1 & c_1 \\ a_2 & b_2 & c_2 \\ 0 & 0 & 1 \end{bmatrix} \begin{bmatrix} x \\ y \\ 1 \end{bmatrix}.$$

We may also prove that any affine transformation on a three-space may be expressed in the form

$$(5\text{-}5) \quad \begin{aligned} x_1' &= a_{11}x_1 + a_{12}x_2 + a_{13}x_3 + a_{14}x_4, \\ x_2' &= a_{21}x_1 + a_{22}x_2 + a_{23}x_3 + a_{24}x_4, \\ x_3' &= a_{31}x_1 + a_{32}x_2 + a_{33}x_3 + a_{34}x_4, \\ x_4' &= \phantom{a_{31}x_1 + a_{32}x_2 + a_{33}x_3 +} x_4, \end{aligned}$$

where the determinant of the coefficients is different from zero (Exercise 12). This system of equations and the corresponding system in the nonhomogeneous coordinates of the points may also be represented by matrix equations (Exercises 13 and 14).

There are two observations that should be made regarding the above representations of affine transformations. First, we note that for affine transformations the representation in terms of homogeneous coordinates is equivalent to the representation in terms of nonhomogeneous coordinates. Homogeneous coordinates are needed only when correspondences among ideal points are to be considered. They are also useful in providing a basis for matrix representations such as that for (5-4) in terms of nonhomogeneous coordinates. The second observation is concerned with identity matrices (Section 4-4).

Any affine transformation may be expressed such that the last row of its square matrix is identical with the last row of the corresponding identity matrix. We shall use square matrices to represent transformations so that we may multiply matrices, as in the matrix equation corresponding to (5–4). The last rows of the square matrices provide a convenient means of recognizing matrices of affine transformations. A square matrix with determinant different from zero may be used to represent an affine transformation if and only if its last row may be obtained from the last row of the corresponding identity matrix by multiplying each element by a non-zero constant. This necessary and sufficient condition illustrates our comment that we shall restrict our transformations by placing conditions upon the elements of the matrices of the transformations.

EXERCISES

1. Indicate which of the following matrices may be used to represent an affine transformation on a line:

(a) $\begin{bmatrix} 1 & 2 \\ 1 & 0 \end{bmatrix}$, (b) $\begin{bmatrix} 2 & 3 \\ 0 & 1 \end{bmatrix}$, (c) $\begin{bmatrix} 2 & 0 \\ 3 & 1 \end{bmatrix}$, (d) $\begin{bmatrix} 0 & 1 \\ 1 & 2 \end{bmatrix}$.

2. Prove that the set of affine transformations on a plane is closed under multiplication.

*3. Prove that the set of affine transformations on a plane forms a group (Section 4–7).

4. Use the methods of synthetic projective geometry and prove that the identity transformation is the only projective transformation on a line that leaves the points (1,0), (1,1), and (0,1) invariant.

5. Repeat Exercise 4 using the methods of analytic projective geometry.

6. Find necessary and sufficient conditions that a transformation (5–3) leave the line $x_3 = 0$ pointwise invariant, i.e., leave every ideal point invariant.

7. Prove algebraically that a transformation (5–3) leaves the points (1,0,0), (0,1,0), (0,0,1), and (1,1,1) invariant if and only if it is the identity transformation.

8. Prove that any affine transformation (5–3) leaving three noncollinear (ordinary) points invariant is the identity transformation.

*9. Prove that there exists an affine transformation transforming any three given noncollinear points A, B, C on a plane respectively into any three noncollinear points A', B', C' on the plane.

10. Find necessary and sufficient conditions that under the transformation (5–3) every line m corresponds to a line parallel to itself.

11. Find necessary and sufficient conditions that under the transformation (5–3) there exists an ordinary point (a,b,c) such that every line on this point is invariant as a line.

†12. Prove that any affine transformation on a three-space may be expressed as a system of equations (5–5) in the homogeneous coordinates of the points, where the determinant of the coefficients is different from zero.

†13. Repeat Exercise 12 using the nonhomogeneous coordinates of the points.

†14. Express the systems of equations obtained in Exercises 12 and 13 as matrix equations.

5–6 Homothetic transformations.   We have seen that every projective transformation on a real projective plane leaves at least one line invariant and at least one point invariant (Section 4–8).   When the ideal line is an invariant line, the transformation is an affine transformation (5–3).   Under an affine transformation of a plane onto itself, each ideal point corresponds to an ideal point but not necessarily to itself.   When each ideal point corresponds to itself, the ideal line is said to be pointwise invariant.   The set of transformations leaving the ideal line invariant forms the group of affine transformations.   The set of transformations leaving the ideal line pointwise invariant forms the group of *homothetic transformations*.

On a projective plane any projective transformation that leaves all points invariant on an invariant line must also leave all lines invariant on an invariant point (principle of planar duality).   Thus every homothetic transformation leaves the points of a pencil of points invariant and the lines of a pencil of lines invariant.   The axis of the pencil of invariant points is called an *axis* of the homothetic transformation; the center of the pencil of invariant lines is called a *center* of the homothetic transformation.   If a homothetic transformation has two distinct axes of invariant points, it is the identity transformation (Exercise 1).   If a homothetic transformation has two distinct centers of invariant lines, it is the identity transformation (Exercise 2).   Then, since every homothetic transformation has at least one axis and at least one center, every homothetic transformation that is not the identity transformation has a unique axis and a unique center.

The above results regarding centers and axes of homothetic transformations will be used with the fact that all homothetic transformations have the ideal line as an axis to define three subsets of the set of homothetic transformations.   If a homothetic transformation has an ordinary line as an axis, then the transformation has at least two axes and is the identity transformation.   If a homothetic transformation has an ideal point as a center, the transformation is called a translation (Section 5–7).   If a homothetic transformation has a unique ordinary point as its center, the transformation is called a dilation (Section 5–8).   Then all homothetic transformations are either translations or dilations.   The identity transformation is a translation but is not a dilation.   These special cases of homothetic transformations will be considered in later sections.   In the present section we shall obtain an algebraic representation for homothetic transformations, i.e., for the set of affine transformations that leave the ideal line pointwise invariant.

Any projective transformation leaving three distinct points on a line invariant is the identity transformation on that line (Section 2–7). Thus a projective transformation is the identity on the ideal line $x_3 = 0$ if and only if the points $(1,0,0)$, $(0,1,0)$ and $(1,1,0)$ are invariant.   Accordingly, we shall equate each of these points to its corresponding point under the general affine transformation (5–3) and obtain

$$(1,0,0) = (a_{11}, a_{21}, 0),$$
$$(0,1,0) = (a_{12}, a_{22}, 0),$$
$$(1,1,0) = (a_{11} + a_{12}, a_{21} + a_{22}, 0).$$

Since these are homogeneous coordinates of the points, any two sets of coordinates for the same point differ by at most a constant factor. Thus $a_{21} = 0 = a_{12}$, and from the third equation $a_{11} + a_{12} = a_{21} + a_{22}$, whence $a_{11} = a_{22}$.   We have now proved that an affine transformation (5–3) is the identity transformation on the line $x_3 = 0$ if and only if $a_{12} = a_{21} = 0$ and $a_{11} = a_{22}$.   In other words, any homothetic transformation has the form

(5–6)
$$\begin{aligned} x_1' &= a_{11}x_1 && + a_{13}x_3, \\ x_2' &= && a_{11}x_2 + a_{23}x_3, \\ x_3' &= && x_3 \end{aligned}$$
$(a_{11} \neq 0)$

or, in terms of the nonhomogeneous coordinates of the points,

$$\begin{aligned} x' &= ax + c_1, \\ y' &= ay + c_2 \end{aligned}$$
$(a \neq 0).$

The set of homothetic transformations forms a group (Exercise 4).

This group is a subgroup of the group of affine transformations and therefore a subgroup of the group of projective transformations on the plane.

Homothetic transformations are sometimes called *similitudes*. This term probably connotes similar figures to most readers and this connotation has some justification.   If p is a line and p corresponds to p' under a homothetic transformation, then p is parallel to p' (Exercise 7).   In euclidean geometry any two triangles ABC and $A'B'C'$ such that AB is parallel to $A'B'$, BC is parallel to $B'C'$, and CA is parallel to $C'A'$ are similar (Exercise 8).   However, two triangles may be similar without having their corresponding sides parallel.   When the corresponding sides of two similar triangles are parallel, the triangles are said to be *homothetic*.   In general, two polygons in euclidean geometry are similar if and only if they can be divided into similarly placed similar triangles.   Two similar polygons whose corresponding sides are parallel are said to be *homothetic*. Furthermore, the lines joining the corresponding vertices of two homothetic polygons meet in the point that is the center of the homothetic transformations of the polygons into each other.   In euclidean geometry this point is called the *center of similitude* or *homothetic center* of the two polygons [5; 77–86].

In affine geometry, if a plane figure F corresponds to a plane figure F' under a homothetic transformation, then the corresponding sides of F and F' are parallel and the lines joining corresponding vertices pass through the center of the homothetic transformation. Any two figures that correspond under a homothetic transformation are called *homothetic figures*.   Furthermore, this definition is equivalent to the definition of homothetic figures in euclidean geometry. In other words, the pointwise invariance of the ideal line under an affine transformation is a necessary and sufficient condition that corresponding figures under the transformation be homothetic figures in euclidean geometry.   Any two homothetic figures in affine geometry correspond to a special case of similar figures in euclidean geometry (i.e., similar figures with corresponding sides parallel).   Thus the pointwise invariance of the ideal line under an affine transformation is a sufficient condition but not a necessary condition for corresponding figures under the transformation to be similar in euclidean geometry (Exercise 9).

EXERCISES

*1. Prove that if two distinct lines are pointwise invariant under a projective transformation of a plane onto itself, then the transformation is the identity transformation.

2. Prove that if the individual lines of two coplanar pencils of lines on distinct centers are invariant under a projective transformation of the plane onto itself, then the transformation is the identity transformation.

3. Prove that the product of any two homothetic transformations (5–6) is a homothetic transformation.

*4. Prove that the set of homothetic transformations (5–6) forms a group.

5. Prove that the invariant point of the homothetic transformation (5–6) is

    (a) unique if and only if $(a_{13}, a_{23}, 1 - a_{11}) \neq (0,0,0)$,

    (b) a unique ideal point if and only if $a_{11} = 1$,

    (c) an ordinary point if and only if $a_{11} \neq 1$.

6. Use the results obtained in Exercise 5 and prove that any homothetic transformation (5–6) is a translation if $a_{11} = 1$, a dilation otherwise.

7. Prove that if a line p corresponds to a line p' under a homothetic transformation, then p is parallel to p'.

8. Prove that if triangle ABC corresponds to triangle $A'B'C'$ under a homothetic transformation, then in euclidean geometry the two triangles are similar.

9. Prove that the pointwise invariance of the ideal line under an affine transformation is a sufficient condition but not a necessary condition for corresponding figures under the transformation to be similar in euclidean geometry.

5–7 Translations. A homothetic transformation (5–6) that has an ideal point as a center is a *translation*. In other words, a translation on a plane is a projective transformation that leaves every point on the ideal line invariant and every line on an ideal point invariant. Intuitively, the effect of a translation upon any figure on a euclidean plane is to slide it a fixed distance in a given direction. The given direction in euclidean geometry corresponds to the ideal point as center in affine geometry.

An algebraic representation for translations may be found by restricting the coefficients of (5–6) such that there is an ideal invariant point. In general, the invariant points of (5–6) satisfy the conditions

$$x_1 = a_{11}x_1 + a_{13}x_3, \quad x_2 = a_{11}x_2 + a_{23}x_3, \quad x_3 = x_3.$$

Accordingly, as in Exercise 5 of Section 5–6, there is a unique invariant point $(a_{13}, a_{23}, 1 - a_{11})$ if and only if this triple is different from $(0,0,0)$. In particular, if $a_{11} = 1$, there is a unique ideal invariant point if $(a_{13}, a_{23}, 0) \neq (0,0,0)$ and all points are invariant [i.e., (5–6) is the identity transformation] if $(a_{13}, a_{23}, 0) = (0,0,0)$. If $a_{11} \neq 1$, there is a unique ordinary invariant point. Thus (5–6) represents a translation if and only if $a_{11} = 1$. In other words, any translation on a plane may be expressed in the form

$$
\begin{aligned}
x_1' &= x_1 & + a_{13}x_3, \\
x_2' &= & x_2 + a_{23}x_3, \\
x_3' &= & x_3
\end{aligned}
$$

or, in terms of the nonhomogeneous coordinates of the points,

(5–7)
$$
\begin{aligned}
x' &= x + a, \\
y' &= y + b.
\end{aligned}
$$

The matrix of any translation on the plane may be expressed in the form

$$
\begin{bmatrix}
1 & 0 & a \\
0 & 1 & b \\
0 & 0 & 1
\end{bmatrix}.
$$

We have obtained an algebraic representation for translations by placing restrictions upon the set of homothetic transformations (5–6). The set of translations on a plane forms a subgroup of the group of affine transformations on a plane (Exercise 14). Later we shall find that all properties that are invariant under the group of translations are also properties of euclidean geometry. In other words, we shall find that the group of translations is also a subgroup of the group of euclidean transformations.

EXERCISES

1. Prove that any affine transformation in which $a_{12} = a_{21} = 0$ and $a_{11} = a_{22} = 1$ is a translation.

2. Given the reference points on a pair of coordinate axes on a plane, construct the points corresponding to the points $(1,0)$, $(1,1)$, $(2,3)$, and $(5,\mathfrak{c})$ under the translation $x' = x + 2$, $y' = y$ when the ideal line is taken as a real line such as m_∞ in Fig. 5–4.

3. Repeat Exercise 2 when the ideal line is taken as the ideal line in the plane of the construction, i.e., when parallel lines are drawn as parallel lines on the euclidean plane on which the construction is made.

4. Repeat Exercises 2 and 3 for the translations:
   (a) $x' = x - 1,\ y' = y + 1$,
   (b) $x' = x + 2,\ y' = y + 3$,
   (c) $x' = x - 2,\ y' = y + 2$.

5. Find the equations of a translation such that the point $(1,2) = (x,y)$ corresponds to $(-1,3)$. We say that $(-1,3)$ is the *image* of $(1,2)$ under the translation and write $T(1,2) = (-1,3)$.

6. Find translations such that:
   (a) $(2,3)$ is the image of $(1,5)$,
   (b) $(-3,2)$ is the image of $(7,3)$,
   (c) $(8,5)$ is the image of $(0,10)$.

7. Does there exist a translation such that $(0,0)$ is its own image? Explain.

*8. Prove that any translation is completely determined by an ordinary point and its image.

*9. Prove that any translation with one ordinary invariant point leaves every ordinary point invariant.

10. Prove that if (x_2,y_2) is the image of (x_1,y_1) under a translation T_1 and (x_3,y_3) is the image of (x_2,y_2) under a translation T_2, then (x_3,y_3) is the image of (x_1,y_1) under the translation T_2T_1.

11. Prove that the product of the following transformations represents a translation:

$$\begin{bmatrix} -1 & 0 & a \\ 0 & -1 & b \\ 0 & 0 & 1 \end{bmatrix},\ \begin{bmatrix} -1 & 0 & c \\ 0 & -1 & d \\ 0 & 0 & 1 \end{bmatrix}.$$

12. Prove that the set of translations on a plane is closed under multiplication.

13. Prove that the set of transformations of the form given in Exercise 11 is not closed under multiplication.

*14. Prove that the set of translations on a plane forms a group.

15. Derive the equations (5–7) by restricting the transformations (5–6) such that every line on an ideal point P_∞: $(p,q,0)$ is invariant as a line.

16. Given a translation T with center P_∞ and $T(A) = A'$, prove that if A, B, P_∞ are not collinear, then $T(B) = B'$ if and only if $ABB'A'$ is a parallelogram.

17. Given a translation T with center P_∞ and $T(A) = A'$, prove that if A, B, P_∞ are collinear, then $T(B) = B'$ if and only if we have the quadrangular set $Q(P_\infty AA',P_\infty B'B)$.

18. Given two triangles A, B, C and A', B', C' such that corresponding sides are parallel, $T(A) = A'$, and $T(B) = B'$ where T is a translation, prove that $T(C) = C'$.

19. Use mathematical induction and prove that the product of any positive integral number of translations is a translation.

5-8 Dilations. A homothetic transformation (5-6) that has a unique ordinary point as its center is called a *dilation*. In other words, a dilation is a projective transformation that leaves every point invariant on the ideal line, leaves every line invariant on a unique ordinary point, and is not the identity transformation. An algebraic representation for dilations may be obtained by restricting the coefficients of (5-6) such that there is a unique ordinary point that is invariant under the transformation. Accordingly, as in Section 5-7, a homothetic transformation (5-6) is a dilation if and only if $a_{11} \neq 1$. Thus any dilation on a plane may be expressed in the form

$$\begin{aligned} x' &= ax + c_1, \\ y' &= ay + c_2 \end{aligned} \qquad (a \neq 0 \text{ or } 1),$$

in terms of nonhomogeneous coordinates, has *center*

$$(5\text{-}8) \qquad x = \frac{c_1}{1-a}, \quad y = \frac{c_2}{1-a},$$

and may be represented by the matrix

$$(5\text{-}9) \qquad \begin{bmatrix} a & 0 & c_1 \\ 0 & a & c_2 \\ 0 & 0 & 1 \end{bmatrix} \qquad (a \neq 0 \text{ or } 1).$$

The transformation (5-6) is

a homothetic transformation   if $a \neq 0$,
a translation   if $a = 1$,
a dilation   if $a \neq 0$ or 1.

The set of homothetic transformations forms a group (Exercise 4, Section 5-6); the set of translations forms a group (Exercise 14, Section 5-7); the set of dilations does not form a group, since it is not closed under multiplication (Exercise 1). Also it does not contain the identity transformation.

A transformation that is distinct from the identity and has the property that it may be applied twice to obtain the identity is called an involution (Section 4-8) or a transformation of *period* two. In algebra, multiplication by -1 has period two. In geometry, a rotation through 180° about a fixed point has period two. Both of these illustrative examples may be visualized as reflections in a point (the origin on the number line and the fixed point respectively). This intuitive concept of a reflection in a point may be formalized as follows: A *point reflection* is a dilation of period two.

The equations of a point reflection

$$(5\text{–}10) \qquad \begin{aligned} x' &= -x + c_1, \\ y' &= -y + c_2 \end{aligned}$$

may be obtained (Exercise 2) from those of a dilation D by requiring that the square of the matrix (5–9) of the dilation be the identity matrix. Under a point reflection with center M every point on the ideal line (axis) is invariant, every line on the point M (center) is invariant, and, if A corresponds to A' under the point reflection, then M bisects the segment AA' (Exercise 11).

Translations and point reflections are special cases of the general euclidean transformation that we seek. Accordingly, their properties are of special interest to us as we determine the concepts underlying our ordinary euclidean geometry. Several of the properties of point reflections and translations are considered in the following set of exercises.

EXERCISES

1. Prove that the set of dilations is not closed under multiplication and therefore does not form a group.

2. Derive the equations (5–10) of any point reflection.

3. Give the coordinates of the center of the point reflection (5–10).

4. Prove that a point reflection is uniquely determined by its center.

5. Prove that any point reflection may be expressed as the product of a point reflection about the origin and a translation.

6. Prove that the product of any positive even number of point reflections is a translation.

7. Prove that any translation is the product of two point reflections one of which is arbitrary.

8. Prove that the product of an odd number of point reflections is a point reflection.

9. Does the set of all point reflections form a group? Explain.

*10. Prove that the set of all point reflections and translations forms a group.

11. Prove that if A' corresponds to A under a point reflection with center M, then M is the mid-point of the line segment AA'.

12. Prove that any dilation that is not a point reflection may be expressed as the product of a dilation whose center is the origin and a point reflection.

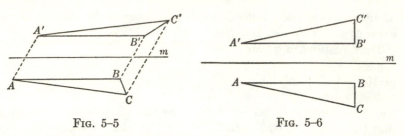

FIG. 5-5   FIG. 5-6

5-9 Line reflections. We now consider the affine plane as a subset
of a projective plane and introduce line reflections as the plane duals
(on the projective plane) of point reflections. We shall find that the
line reflections are not homothetic transformations.

Intuitively a figure F may be reflected in a line m to obtain a
figure F' if the line segments joining corresponding points of F and F'
are all parallel and have their mid-points on the line m (Fig. 5-5).
If, in addition, the segments are perpendicular to the line m (Fig. 5-6),
the line reflection is called an orthogonal line reflection. Since perpendicularity has not yet been defined, we shall be concerned in this
chapter with line reflections such as that in Fig. 5-5 where any visual
appearance of perpendicularity will be coincidental. Formally, a
line reflection may be defined as the plane dual of a point reflection.
A point reflection is a projective transformation of period two leaving
every point on the ideal line (axis) invariant and every line on an
ordinary point (center) invariant where the center is not on the axis.
A *line reflection* is a projective transformation of period two leaving
every line on an ideal point (center) invariant and every point on an
ordinary line (axis) invariant where the axis is not on the center.

We shall develop an algebraic representation for line reflections by
restricting the coefficients of the general affine transformation (5-3).
Any line reflection is an affine transformation, since the ideal line is
on the center of the line reflection and every line on the center is
invariant. If the ideal line were pointwise invariant under a line
reflection, there would be two pointwise invariant lines on the plane,
all lines on the plane would be invariant, all points on the plane would
be invariant, and the transformation would be the identity transformation (Exercise 1, Section 5-6), contrary to the assumption that
line reflections are of period two. Thus under any line reflection the
ideal line is invariant as a line but is not pointwise invariant. However, there are two distinct invariant ideal points under any line

reflection — the center and the ideal point on the axis. Accordingly, since a line reflection is of period two, it must effect a transformation on the ideal line that is of period two and leaves two distinct points invariant. In other words, assuming that we are concerned with a real affine plane, a line reflection must effect a hyperbolic involution on the ideal line.

The general affine transformation (5–3) effects the transformation

$$x_1' = a_{11}x_1 + a_{12}x_2, \qquad (a_{11}a_{22} - a_{12}a_{21} \neq 0)$$
$$x_2' = a_{21}x_1 + a_{22}x_2$$

on the ideal line $x_3 = 0$ when x_1 and x_2 are considered as homogeneous coordinates on the ideal line. This transformation has period two (i.e., is an involution) on the ideal line if and only if $a_{11} = - a_{22}$ (Exercise 3, Section 4–8). Then as in (4–28) the invariant points under the involution satisfy the relation

$$(5\text{–}11) \qquad a_{21}x_1^2 - 2a_{11}x_1x_2 - a_{12}x_2^2 = 0$$

and there are two distinct invariant points if and only if $a_{11}^2 + a_{12}a_{21} > 0$. Thus the effect of the transformation (5–3) on the ideal line is an involution with distinct, real, invariant points if and only if $a_{11} = - a_{22}$ and $a_{11}^2 + a_{12}a_{21} > 0$.

We next consider ordinary points and the transformation

$$(5\text{–}12) \qquad \begin{aligned} x' &= a_{11}x + a_{12}y + a_{13}, \\ y' &= a_{21}x - a_{11}y + a_{23} \end{aligned} \qquad (a_{11}^2 + a_{12}a_{21} > 0)$$

obtained from (5–3) under the above conditions. It will be sufficient (Exercise 1) to find conditions that (5–12) have period two and leave an ordinary line pointwise invariant.

The transformation (5–12) has period two if it is not the identity, and the square of its matrix (Exercise 2)

$$\begin{bmatrix} a_{11}^2 + a_{12}a_{21} & 0 & a_{11}a_{13} + a_{12}a_{23} + a_{13} \\ 0 & a_{11}^2 + a_{12}a_{21} & a_{21}a_{13} - a_{11}a_{23} + a_{23} \\ 0 & 0 & 1 \end{bmatrix}$$

is the identity matrix. Thus (5–12) has period two if and only if we have the three equations

$$(5\text{–}13) \qquad \begin{aligned} a_{11}^2 + a_{12}a_{21} &= 1, \\ a_{12}a_{23} + (a_{11} + 1)a_{13} &= 0, \\ a_{21}a_{13} + (1 - a_{11})a_{23} &= 0. \end{aligned}$$

Finally, the coordinates of the invariant points under the transformation (5–12) satisfy the equations

$$(a_{11} - 1)x \qquad + a_{12}y + a_{13} = 0,$$
$$a_{21}x - (a_{11} + 1)y + a_{23} = 0.$$

Thus there is a line of invariant points if and only if these equations represent the same line (4–9), i.e., if and only if the triples of coefficients belong to the same class. In other words, the above equations are dependent, and the transformation (5–12) leaves a line invariant, if and only if there exists a number $k \neq 0$ such that

(5–14)
$$a_{11} - 1 = ka_{21},$$
$$a_{12} = -k(a_{11} + 1),$$
$$a_{13} = ka_{23}.$$

The line is an ordinary line, since $a_{11} - 1$ and $a_{11} + 1$ cannot both be zero. Furthermore, since the conditions (5–14) are a consequence of the equations (5–13) (Exercise 3), we have proved that if the transformation (5–12) has period two, it leaves an ordinary line pointwise invariant. In other words, the transformation (5–12) is a line reflection if and only if the equations (5–13) hold.

We have proved that any line reflection may be expressed in the form (5–12) with the conditions (5–13) upon its coefficients. Then the axis of the line reflection is the line

(5–15) $$a_{21}x - (a_{11} + 1)y + a_{23} = 0$$

with ideal point $(a_{11} + 1, a_{21}, 0)$, and the center of the line reflection is

(5–16) $$(-a_{12}, a_{11} + 1, 0),$$

which may be obtained from (5–11) as the second invariant point on the ideal line (Exercise 5). Since a line reflection has period two, the product of any two line reflections with coincident centers and parallel axes is a homothetic transformation (Exercise 6). Since a homothetic transformation cannot be a line reflection (Exercise 7), the set of line reflections does not form a group. The importance of line reflections is based upon their having period two and therefore effecting involutions on the ideal line. The significance of these involutions on the ideal line will be more evident after we have used an elliptic involution on the ideal line to define perpendicularity in euclidean geometry (Section 6–1).

EXERCISES

1. Prove that an affine transformation of period two leaving an ordinary line pointwise invariant and two distinct ideal points invariant is a line reflection.

2. Find the square of the matrix of the transformation (5–12).

3. Prove that the equations (5–14) are a consequence of the equations (5–13).

4. Find and plot the images of the points (0,0), (1,0), (1,1), (1,2), (0,1), and $(-1,-1)$ under the line reflection

$$x' = 2x + 3y - 4, \quad y' = -x - 2y + 4.$$

*5. Prove that if the conditions (5–13) hold for a transformation (5–12) and the ideal point on the line (5–15) satisfies the equation (5–11), then the point (5–16) is the other invariant ideal point under (5–12).

6. Prove that the product of any two line reflections with coincident centers and parallel axes is a homothetic transformation. Discuss the relative advantages of synthetic and algebraic methods for this proof.

7. Prove that if a transformation is a homothetic transformation, it is not a line reflection.

*8. Prove that if P and P' are corresponding points under a line reflection with axis m, then the mid-point of the segment PP' is on m.

5–10 Equiaffine and equiareal transformations. An affine transformation under which any triangle corresponds to a triangle of the same measure or "signed area" is called an equiaffine transformation. An affine transformation under which any triangle corresponds to a triangle of the same "area" is called an equiareal transformation. We shall conclude our study of affine geometry by defining the measure of a triangle and considering a few of the properties of the sets of equiaffine and equiareal transformations. Each set of transformations forms a subgroup of the group of affine transformations.

We have seen (Exercise 10, Section 4–2) that any three ordinary points $A:(a_1,a_2,1)$, $B:(b_1,b_2,1)$ and $C:(c_1,c_2,1)$ are collinear if and only if the determinant

$$m(ABC) = \begin{vmatrix} a_1 & a_2 & 1 \\ b_1 & b_2 & 1 \\ c_1 & c_2 & 1 \end{vmatrix}$$

vanishes. We now define this determinant as the *measure* of the triangle determined by the noncollinear points A, B, and C. Since

$m(ABC) = -m(ACB)$ (Exercise 1), the sign of the measure of a
triangle depends upon the order in which the vertices are named. In
this sense we have defined the measure of an *ordered point triad* obtained by taking the vertices of the triangle in the order in which they
are stated. However, we shall continue to refer to the measure of a
triangle and mean thereby the measure of the ordered point triad
obtained by taking the vertices of the triangle in the order in which
they are stated. Three points A, B, C form a triangle if and only if
$m(ABC) \neq 0$. The significance of the sign and magnitude of $m(ABC)$
will become evident as we consider the measure of triangles in more
detail.

The significance of the sign of $m(ABC)$ is implied by the relationships (Exercise 2)

$$m(ABC) = m(BCA) = m(CAB),$$
$$m(ACB) = m(BAC) = m(CBA),$$

where $m(ABC) = -m(ACB)$. From these relationships we may observe that the sign of the measure of a triangle depends upon the
cyclic order in which the vertices are stated. For example, if we
consider the points A, B, C as though they were on a circle (Fig. 5–7),
then the sets of ordered points ABC, BCA, and CAB obtained by
traversing the circle in the same sense (counterclockwise in Fig. 5–7)
are said to be in the same *cyclic order*. Similarly, the sets of ordered
points ACB, BAC, and CBA obtained by starting at each of the given
points and traversing the circle in Fig. 5–7 in a clockwise sense are
said to be in the same cyclic order. These results could also be obtained by traversing the triangle instead of the circle. Thus the sign
of the measure of an ordered point triad, or triangle, depends upon
the sense (clockwise or counterclockwise) in which the triangle is
traversed when the points are joined in the order stated. When the
coordinate axes on the euclidean plane are taken in their usual position (positive sense to the right on the x-axis and up on the y-axis),
the measure of an ordered point triad is positive whenever the triangle
is traversed in a counterclockwise sense and negative whenever the
triangle is traversed in a clockwise sense.

The magnitude or numerical value of $m(ABC)$ is related to the
area (at present undefined) of the triangle ABC. After perpendicularity has been defined, we shall consider the numerical value of
$m(ABC)$ as twice the area of the triangle in terms of a unit square.

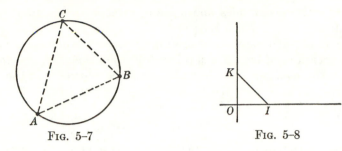

FIG. 5–7   FIG. 5–8

In other words, the numerical value of $m(ABC)$ is the area of triangle ABC in terms of a unit triangle.

Given any real affine plane with a coordinate system, we may take triangle OIK determined by the points with homogeneous coordinates $(0,0,1)$, $(1,0,1)$, and $(0,1,1)$ respectively and with measure 1 (Exercise 3) as our *unit triangle* (Fig. 5–8). Then the measure $m(ABC)$ of any triangle ABC is positive if triangles ABC and OIK are traversed in the same sense (i.e., if they are *similarly oriented*) and negative if the triangles are *oppositely oriented*. Triangles ABC and $A'B'C'$ in Fig. 5–6 are oppositely oriented. The numerical value of $m(ABC)$ represents the area of triangle ABC relative to that of the unit triangle OIK.

An affine transformation (5–3) transforms the unit triangle OIK into a triangle of measure $+1$ if and only if the determinant of the transformation is $+1$ (Exercise 4). In general (Exercise 5), any triangle ABC with measure $m(ABC)$ is transformed under an affine transformation with determinant D into a triangle $A'B'C'$ with measure $D[m(ABC)]$. Thus an affine transformation preserves the measure of triangles if and only if it has determinant $+1$; it preserves the magnitude of the measures of triangles if and only if it has determinant of numerical value 1. The set of affine transformations that preserve the measures of triangles is called the set of *equiaffine transformations*. Any equiaffine transformation may be expressed in the form

$$(5\text{–}17) \qquad \begin{aligned} x' &= a_1x + b_1y + c_1, \\ y' &= a_2x + b_2y + c_2 \end{aligned} \qquad (a_1b_2 - b_1a_2 = 1).$$

The set of equiaffine transformations forms a group (Exercise 6) and has many of the properties of the set of euclidean transformations (Exercises 7 through 11). Intuitively, any affine transformation that preserves the magnitudes and signs of areas is an equiaffine transformation.

The set of affine transformations that preserve the numerical values of the measures of triangles is called the set of *equiareal transformations*. Any equiareal transformation may be expressed in the form (Exercise 12)

(5–18)
$$x' = a_1x + b_1y + c_1, \qquad (a_1b_2 - b_1a_2 = \pm 1).$$
$$y' = a_2x + b_2y + c_2$$

The set of equiareal transformations forms a group (Exercise 13) which has the equiaffine group as a subgroup (Exercise 14). Intuitively, any affine transformation that preserves areas is an equiareal transformation. Thus any two figures that correspond under an equiareal transformation are said to have the same *area*. From this point of view the area of any triangle is a positive number, the numerical value of the measure of a corresponding ordered point triad. In Chapter 6 we shall find that the group of euclidean transformations is composed of the set of equiareal transformations that preserve perpendicularity.

Exercises

1. Prove that $m(ABC) = -m(ACB)$.

2. Prove that $m(ABC) = m(BCA) = m(CAB)$ and $m(ACB) = m(BAC) = m(CBA)$.

3. Prove that the unit triangle OIK in Fig. 5–8 has measure $+1$.

4. Prove that an affine transformation transforms the unit triangle into a triangle with measure $+1$ if and only if the determinant of the transformation is $+1$.

*5. Prove that if $ABC \overset{\sim}{\to} A'B'C'$ under an affine transformation with determinant D, then $m(A'B'C') = D[m(ABC)]$.

*6. Prove that the set of equiaffine transformations forms a group.

7. Indicate which of the following sets of transformations on a plane are subsets of the set of equiaffine transformations and state the reason for your answer:

    (a) homothetic transformations,   (b) line reflections,
    (c) translations,   (d) dilations,
    (e) point reflections,   (f) identity transformation,
    (g) affine transformations,   (h) equiaffine transformations.

8. Prove that if two triangles can be designated by ABC and BCA respectively, they have the same measure.

9. Prove that if triangle $A'B'C'$ corresponds to triangle ABC under a point reflection or a translation, then $m(A'B'C') = m(ABC)$.

10. Given triangle ABC and the mid-point D of the segment BC, prove that $m(ABD) = m(ADC)$.

11. Given triangle ABC and a point D such that CD is parallel to AB, prove that $m(ABC) = m(ABD)$.

*12. Prove that any equiareal transformation may be represented in the form (5–18).

*13. Prove that the set of equiareal transformations forms a group.

14. Indicate which of the sets of transformations listed in Exercise 7 are subsets of the set of equiareal transformations.

5–11 Survey.　The concept of euclidean geometry as a special case of projective geometry is based upon the specialization of projective transformations to obtain euclidean transformations.　The special cases of projective transformations on a plane that we have considered in the present chapter are indicated in the following array.　The lines are used to indicate that a set of transformations is a subset of a set above it; an asterisk indicates that the set forms a group.

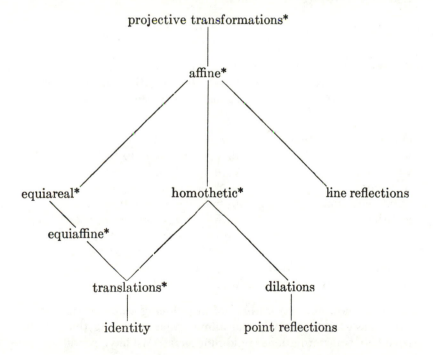

The restrictions upon the projective transformations on a plane may be summarized as follows: A projective transformation is

affine if the ideal line $x_3 = 0$ is invariant,
homothetic if every point on the ideal line is invariant,
equiareal if its determinant is ± 1,
equiaffine if its determinant is $+1$.

A homothetic transformation is a

translation if its center is an ideal point,
dilation if its center is an ordinary point,
point reflection if it is a dilation of period two.

The plane dual of a point reflection is a line reflection.

The relations among the above sets of transformations may also be clearly stated in terms of the coefficients of the equations representing them. A projective transformation on a plane (4–11) is an affine transformation (5–3) if $a_{31} = a_{32} = 0$. Any affine transformation may be expressed in the form (5–4):

$$x' = a_1 x + b_1 y + c_1,$$
$$y' = a_2 x + b_2 y + c_2 \qquad (a_1 b_2 - b_1 a_2 \neq 0)$$

in terms of the nonhomogeneous coordinates of the points. The above subsets of the set of affine transformations may be specified in terms of the following conditions upon the coefficients of the equations (5–4):

affine transformation: $a_1 b_2 - b_1 a_2 \neq 0,$
homothetic transformation: $a_1 = b_2 \neq 0, b_1 = a_2 = 0,$
translation: $a_1 = b_2 = 1, b_1 = a_2 = 0,$
dilation: $a_1 = b_2 \neq 0$ or $1, b_1 = a_2 = 0,$
point reflection: $a_1 = b_2 = -1, b_1 = a_2 = 0,$
identity: $a_1 = b_2 = 1, b_1 = a_2 = c_1 = c_2 = 0,$
line reflection: $\begin{cases} a_1 b_2 - b_1 a_2 = -1, a_1 + b_2 = 0, \\ b_1 c_2 + (a_1 + 1)c_1 = 0 = a_2 c_1 + (1 - a_1)c_2, \end{cases}$
equiareal transformation: $a_1 b_2 - b_1 a_2 = \pm 1,$
equiaffine transformation: $a_1 b_2 - b_1 a_2 = 1.$

We are now on the threshold of introducing a new restriction that will enable us to identify the affine transformations that are also euclidean transformations (rigid motions). We have postulated projective three-space and projective transformations. We have selected an ideal plane and defined the set of projective transformations that leave the ideal plane invariant as a set of points to be affine transformations. In the present chapter we have considered primarily affine transformations on a real plane. In the next chapter we shall select an involution on the ideal line associated with the real plane. The equiareal transformations that leave this involution invariant will be called euclidean transformations. Perpendicular lines will be defined in terms of this fixed involution. In this way we shall obtain the set of euclidean transformations as a subgroup of the set of projective transformations and consider euclidean geometry as a special case of projective geometry. We shall endeavor to gain insight into the fundamental concepts of euclidean geometry by considering their dependence upon the postulates and concepts of projective and affine geometries. We could have postulated euclidean geometry directly without considering projective and affine geometries. The longer path has been taken with the expectation of finding the experience rewarding both in the fundamental relationships discovered and in the insight gained into some of the fundamental concepts underlying euclidean geometry.

REVIEW EXERCISES

1. Prove that the product of any two line reflections is an equiaffine transformation.

2. Prove that the product of any positive even number of line reflections is an equiaffine transformation.

3. Define the asymptotes of a hyperbola on an affine plane.

4. Prove that if a parallelogram is inscribed in a conic, its diagonals are diameters of the conic.

5. Given a central conic with a diameter p, let P be the pole of p, let P' be the ideal point of p, and let p' be the polar of P'. Prove that p' is on P and that p' is a diameter of the conic.

6. Prove that the pairing of ideal points P, P' in Exercise 5 is an involution.

7. Prove that the pairing of diameters p, p' in Exercise 5 is an involution. The two diameters of each pair under this involution are called *conjugate diameters* under the polarity defining the conic.

8. If T is a translation and has period k for some positive integer k, prove that T is the identity transformation.

9. If under a projective transformation of an affine plane onto itself triangle ABC corresponds to triangle $A'B'C'$ such that A corresponds to A',

B corresponds to B', C corresponds to C', and the mid-point of each side of the triangle ABC corresponds to the mid-point of the corresponding side of triangle $A'B'C'$, prove that the transformation is an affine transformation.

10. If corresponding sides of the triangles are parallel under the transformation described in Exercise 9, prove that the transformation is a homothetic transformation.

11. If AA' is parallel to BB' under the transformation described in Exercise 10, prove that the transformation is a translation.

12. If $A \neq A'$ and $AA' \cdot BB'$ is an ordinary point under the transformation described in Exercise 10, prove that the transformation is a dilation.

13. If AB' is parallel to $A'B$ under the transformation described in Exercise 10, prove that the transformation is a point reflection.

14. Given distinct points A, B, prove that the pairs of points C, C' satisfying relations of the form $H(AB, CC')$ are pairs of an involution.

15. Given any quadrangular set of points, prove that the pairs of points on opposite sides of the quadrangle are pairs of an involution.   (*Hint:* Use Review Exercise 15 of Chapter 4.)

CHAPTER 6

EUCLIDEAN PLANE GEOMETRY

We are now ready to discuss euclidean plane geometry as a special case of projective geometry. In accordance with Felix Klein's concept of a geometry as a study of properties that are invariant under a group of transformations, we shall obtain the group of euclidean transformations as a special subset of the group of affine transformations. In particular, we shall define perpendicular lines in terms of an elliptic involution, I_∞, on the ideal line (Section 6-1) and define the affine transformations that leave I_∞ invariant to be similarity transformations. Then the set of similarity transformations that are also equiareal transformations (Section 5-10) forms the group of euclidean transformations (Section 6-4). Under the group of euclidean transformations parallelism, areas, distances (Section 6-5), and angles (Sections 6-6 and 6-7) are invariant. We conclude our development of euclidean geometry with a brief study of the dependence of several common figures of euclidean geometry upon these invariants (Section 6-8) and a survey of our development of euclidean geometry as a special case of projective geometry.

6-1 Perpendicular lines. The *euclidean plane* considered as a set of points is isomorphic to the set of real points on an affine plane. From an algebraic point of view, a euclidean plane is isomorphic to the set of all pairs of real numbers. From a synthetic point of view, a euclidean plane may be obtained by defining a projective plane, selecting one line on the plane as the ideal line, and postulating an isomorphism between the set of real numbers and the set of ordinary points on an ordinary line. Although we define a euclidean plane to be a real affine plane, we shall consider a euclidean plane as a subset of projective three-space. Then, since a euclidean plane is a subset of a three-space, we may consider a "rotation" of the plane about a line on the plane. Also, since the plane is a subset of projective space, we may refer to the ideal and complex points associated with a euclidean plane and mean thereby the ideal and complex points associated with an affine plane on which the real ordinary points are isomorphic to the euclidean plane.

Consider a projective plane π on which an ideal line m_∞ has been selected. In Chapter 5 we considered the transformations (affine transformations) that left m_∞ invariant as a line. We now place a further limitation upon the set of transformations under consideration by requiring that the pairs of an elliptic involution on m_∞ be invariant. In other words, given an elliptic involution on m_∞ with pairs of corresponding points R and R', S and S', T and T', ..., we shall consider the set of transformations under which each pair R, R' corresponds to a pair (not necessarily the same pair) of the involution.

Let I_∞ be an arbitrary elliptic involution on m_∞. We shall call I_∞ the *absolute* (or orthogonal) *involution*. The set of projective transformations leaving I_∞ invariant (i.e., leaving invariant the pairing of ideal points determined by the absolute involution) forms a subgroup of the group of affine transformations and is called the group of *similarity transformations* (Exercise 6, Section 6–2). If a plane figure F corresponds to a figure F' under a similarity transformation, the two figures are said to be *similar;* that is, $F \sim F'$ (Section 6–2). If m and m' are two lines whose ideal points are a pair of the absolute involution, then the lines are said to be *perpendicular* or orthogonal; that is, $m \perp m'$. Thus a projective transformation that preserves parallelism is an affine transformation; an affine transformation that preserves perpendicularity is a similarity transformation.

Any line on a projective plane could be selected and fixed to obtain affine geometry. The two affine geometries associated with two different selections of the ideal line are equivalent (isomorphic) under a projective transformation (a change of coordinates), since any two lines are equivalent under a projective transformation. Similarly, any elliptic involution on the ideal line may be selected and fixed to obtain the geometry of the group of similarity transformations (technically, a *parabolic metric geometry*). Since any two elliptic involutions on a line are equivalent under a real projective transformation (Exercise 10, Section 4–8), any two selections of the absolute involution give rise to equivalent geometries. As in the case of the selection of the ideal line, we shall select a particular involution I_∞ as the absolute involution and keep this involution invariant.

Any projective involution on a line (Section 4–8) may be expressed in the form

(6–1)
$$\begin{aligned} x_1' &= a_{11}x_1 + a_{12}x_2, \\ x_2' &= a_{21}x_1 - a_{11}x_2 \end{aligned} \qquad (a_{11}^2 + a_{12}a_{21} \neq 0),$$

with invariant points satisfying the equation

$$a_{21}x_1^2 - 2a_{11}x_1x_2 - a_{12}x_2^2 = 0.$$

By Theorem 4–4 the involution is elliptic if and only if $a_{11}^2 + a_{12}a_{21}$ < 0. Since any two elliptic involutions are equivalent under a projective transformation, we may use (Exercise 1) a projective transformation (change of coordinates) to express the absolute involution I_∞ on the ideal line in the form

$$(6\text{–}2) \qquad \begin{aligned} x_1' &= x_2, \\ x_2' &= -x_1, \end{aligned}$$

where x_1 and x_2 are homogeneous coordinates on the ideal line $x_3 = 0$. For example, if I_∞ is given with the equations (6–1), we may use the projective transformation

$$(6\text{–}3) \qquad \begin{aligned} x_1' &= a_{21}x_1 - a_{11}x_2, \\ x_2' &= -a_{11}x_1 - a_{12}x_2, \\ x_3' &= (a_{11}^2 + a_{12}a_{21})x_3 \end{aligned}$$

of the euclidean plane onto itself. Then on the ideal line the transformation obtained when (6–1) is followed by (6–3) has matrix

$$\begin{bmatrix} a_{21} & -a_{11} \\ -a_{11} & -a_{12} \end{bmatrix} \begin{bmatrix} a_{11} & a_{12} \\ a_{21} & -a_{11} \end{bmatrix} = \begin{bmatrix} 0 & a_{11}^2 + a_{12}a_{21} \\ -(a_{11}^2 + a_{12}a_{21}) & 0 \end{bmatrix}$$

and is therefore an involution of the form (6–2). We have proved that given any involution (6–1) on the ideal line, there exists a projective transformation (6–3) such that the given involution is equivalent to the involution (6–2). Accordingly, we shall hereafter assume that on the ideal line the absolute involution I_∞ has the form (6–2). This involution may also be expressed (Exercise 2) in the form

$$x' = -1/x$$

in terms of the nonhomogeneous coordinates of the points on the line $x_3 = 0$.

Any line r on a euclidean plane has an associated ideal point, say $R_\infty : (x_1, x_2, 0)$. Under the absolute involution (6–2) the point R_∞ corresponds to the point $R_\infty' : (x_2, -x_1, 0)$. All lines perpendicular to the given line r must be on the point R_∞' and any line r' through R_∞' is perpendicular to the given line r. When the absolute involution is invariant under a projective transformation T, any pair of points

R_∞, R_∞' of the absolute involution corresponds under the transformation T to a pair of points S_∞, S_∞' of the involution and, accordingly, any pair of perpendicular lines r, r' corresponds to a pair of perpendicular lines s, s'.

The ideal points $R_\infty:(x_1,x_2,0)$ and $R_\infty'(x_2,-x_1,0)$ of any pair of perpendicular lines have nonhomogeneous coordinates $x = x_1/x_2$ and $x' = -1/x$ respectively. Intuitively, the nonhomogeneous coordinates of the ideal point of any line may be taken as the reciprocal of the slope of the line. In this sense the absolute involution (6–2) provides a pairing of lines whose slopes are negative reciprocals, i.e., a pairing of perpendicular lines. Also, as in your previous work in geometry, all lines perpendicular to a given line $ax + by + c = 0$ may be expressed in the form $bx - ay + d = 0$, and, conversely, all lines that may be expressed in this form are perpendicular to the given line (Exercise 14). The above definition of perpendicular lines enables us to define such common concepts and figures as rectangle, right triangle, perpendicular bisector of a line segment, altitude of a triangle, etc., and to prove some of the common properties of these figures (Exercises 4 to 17).

The concept of euclidean geometry as a parabolic metric geometry based upon an elliptic involution on the associated ideal line may cause concern to some readers because of the presence of the two adjectives: elliptic and parabolic. In Section 5–4 we found that a hyperbola had two distinct real ideal points, a parabola had one distinct real ideal point, and an ellipse had two imaginary ideal points. This association of hyperbolic with two distinct real, parabolic with one distinct real, and elliptic with two imaginary was also used in the classification of projectivities on a line (Section 4–8) and in the discussion of involutions on a line (Section 4–11). The same association is used here. Euclidean geometry is a parabolic geometry because each line has one distinct real ideal point. It is associated with an elliptic involution on the ideal line because on a euclidean plane no line is perpendicular to itself; i.e., the invariant points of the absolute involution have imaginary nonhomogeneous coordinates.

Exercises

1. Prove that any elliptic involution on a line may be expressed in the form (6–2).

2. Prove that the absolute involution (6–2) has the form $x' = -1/x$ in terms of the nonhomogeneous coordinates of the points on the line $x_3 = 0$.

3. Prove that lines parallel to the same or parallel lines are parallel to each other.

4. Prove that lines perpendicular to the same or parallel lines are parallel.

5. Prove that a line that is perpendicular to one of two parallel lines is perpendicular to the other also.

6. Prove that if two given lines are perpendicular to perpendicular lines, the given lines are perpendicular to each other.

7. Define a *right triangle.*

8. Define a *rectangle.*

9. Prove that the diagonals of a rectangle bisect each other.

10. Prove that the pairs of perpendicular lines through any point are pairs of an involution.

11. Define the *perpendicular bisector* of a line segment.

12. Define an *altitude* of a triangle.

13. Prove that the coordinate axes $x_1 = 0$ and $x_2 = 0$ are perpendicular to each other when (6–2) is taken as the absolute involution.

14. Prove that all lines perpendicular to the line $ax + by + c = 0$ may be expressed in the form $bx - ay + d = 0$ and conversely.

15. Prove that if two pairs of opposite sides of a complete quadrangle intersect the ideal line in pairs of points of the absolute involution, then the third pair of opposite sides of the quadrangle also intersects the ideal line in a pair of the absolute involution.

16. Prove that the altitudes of any triangle are concurrent, i.e., pass through a common point.

17. Prove that the perpendicular bisectors of the sides of any triangle are concurrent.

6–2 Similarity transformations. Similarity transformations have been defined (Section 6–1) as projective transformations that leave the absolute involution invariant. Thus any similarity transformation leaves the ideal line invariant as a line and is an affine transformation

$$(6\text{–}4) \qquad \begin{bmatrix} a_{11} & a_{12} & a_{13} \\ a_{21} & a_{22} & a_{23} \\ 0 & 0 & 1 \end{bmatrix} \qquad (a_{11}a_{22} - a_{12}a_{21} \neq 0).$$

We shall obtain the equations of similarity transformations by selecting the subset of the affine transformations under which the absolute involution is invariant.

The transformation (6–4) leaves the absolute involution (6–2) invariant if and only if every pair of points $(x_1,x_2,0)$, $(x_2,-x_1,0)$ of the involution corresponds under the transformation (6–4) to a pair of the involution, i.e., if and only if the points

(6-5) $(a_{11}x_1 + a_{12}x_2,\ a_{21}x_1 + a_{22}x_2,\ 0),$
$(a_{11}x_2 - a_{12}x_1,\ a_{21}x_2 - a_{22}x_1,\ 0)$

are a pair of the absolute involution. Since we are concerned with homogeneous coordinates, these points form a pair of the absolute involution if and only if there exists a real number $k \neq 0$ such that for all x_1 and x_2

$$a_{11}x_1 + a_{12}x_2 = k(a_{21}x_2 - a_{22}x_1),$$
$$a_{21}x_1 + a_{22}x_2 = -k(a_{11}x_2 - a_{12}x_1),$$

that is, if and only if

$$(a_{11} + ka_{22})x_1 + (a_{12} - ka_{21})x_2 = 0,$$
$$(a_{21} - ka_{12})x_1 + (a_{22} + ka_{11})x_2 = 0.$$

Since these relations must hold for all x_1 and x_2, all four coefficients must vanish. The vanishing of the coefficients

$$a_{11} + ka_{22} \quad \text{and} \quad a_{22} + ka_{11}$$

implies that $k^2 = 1$ and $a_{11}^2 = a_{22}^2$ (Exercise 2). Similarly, the vanishing of the coefficients

$$a_{12} - ka_{21} \quad \text{and} \quad a_{21} - ka_{12}$$

implies that $a_{12}^2 = a_{21}^2$. Finally, the existence of a single value of $k \neq 0$ such that all four of the coefficients vanish implies that $a_{11}a_{12} + a_{21}a_{22} = 0$. In other words, the points (6-5) are a pair of the absolute involution and the transformation (6-4) leaves the absolute involution invariant if and only if

(6-6) $a_{11}^2 = a_{22}^2, \quad a_{12}^2 = a_{21}^2, \quad \text{and} \quad a_{11}a_{12} + a_{21}a_{22} = 0.$

This result may also be expressed as follows (Exercise 5): Any similarity transformation may be expressed in the form

(6-7) $x' = ax - by + c,$ $(a^2 + b^2 \neq 0,\ e^2 = 1).$
$y' = bex + aey + d$

The set of similarity transformations forms a group (Exercise 6).

We have seen that the set of similarity transformations is a subgroup of the group of affine transformations. Soon we shall find that the group of euclidean transformations is a subgroup of the group of similarity transformations. We could define the set of euclidean transformations as the set of similarity transformations that were also equiareal transformations. Any two figures that correspond under an equiareal transformation have the same area. Any two figures that correspond under a similarity transformation are said to be *similar* and to have the same *shape*. Thus the possibility of defining a euclidean transformation as a similarity transformation that is also an equiareal transformation implies that any two figures that have the same shape and the same area are "congruent" in euclidean geometry. We shall define euclidean transformations in terms of orthogonal line reflections (Section 6–3) and show that such a definition is equivalent to the one suggested above.

EXERCISES

1. Prove that any two homothetic figures (Section 5–6) are similar figures and have corresponding sides parallel.

2. Prove that $a_{11} + ka_{22} = 0$ and $a_{22} + ka_{11} = 0$ imply $k^2 = 1$ and $a_{11}^2 = a_{22}^2$.

3. Prove that if the absolute involution is invariant under an affine transformation (6–4), then the conditions (6–6) hold.

4. State and prove the converse of Exercise 3.

5. Prove that any similarity transformation may be expressed in the form (6–7).

6. Prove that the set of similarity transformations forms a group.

7. Prove that if the corresponding sides of two triangles are parallel, the triangles are similar. (*Hint:* See Exercise 9, Section 5–5.)

8. Prove that if the corresponding sides of two triangles are perpendicular, the triangles are similar.

9. Prove that any given triangle ABC is similar to the triangle $AB'C'$ obtained by taking B' and C' as the mid-points of the segments AB and AC respectively.

10. Prove that any diagonal of a parallelogram divides the parallelogram into similar triangles.

11. Prove that if $A \neq B$ and $C \neq D$, the segments AB and CD are similar.

12. Prove that each of the following sets of transformations are subsets of the set of similarity transformations

    (a) translations,   (b) dilations,

    (c) homothetic transformations,   (d) point reflections.

6–3 Orthogonal line reflections. A line reflection was defined in Section 5–9 as a projective transformation of period two that leaves every point invariant on an ordinary line (axis) and every line invariant on an ideal point (center) that is not on the axis. If P_∞ is the center of a line reflection, then every pair of corresponding points P and P' are collinear with P_∞ and each segment PP' has its mid-point on the axis of the line reflection (Exercise 8, Section 5–9). The axis is perpendicular to the line segment PP' if and only if P_∞ and the ideal point P'_∞ on the axis are a pair of the absolute involution. A line reflection whose axis is perpendicular to the line segments joining corresponding points is called an orthogonal line reflection. In other words, we define a line reflection having its center P_∞ and the ideal point P'_∞ on its axis as a pair of points of the absolute involution to be an *orthogonal line reflection*.

Any line reflection (Section 5–9), not necessarily orthogonal, may from (5–12) and (5–13) be expressed algebraically in the form

$$\begin{aligned} x'_1 &= a_{11}x_1 + a_{12}x_2 + a_{13}x_3, \\ x'_2 &= a_{21}x_1 - a_{11}x_2 + a_{23}x_3, \\ x'_3 &= \qquad\qquad\qquad x_3, \end{aligned}$$

(6–8)

where

$$\begin{aligned} a_{11}^2 + a_{12}a_{21} &= 1, \\ a_{12}a_{23} + (a_{11} + 1)a_{13} &= 0, \text{ and} \\ a_{21}a_{13} + (1 - a_{11})a_{23} &= 0. \end{aligned}$$

The line reflection (6–8) has center

$$(-a_{12}, a_{11} + 1, 0)$$

(Exercise 5, Section 5–9) and the point

$$(a_{11} + 1, a_{21}, 0)$$

as the ideal point on its axis (Section 5–9). The line reflection (6–8) is an orthogonal line reflection if and only if these two ideal points are a pair under the absolute involution (6–2), i.e., if and only if $a_{12} = a_{21}$ (Exercise 1). Then in nonhomogeneous coordinates any orthogonal line reflection may (Exercise 2) be expressed in the form

(6–9)

$$\begin{aligned} x' &= ax + by + c, \\ y' &= bx - ay + d, \end{aligned}$$

where

$$a^2 + b^2 = 1, \quad (a + 1)c + bd = 0 = bc + (1 - a)d.$$

The matrix equation for (6–9) in terms of the homogeneous coordinates of the points is

$$(6\text{–}10) \qquad \begin{bmatrix} x_1' \\ x_2' \\ x_3' \end{bmatrix} = \begin{bmatrix} a & b & c \\ b & -a & d \\ 0 & 0 & 1 \end{bmatrix} \begin{bmatrix} x_1 \\ x_2 \\ x_3 \end{bmatrix}.$$

Intuitively, there exists a unique orthogonal line reflection taking any given point P into any other given point Q, namely, the reflection in the perpendicular bisector of the line segment PQ. This result may also be proved algebraically (Exercise 5). When $P = Q$, the orthogonal line reflection taking P into Q is not unique, since any axis through P may be used. These results may be used to prove that the origin $(0,0)$ may be transformed into any point $Q:(c,d)$ and therefore that the numbers c and d in (6–9) may be taken as arbitrary numbers (Exercise 6).

Any orthogonal line reflection is a similarity transformation (Exercise 7) and also an equiareal transformation (Exercise 8). Any product of orthogonal line reflections is a similarity transformation (preserves shape) and also an equiareal transformation (preserves area) (Exercise 9). Any product of an even number of orthogonal line reflections preserves shape, area, and the signs of the measures of ordered point triads (i.e., is an equiaffine transformation) (Exercise 10). Finally, any transformation that is both a similarity transformation and an equiareal transformation may be expressed as a product of orthogonal line reflections; any transformation that is both a similarity transformation and an equiaffine transformation may be expressed as a product of an even number of orthogonal line reflections (Exercise 11). This property establishes the equivalence of the definitions of a euclidean transformation as a product of orthogonal line reflections and as a transformation that is both a similarity transformation and an equiareal transformation (i.e., as an affine transformation that preserves shape and area).

Intuitively, an orthogonal line reflection may be visualized in three-space as a 180° rotation of the plane about a line on the plane (the axis of the line reflection). This transformation can be performed as a rigid motion in space but cannot be performed as a rigid motion when the plane is considered by itself. In general, the similarity transformations that are also equiareal transformations are rigid motions when the plane is considered as a subset of three-space; the similarity transformations that are also equiaffine transformations are rigid motions when the plane is considered by itself.   As indicated in Section 5–10, all equiaffine transformations are also equiareal transformations (i.e., all transformations that are rigid motions when the plane is considered by itself are rigid motions when the plane is considered as a subset of three-space).

EXERCISES

1. Prove that the center and the ideal point on the axis of the line reflection (6–8) are a pair of the absolute involution (6–2) if and only if $a_{12} = a_{21}$.

2. Prove that any orthogonal line reflection may be expressed in the form (6–9) with the conditions stated on the coefficients.

3. Prove that the product of two orthogonal line reflections with parallel axes is a translation perpendicular to the axes.

4. Prove that the product of two orthogonal line reflections whose axes are perpendicular is a point reflection.

5. Prove that for any two given distinct points P and Q, there exists a unique orthogonal line reflection T such that $T(P) = Q$.

6. Prove that there exist orthogonal line reflections (6–9) for any given pair of numbers c, d.

7. Prove that the set of orthogonal line reflections is a subset of the set of similarity transformations.

8. Prove that any orthogonal line reflection is an equiareal transformation.

*9. Prove that any product of orthogonal line reflections preserves both shape and area.

10. Prove that any product of an even number of orthogonal line reflections is an equiaffine transformation.

11. Prove that any transformation that is both a similarity transformation and an equiareal transformation may be expressed as a product of orthogonal line reflections; any transformation that is both a similarity transformation and an equiaffine transformation may be expressed as a product of an even number of orthogonal line reflections.

6–4 Euclidean transformations.   We now define any product of a finite number of orthogonal line reflections to be a *euclidean transformation.*   These transformations will be classified as *odd* or *even* according as they are obtained as a product of an odd or an even number of orthogonal line reflections.   Since the matrix of any orthogonal line reflection (6–9) has determinant -1, every odd euclidean transformation has determinant -1 and every even euclidean transformation has determinant $+1$ (Exercise 1). An even euclidean transformation is frequently called a *displacement*. From the point of view of similarity, equiareal, and equiaffine transformations, any similarity transformation that is also an equiaffine transformation is a displacement; any similarity transformation that is also an equiareal transformation and is not an equiaffine transformation is an odd euclidean transformation.

The matrix of the product of any two orthogonal line reflections (6–9) with centers $(-b, a+1, 0)$ and $(-b', a'+1, 0)$, respectively, has the form

$$\begin{bmatrix} a & b & c \\ b & -a & d \\ 0 & 0 & 1 \end{bmatrix} \begin{bmatrix} a' & b' & c' \\ b' & -a' & d' \\ 0 & 0 & 1 \end{bmatrix}$$
$$= \begin{bmatrix} aa' + bb' & ab' - a'b & ac' + bd' + c \\ a'b - ab' & aa' + bb' & bc' - ad' + d \\ 0 & 0 & 1 \end{bmatrix}$$

and determinant $+1$. If the two line reflections have the same centers, then $a = ka'$, $b = kb'$, and the product represents a translation (Exercise 2). If the line reflections have distinct centers, then their axes intersect in an ordinary point (s_1, s_2) and their product may be expressed in the form (Exercise 3)

$$(6\text{–}11) \qquad \begin{aligned} x' &= a(x - s_1) - b(y - s_2) + s_1, \\ y' &= b(x - s_1) + a(y - s_2) + s_2 \end{aligned} \qquad (a^2 + b^2 = 1).$$

Then any product of an even number of orthogonal line reflections (i.e., any displacement) may be expressed in the form (Exercise 4)

$$(6\text{–}12) \qquad \begin{aligned} x' &= ax - by + c, \\ y' &= bx + ay + d \end{aligned} \qquad (a^2 + b^2 = 1).$$

The group (Exercise 5) of displacements (even euclidean transformations) is often called the group of rigid motions on the plane when the plane is considered by itself rather than as a subset of three-space. Euclidean geometry on a plane considered by itself is therefore the geometry of the group of even euclidean transformations. When the plane is considered as a subset of three-space, the odd euclidean transformations must also be considered. In your previous study of euclidean geometry you have probably heard that any rigid motion on a plane may be expressed in terms of translations and rotations about a point. We now define rotations about a point and prove that rigid motions (displacements) on a plane considered by itself do have this property.

Translations and point reflections are displacements since any translation or point reflection may be expressed as a product of two orthogonal line reflections (Exercises 6 and 7). The displacement (6–12) is a translation if $a = 1$ and is a point reflection if $a = -1$. We now define the product of two orthogonal line reflections whose axes have an ordinary point O in common to be a *rotation* about the point O. The point O is called the *center* of the rotation. If the axes of the two line reflections are perpendicular, the rotation is a point reflection (Exercise 7). In general, the product of any two orthogonal line reflections is a rotation if the axes intersect, a translation if the axes are parallel. Then since any displacement is a product of an even number of orthogonal line reflections, any displacement is a product of translations and rotations (Exercise 15). Note that the identity transformation (the product of an orthogonal line reflection with itself) may be considered either as a translation or a rotation since the axis of the line reflection is parallel to itself and also has an ordinary point in common with itself.

An odd euclidean transformation has been defined as the product of an odd number of orthogonal line reflections. Then, since the identity transformation is a displacement, any odd euclidean transformation may be expressed as the product of a displacement and an orthogonal line reflection. Algebraically, any odd euclidean transformation may be expressed (Exercise 16) in the form

$$(6\text{–}13) \qquad \begin{aligned} x' &= ax + by + c, \\ y' &= bx - ay + d \end{aligned} \qquad (a^2 + b^2 = 1).$$

Odd euclidean transformations are often called *symmetries*, and two figures that correspond under an odd euclidean transformation are said to be symmetric. We shall, however, avoid this terminology, since it conflicts with common usage in elementary mathematics. For example, figures that are symmetric with respect to a line are symmetric in the above sense, but figures that are symmetric with respect to a point are not symmetric in the above sense, since they correspond under a point reflection, i.e., under an even euclidean transformation. In view of this difficulty in the usage of the word "symmetry," we shall use the longer term "odd euclidean transformation."

Euclidean geometry is often referred to as the geometry of rigid motions. In this sense the set of euclidean transformations is the set of rigid motions. An orthogonal line reflection and, in general, any odd euclidean transformation may be considered as a rigid motion when the plane is considered as a subset of a three-space but cannot be considered as a rigid motion when the plane is considered by itself. Any displacement (even euclidean transformation) may be considered as a rigid motion whether the plane is considered by itself or as a subset of three-space. This distinction between even and odd euclidean transformations is precisely the distinction between the set of similarity transformations that are also equiaffine transformations and the set of similarity transformations that are also equiareal transformations but are not equiaffine transformations. It provides one basis for our assumption (Section 6–1) that the euclidean plane may be considered as a subset of a three-space. Under this assumption the set of all euclidean transformations, odd and even, constitutes the set of *rigid motions* of a plane onto itself in three-space. The group of even euclidean transformations (Exercise 5) is the set of rigid motions of a plane onto itself on the plane; the group of all euclidean transformations (Exercise 19) is the set of rigid motions for euclidean plane geometry when the plane is considered as a subset of three-space, i.e., when the rigid motions are not restricted to the plane and therefore may include a rotation about a line of the plane. We shall hereafter use the words "rigid motion" and "euclidean transformation" interchangeably.

Any rigid motion is either a displacement (6–12) or an odd euclidean transformation (6–13). Accordingly (Exercise 18), any rigid motion may be represented in the form

$$(6\text{–}14) \qquad \begin{aligned} x' &= ax - by + c, \\ y' &= bex + aey + d \end{aligned} \qquad (a^2 + b^2 = e^2 = 1).$$

The transformation (6–14) represents an even transformation when $e = +1$, an odd transformation when $e = -1$. From the point of view of Klein, euclidean geometry is a study of the properties that are invariant under the group of euclidean transformations (6–14). We shall continue to consider this point of view and shall study some of the invariants under the group of transformations (6–14).

Any two figures F_1 and F_2 that correspond under a rigid motion (6–14) are said to be *congruent*, written $F_1 \cong F_2$. Thus euclidean geometry may be considered as the study of the common properties of congruent figures. From Exercise 9, Section 6–3, and the definition of euclidean geometry, any two congruent figures have the same area and the same shape; i.e., area and shape are invariants in euclidean geometry. In the next section we shall define distance and prove that distance is also invariant under the group of euclidean transformations.

Exercises

1. Prove that every odd euclidean transformation has determinant -1 and every even euclidean transformation has determinant $+1$.

2. Prove that any product of two orthogonal line reflections with parallel axes is a translation.

3. Prove that any product of two orthogonal line reflections with intersecting axes may be expressed in the form (6–11).

4. Prove that any displacement may be expressed in the form (6–12).

5. Prove that the set of all displacements forms a group.

6. Prove that any translation may be expressed as a product of two orthogonal line reflections.

7. Prove that any point reflection may be expressed as a product of two orthogonal line reflections with perpendicular axes, and conversely.

8. Does the set of rotations form a group? Explain.

9. Prove that the set of all rotations having the origin as center forms a group.

10. Prove that the set of all orthogonal line reflections forms a group.

11. Find the general coordinates of the center of the transformation (6–12) when it represents a rotation.

12. Find an algebraic representation for a rotation with center (h,k).

13. Prove that any displacement leaving exactly one ordinary point invariant is a rotation.

14. Prove that any displacement leaving at least two ordinary points invariant is the identity transformation.

15. Prove that any displacement may be expressed as a translation or as the product of a rotation about the origin and a translation.

16. Prove that any odd euclidean transformation may be expressed in the form (6–13).

17. Prove that the product of any two odd euclidean transformations is an even euclidean transformation.

18. Prove that any rigid motion may be expressed in the form (6–14).

19. Prove that the set of rigid motions (6–14) forms a group.

20. Consider the unit square $OIJK$ with vertices having nonhomogeneous coordinates (0,0), (1,0), (1,1), and (0,1) respectively. Give a euclidean transformation such that $OIK \cong JIK$.

21. Prove that any diagonal of a parallelogram divides it into congruent triangles.

6–5 Distances. In this section we shall use an algebraic approach and introduce the concept of a distance AB between any two points on a euclidean plane. The invariance of distances under the group of euclidean transformations will be established, and indeed may be used to define euclidean geometry as a special case of affine geometry. This invariance of distances as a distinguishing characteristic of euclidean geometry provides a basis for the intuitive concept of euclidean transformations as "rigid" motions. We shall use the distance function to define circles and several other common figures (Exercise 10), to find formulas for areas of several polygons (Exercises 13 and 14), and to prove that two triangles are congruent if their corresponding sides are equal (Exercise 17).

Given any three points A, B, C, distance should be defined such that the distances AB, BA, BC, and AC have the following properties:

   (i) $AB = 0$ if and only if $A = B$,
   (ii) $AB = BA$, and
   (iii) $AB + BC \geq AC$.

We define the distance AB between any two points $A:(x_1, y_1)$ and $B:(x_2, y_2)$ to be the nonnegative square root of the expression

$$(x_2 - x_1)^2 + (y_2 - y_1)^2,$$

where the x_j and y_j are nonhomogeneous coordinates of the points. In other words,

(6–15) $AB = \sqrt{(x_2 - x_1)^2 + (y_2 - y_1)^2}.$

This *euclidean distance function* has the above three properties (Exercise 1). Also, it is invariant under any euclidean transformation (6–14), since under such a transformation $a^2 + b^2 = e^2 = 1$ and

$$(x_1', y_1') = (ax_1 - by_1 + c,\ bex_1 + aey_1 + d),$$
$$(x_2', y_2') = (ax_2 - by_2 + c,\ bex_2 + aey_2 + d),$$

whence
$$x_2' - x_1' = a(x_2 - x_1) - b(y_2 - y_1),$$
$$y_2' - y_1' = be(x_2 - x_1) + ae(y_2 - y_1),$$

and

$$(x_2' - x_1')^2 + (y_2' - y_1')^2 = (a^2 + b^2e^2)(x_2 - x_1)^2$$
$$+ 2ab(e^2 - 1)(x_2 - x_1)(y_2 - y_1) + (a^2e^2 + b^2)(y_2 - y_1)^2$$
$$= (x_2 - x_1)^2 + (y_2 - y_1)^2.$$

Furthermore, if the distance function (6–15) is invariant under an affine transformation (6–4), then the transformation is a euclidean transformation (6–14) (Exercise 2).

The above discussion of the euclidean distance function provides a third basis for defining euclidean transformations. A euclidean transformation may now be considered as

(i) a finite product of orthogonal line reflections,
(ii) an equiareal transformation that is also a. similarity transformation, or
(iii) an affine transformation that leaves the euclidean distance function invariant.

In other words, a euclidean transformation is an affine transformation (preserves parallelism) which

(i) may be expressed as a product of translations and rotations (including rotations about a line),
(ii) preserves area and shape, or
(iii) preserves distances.

These three concepts of euclidean transformations are equivalent. They can help us gain a better understanding of the relationships among the concepts of parallelism, perpendicularity, shape, area, and distance. If a transformation that takes lines into lines and preserves the incidence of points and lines on a euclidean plane also preserves the euclidean distance relation, then it must preserve shape and area. If the transformation preserves both shape and area, then it must preserve distance. It may preserve area (be an equiareal transformation) without preserving shape or distance; it may preserve shape (be a similarity transformation) without preserving area or distance.

The above definition of distance is consistent with our assumption in Exercise 8 of Section 5–3 that the opposite sides of a parallelogram are equal (Exercise 3). This earlier assumption was used to compare distances on parallel lines.   The definition (6–15) may be used to associate nonnegative numbers with all distances on a euclidean plane and to compare distances on any two lines (parallel or intersecting). It may also be used to obtain an equation for a circle.   Given any point O and any distance r, the set of points P such that $OP = r$ is called a *circle* with *center* O and *radius* r.   When the center has coordinates (h,k), the circle consists of the points (x,y) satisfying the equation $(x - h)^2 + (y - k)^2 = r^2$ (Exercise 5).   Several other figures whose definitions involve the equality of distances are considered in Exercise 10.

Finally, we shall consider the use of distances in obtaining formulas for areas.   Even though our definition of the area of a triangle does not involve distances directly, distances and areas have very similar properties.   We have defined distances such that for any two points A and B we have

(i) the distance AB is a nonnegative number, and

(ii) if C is an interior point on the line segment AB, then $AB = AC + CB$.

We now define the *area of any triangle ABC* to be $\frac{1}{2} \mid m(ABC) \mid$ square units.   Then (Exercise 11)

(i) the area of any triangle ABD is a nonnegative number, and

(ii) if C is an interior point of the line segment AB, the area of triangle ABD is equal to the sum of the areas of triangles ACD and CBD.

In other words, the area of any triangle is a nonnegative number, and, if a given triangle is subdivided so that it "consists" of two "non-overlapping" triangles, the area of the given triangle is equal to the sum of the areas of the two component triangles.   We shall assume that any polygon may be divided into component triangles and that the area of the polygon is the sum of the areas of its component triangles.   Although this intuitive idea may be formalized, we shall, without further discussion, use it for triangles and quadrilaterals and obtain the usual area formulas for several common figures (Exercises 12, 13, and 14).   We shall identify each number representing an area as a number of square units in order to distinguish numbers representing areas from numbers representing distances and to connote the intuitive concept of the area of a figure as the number of unit squares needed to cover the figure exactly.

EXERCISES

1. Prove that the distance function (6–15) has the three properties stated at the beginning of this section.

2. Prove that the distance function (6–15) is invariant under an affine transformation (6–4) if and only if the transformation has the form (6–14).

3. Prove that the opposite sides of a parallelogram are equal.

4. Prove that the diagonals of a rectangle are equal.

5. Prove that any circle with center (h,k) and radius r consists of the points (x,y) satisfying the equation $(x - h)^2 + (y - k)^2 = r^2$.

6. Prove that any two circles are similar.

7. Prove that any two circles with equal radii are congruent.

8. Prove that a circle with center O and radius OQ may also be defined as the set of points obtainable from Q using rotations about O.

9. Prove that if $AB \perp BC$, then $\overline{AB}^2 + \overline{BC}^2 = \overline{AC}^2$.

10. Define the following figures:
   (a) square,   (b) isosceles triangle,   (c) rhombus,
   (d) equilateral triangle,   (e) isosceles trapezoid.

11. Prove that the area of any triangle ABD is a nonnegative number and that, if C is an interior point of the line segment AB, the area of triangle ABD is equal to the sum of the areas of triangles ACD and CBD.

12. Prove that the square with vertices $(0,0)$, $(1,0)$, $(1,1)$, and $(0,1)$ has an area of 1 square unit.

13. Prove that the area of a triangle ABC is equal to one half of the product of the base AB and the length of the altitude through C when triangle ABC has vertices
   (a) $(0,0)$, $(a,0)$, $(0,b)$,   (b) (h,k), $(h + a, k)$, $(h, k + b)$,
   (c) (a,b), (c,d), (e,f).

14. Use the area of a triangle to define the area of an arbitrary (a) square, (b) rectangle, (c) parallelogram, (d) trapezoid.

15. Prove that $\triangle ABC \cong \triangle ABC'$ if $AC = AC'$ and $BC = BC'$.

16. Prove that $\triangle ABC \cong \triangle AB'C'$ if the corresponding sides of the two triangles are equal.

17. Prove that if the corresponding sides of two triangles are equal, the two triangles are congruent (s.s.s. = s.s.s.).

18. Define two diameters of a central conic (Section 5–4) to be conjugate diameters if each is on the pole of the other. Then prove that a central conic is a circle if and only if its pairs of conjugate diameters have ideal points that are pairs of the absolute involution.

19. A *chord* of a conic may be defined as a line segment AB where A and B are points of the conic. A diameter that bisects all the chords that are perpendicular to that diameter, is called an *axis* of the conic. Prove that a parabola has exactly one axis.

20. Prove that all diameters of a circle are axes of the circle.

6–6 Directed angles. The set of points on the x-axis with coordinates x where $0 \leq x$ (i.e., the origin and the positive x-axis) may be designated by the symbol $\{OA\}$, where O is the origin and A is any point $(a,0)$ where $0 < a$. Any image $\{O'A'\}$ of $\{OA\}$ under a euclidean transformation is called a half-line or a *ray* with initial point O'. If $\{OA\}$ and $\{OB\}$ are any two rays with the origin as initial point, there exists a unique rotation about O

$$(6\text{--}16) \qquad \begin{aligned} x' &= ax - by, \\ y' &= bx + ay \end{aligned} \qquad (a^2 + b^2 = 1)$$

of the ray $\{OA\}$ onto the ray $\{OB\}$ (Exercise 1). In general, if $\{SR\}$ and $\{ST\}$ are two rays with a common initial point $S:(s_1,s_2)$, there exists a unique rotation about S

$$(6\text{--}17) \qquad \begin{aligned} x' &= a(x - s_1) - b(y - s_2) + s_1, \\ y' &= b(x - s_1) + a(y - s_2) + s_2 \end{aligned} \qquad (a^2 + b^2 = 1)$$

of the ray $\{SR\}$ onto the ray $\{ST\}$ (Exercise 2).

The figure composed of two rays $\{SR\}$ and $\{ST\}$ with a common initial point S is called an *angle* $\angle RST$ with *vertex* S and *sides* $\{SR\}$ and $\{ST\}$. Given any angle RST, there exists a unique rotation of the side $\{SR\}$ onto the side $\{ST\}$ and a second unique rotation (the inverse of the first) of the side $\{ST\}$ onto the side $\{SR\}$ (Exercise 3). We may select one of these rotations by specifying the rays in a particular order and requiring that the first ray be rotated onto the second. The figure composed of an ordered pair of rays $\{SR\}$ and $\{ST\}$ with a common initial point S is called a *directed angle RST* with *vertex S, initial side* $\{SR\}$, and *terminal side* $\{ST\}$. Then there is a unique rotation (6–17) associated with any directed angle RST. We shall use this correspondence between directed angles and rotations to determine measures of directed angles and of rotations.

Given any directed angle RST with vertex $S:(s_1,s_2)$, there is a unique translation

$$x' = x - s_1, \quad y' = y - s_2$$

of $\angle RST$ onto a directed angle AOB where $O:(0,0)$. The rotation associated with $\angle RST$ has the form (6–17); the rotation associated with $\angle AOB$ has the form (6–16). Furthermore, since the two directed angles are related by a translation, the two rotations are the same except for their constant terms (Exercise 4). We shall call any two such rotations (i.e., rotations with the same coefficients a and b)

equivalent rotations.   Intuitively, equivalent rotations are rotations in the same sense (clockwise or counterclockwise) and of the same amount.   Equivalent rotations may have the same or different centers.   Two directed angles are said to be *equal* if and only if they are associated with equivalent rotations.

Any directed angle with rotation (6–17) corresponds under a translation to a directed angle with its vertex at the origin and an equivalent rotation (6–16).   Then since (6–16) takes the point U:(1,0) into the point P:(a,b), we may associate any directed angle RST with a rotation having matrix

$$\begin{bmatrix} a & -b & c \\ b & a & d \end{bmatrix} \qquad (a^2 + b^2 = 1)$$

and with a directed angle UOP involving the points U:(1,0), O:(0,0), and P:(a,b).   In other words, given any directed angle RST, we may find an equal directed angle with its vertex at the origin and with its initial side along the positive x-axis.   Thus a measure defined for directed angles of the form $\angle UOP$ could be used to obtain a measure for any directed angles.   Also, since equal directed angles are associated with equivalent rotations, we may use either the angles or the rotations (6–16) to define a measure for both directed angles and rotations.

We have defined two directed angles to be equal if their associated rotations are equivalent.   We next consider the set of transformations under which directed angles correspond to equal directed angles.   Consider a general directed angle of the form $\angle UOP$ with associated rotation (6–16).   Under any affine transformation,

$$\begin{bmatrix} a_{11} & a_{12} & a_{13} \\ a_{21} & a_{22} & a_{23} \\ 0 & 0 & 1 \end{bmatrix} \qquad (a_{11}a_{22} - a_{12}a_{21} \neq 0),$$

the points O, U, and P correspond respectively to O':(a_{13}, a_{23}), U':($a_{11} + a_{13}, a_{21} + a_{23}$), and P':($aa_{11} + ba_{12} + a_{13}, aa_{21} + ba_{22} + a_{23}$). The equivalent rotation about O' has the form

$$x' = a(x - a_{13}) - b(y - a_{23}) + a_{13},$$
$$y' = b(x - a_{13}) + a(y - a_{23}) + a_{23}$$

and takes U' into P'':($aa_{11} - ba_{21} + a_{13}, ba_{11} + aa_{21} + a_{23}$).   The above affine transformation takes each directed angle UOP into an equal directed angle $U'O'P'$ if and only if P'', O', and P' are

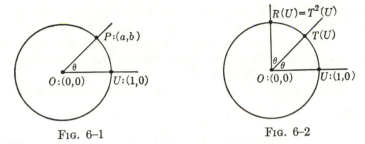

FIG. 6–1   FIG. 6–2

collinear for all values of a and b such that $a^2 + b^2 = 1$, i.e., if and only if $a_{22} = a_{11}$ and $a_{12} = -a_{21}$ (Exercise 6). In other words, an affine transformation takes directed angles into equal directed angles if and only if it is a similarity transformation (6–7) with $e = +1$, that is, a displacement. Thus two directed angles that correspond under a displacement are equal, and conversely, if two directed angles are equal, they correspond under a displacement (Exercise 7).

We now tentatively assume that there exists a measure for directed angles, consider some of its properties, and use these properties in selecting a definition of the measure. Let θ be the measure of any directed angle UOP as in Fig. 6–1 and let T be the associated rotation (6–16). Then $T(U) = P$. If $a = 1$, then $b = 0$, since $a^2 + b^2 = 1$, and T is the identity transformation. If $a = -1$, then $b = 0$, T is a point reflection, and T^2 is the identity transformation. If $b = 1$, then $a = 0$ and T has the form

$$(6\text{–}18) \qquad\qquad x' = -y, \quad y' = x.$$

When T is of the form (6–18), every line m through the origin is transformed into a line m' such that $m' \perp m$ (Exercise 8) and T^4 is the identity transformation (Exercise 9). If $b = -1$, then $a = 0$, T has the form

$$(6\text{–}19) \qquad\qquad x' = y, \quad y' = -x,$$

T transforms every line m through the origin into a line m' where $m' \perp m$, and again T^4 is the identity transformation (Exercise 10). The transformations (6–18) and (6–19) each take any line m through the origin into a line m' perpendicular to m and each have period four. Also they are inverse transformations, since any point (x,y) is transformed into $(-y,x)$ by (6–18) and that point is transformed into (x,y) by (6–19); i.e., their product is the identity transformation.

However, when T is the rotation (6–18), $T^3T = T^4$ is the identity transformation, T^3 is the inverse of T, and therefore T^3 is (6–19). This property of (6–18) may also be verified from the equations of the transformations. Similarly, if T is the transformation (6–19), then T^3 is (6–18) (Exercise 11).

If $a = b = 1/\sqrt{2}$, the rotation (6–16) may be expressed in the form

$$(6\text{–}20) \qquad\qquad x'\sqrt{2} = x - y, \quad y'\sqrt{2} = x + y.$$

This transformation takes (1,0) into $(1/\sqrt{2}, 1/\sqrt{2})$ and takes that point into (0,1). Thus if T has the form (6–20), then T^2 has the form (6–18) and T^8 is the identity transformation (Exercise 12). We next use these special cases ($a = 1$, $a = -1$, $b = 1$, $b = -1$, and $a = b = 1/\sqrt{2}$) of the rotation (6–16) to illustrate the properties of measures of directed angles and of rotations.

Let T be the rotation (6–20) and let R be the rotation (6–18). Then for $U{:}(1,0)$ we have $T^2(U) = (0,1) = R(U)$ as in Fig. 6–2. Thus, if θ represents the measure of (6–20), we would like 2θ to represent the measure of (6–18). Similarly, the measure of a point reflection should be twice the measure of (6–18) and the measure of the identity transformation should be the same as

    (i) twice the measure of a point reflection,
   (ii) four times the measure of (6–18),
  (iii) four times the measure of (6–19), and
  (iv) eight times the measure of (6–20).

Also, the representations for the measures of these rotations should be such that the sum of the measures of (6–18) and (6–19) represents the measure of the identity transformation.

We now define the measure of the rotation (6–18) and of its associated directed angle UOP to be $+1$ *right angles*. We say that the corresponding directed angle UOP is a positive right angle, or a right angle in the positive sense. The inverse, or opposite, relationship between the rotations (6–18) and (6–19) is indicated by defining the measure of (6–19) as -1 right angles. Then, since both (6–18) and (6–19) have period four, the measure of the identity transformation may be expressed either as $+4$ right angles or as -4 right angles. Also, since the product of the identity transformation with itself is the identity transformation, the measure of the identity transformation may be expressed as 0 right angles, $+8$ right angles, -8 right angles, +12 right angles, or, in general, $4n$ right angles where n is any integer (positive, zero, or negative). This property of the identity transformation implies that if the measure of any rotation (or directed angle) may be represented as θ right angles, then that measure may also be represented by $(\theta + 4n)$ right angles for any integer n. In other words, every directed angle and every rotation (6–16) has many representations of its measure. Among these representations there is exactly one, say θ right angles, such that $0 \leq \theta < 4$ right angles. This representation is called the *principal value* of the measure of the rotation and of the directed angle. The measure of any given rotation has a unique principal value; the measure of any given directed angle has a unique principal value. In both cases the principal value is nonnegative and, when expressed in terms of right angles, is less than 4. Measures of rotations and directed angles may also be expressed in terms of straight angles (+1 *straight angles* = +2 right angles), revolutions (+1 rev. = +4 right angles), and other units of angular measurement.

Given any rotation T of the form (6–16), we might assert that there always exists a least nonnegative real number k such that T^k is the identity transformation and therefore that the measure of T is $(1/k)$ revolutions. However, T^k has only been defined for positive integers k. We could define T^k for other real numbers k, but this would require a procedure analogous to a development of the positive real numbers from the positive integers. Accordingly, we shall define the principal value θ of the measure of each rotation and each associated directed angle to be

$$\text{arc cos } a \quad \text{when } 0 \leq b$$
$$1 \text{ rev.} - \text{arc cos } a \quad \text{when } b < 0$$

where "arc cos a" is read "the angle whose cosine is a" and designates a nonnegative number of revolutions that is less than or equal to $\frac{1}{2}$. Thus for any given real number a such that $-1 \leq a \leq 1$, we use "arc cos a" to designate an angle θ such that $0 \leq \theta \leq 180°$. The number θ may then be found using tables of values of cosines of angles. The tables may be in terms of degrees ($360° = 1$ rev.), radians (2π radians = 1 rev.), or possibly other units. For each value of θ the formulation of the tables is related to the problem of finding a real number k such that the corresponding rotation is $(1/k)$ revolutions.

We may summarize our discussion of directed angles as follows: A directed angle is the figure formed by an ordered pair of rays with a common initial point. Associated with any given directed angle there is a unique rotation of the initial side onto the terminal side, a unique equivalent rotation about the origin (equivalent rotations differ at most in their constant terms), and a unique equal directed angle with its vertex at the origin and its initial side along the positive x-axis. Two directed angles are defined to be equal if they are associated with equivalent rotations. This definition implies that two directed angles are equal if and only if they correspond under a displacement. After considering several special rotations, we defined right angles in the positive and negative senses. When the coordinate axes are taken in their usual positions, a positive angle is associated with a counterclockwise rotation and a negative angle is associated with a clockwise rotation. We found that the measures of rotations and directed angles may be represented in many ways in terms of the same units of measure (for example, right angles), but that the measure of each rotation and each directed angle has a unique principal value. Finally, we observed that the problem of expressing the principal value of the measure of any given directed angle in terms of a given unit of angular measure could be most easily solved by using tables. Accordingly, the principal value of the measure of any directed angle with associated rotation (6–17) was defined in terms of "arc cos a." Then the principal value of the measure of any directed angle can be found and is invariant under the group of displacements (Exercise 14). The common relations among directed angles may now be developed. Some of these are considered in the following exercises; others will be considered when we discuss undirected angles (Section 6–7).

EXERCISES

1. Given rays $\{OA\}$ and $\{OB\}$ where O is the origin, prove that there exists a unique rotation (6–16) of $\{OA\}$ onto $\{OB\}$.

2. Given rays $\{SR\}$ and $\{ST\}$, prove that there exists a unique rotation (6–17) about $S:(s_1,s_2)$ of $\{SR\}$ onto $\{ST\}$.

3. Prove that the inverse of any rotation (6–16) or (6–17) may be obtained by replacing "b" by "$-b$" in the equations of the rotation.

4. If the points R, S, and T correspond under a translation to A, $O:(0,0)$, and B respectively, prove that the rotation (6–17) takes $\{SR\}$ onto $\{ST\}$ and

the rotation (6–16) takes $\{OA\}$ onto $\{OB\}$ if and only if the two rotations are equivalent under the translation.

5. Prove that if two directed angles are related by a displacement, their associated rotations are equivalent.

6. Prove that an affine transformation takes any directed angle into an equal directed angle if and only if it is a similarity transformation with a positive determinant.

7. Prove that if two angles are equal, they correspond under a displacement and, conversely.

8. Prove that the rotation (6–18) takes any line m through the origin into a line m' that is perpendicular to m.

9. Prove that the rotation (6–18) has period four, that is, T^4 is the identity transformation and no one of the transformations T, T^2, T^3 is equal to the identity transformation.

10. Repeat Exercises 8 and 9 for the rotation (6–19).

11. Prove that if T is the rotation (6–19), then T^3 is the rotation (6–18).

12. Prove that the rotation (6–20) has period eight.

13. Prove that any directed angle corresponds under a displacement to a unique angle UOP where we have U:(1,0), O:(0,0), and P:(a,b) and the co-ordinates of P satisfy the relation $a^2 + b^2 = 1$.

14. Prove that the principal values of the measures of directed angles are invariant under similarity transformations with positive determinants.

15. Prove that if the measure of a rotation may be expressed as θ rev., then the measure of its inverse rotation may be expressed as $-\theta$ rev.

16. Prove that for directed angles $\angle AOB = -\angle BOA$.

17. Prove that any odd euclidean transformation takes a directed angle of measure θ into a directed angle of measure $-\theta$.

6–7 Angles. A directed angle has been defined as the figure formed by an ordered pair of rays with a common initial point. The directed angle AOB is associated with the (unique) rotation of $\{OA\}$ onto $\{OB\}$. An undirected angle or simply an angle has been defined as the figure formed by a pair (no order specified) of rays with a common initial point. The angle AOB may be associated with both the rotation of $\{OA\}$ onto $\{OB\}$ and the rotation of $\{OB\}$ onto $\{OA\}$. If one of these rotations has measure θ, then the other has measure $-\theta$. Thus $\angle AOB$ is associated with an amount of rotation without regard to the sense (clockwise or counterclockwise) of the rotation. Accordingly, we assume that hereafter "angle" refers to undirected angle unless otherwise specified, and we define the principal value of the measure of an angle AOB to be $|\theta|$ where the measure of the

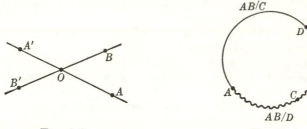

FIG. 6–3   FIG. 6–4

directed angle AOB may be expressed as θ right angles and $0 \leq \theta \leq 2$ (Exercise 1). Then any angle may be associated with a rotation (6–17) and has a corresponding measure "arc cos a." Two angles with measures θ and ϕ respectively are equal if and only if $\theta = (\phi + n)$ rev. or $\theta = (n - \phi)$ rev. (Exercise 2).

The usual relations among angles may now be developed. For example, complementary angles, supplementary angles, acute angles, and obtuse angles may be defined (Exercise 3). Two angles are equal if and only if they correspond under a euclidean transformation (Exercise 4). If two angles correspond under a similarity transformation, then they are equal (Exercise 5). Any two right angles are equal (Exercise 6) and, by definition, $\angle AOB = \angle BOA$.

We have defined and discussed angles formed by two rays or half-lines. Let us now consider the figure formed by any two intersecting lines $A'OA$ and $B'OB$ (Fig. 6–3). Each line contains two rays with initial point O (Exercise 7). The two lines determine four angles ($\angle AOB$, $\angle BOA'$, $\angle A'OB'$, and $\angle B'OA$) and eight directed angles. Since $\angle AOB$ is transformed into $\angle A'OB'$ and $\angle BOA'$ is transformed into $\angle B'OA$ under a point reflection with center O, we have $\angle AOB = \angle A'OB'$ and $\angle BOA' = \angle B'OA$ from Exercise 4. The special relationship between the equal angles of such pairs is indicated by calling them vertical angles. In general, whenever two angles with a common vertex correspond under a point reflection, the angles are said to be *vertical angles*. As in Fig. 6–3, any two intersecting lines determine two pairs of vertical angles and any two vertical angles are equal. Several other properties of angles formed by intersecting lines are considered in the exercises at the end of this section.

We conclude our discussion of angles with a comparison of the properties of the set of angles formed by lines of a pencil of lines and

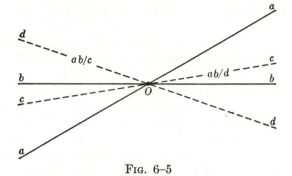

Fig. 6–5

the properties of the set of line segments formed by points of a pencil of points.

(i) In projective geometry, any two points determine a unique line; any two lines determine a unique point.

(ii) Two points A, B on a line m may be used to form two line segments AB/D and AB/C (Fig. 6–4) where C, D are points on m and the pairs A, B and C, D separate each other (Section 3–4); two lines a, b on a point O may be used to form two pairs of vertical angles ab/d and ab/c where c, d are lines on O and the pairs a, b and c, d separate each other (Fig. 6–5).

(iii) In affine geometry, we selected a point P_∞ on each line and defined the segment AB/P_∞ to be the segment AB; in euclidean geometry, we may define the angle ab between two lines to be a zero angle if a and b coincide, a right angle if $a \perp b$, and an acute angle otherwise.

(iv) The distance (linear measure) AB depends upon the unit of measure (choice of coordinates); the angular measure $\angle ab$ depends upon the unit of angular measure.

(v) The distance AB is a nonnegative number; the principal value of the measure of the angle ab is a nonnegative number.

(vi) The distance $AB = 0$ if and only if the points coincide; the principal value of the measure of the angle ab is zero if and only if the lines coincide.

(vii) Given any real number r, there exists a distance $OR = r$; given any real number θ, there exists a directed angle with measure of principal value θ right angles if and only if $0 \leq \theta < 4$, an angle with measure of principal value θ right angles if and only if $0 \leq \theta \leq 2$, and two lines a, b such that $\angle ab = \theta$ right angles, if and only if $0 \leq \theta \leq 1$.

(viii) Given a point O on a line m and any real number r, there exist exactly two points R and R' on m such that $OR = r = OR'$; given any line m on a point O and any real number θ, there exist two lines r and r' on O such that $\angle mr = \theta = \angle mr'$ if $0 < \theta < 1$ right angle, a unique line r such that $\angle mr = \theta$ if $\theta = 0$ or 1 right angle, and no lines r such that $\angle mr = \theta$ if $\theta < 0$ or $\theta > 1$ right angle.

(ix) Any line segment AB has a unique associated distance AB; the angle ab between two intersecting lines has a unique associated measure, say ϕ rev. where $0 \leq \phi \leq \frac{1}{4}$. If the measure of a directed angle has principal value θ rev., it may also be expressed as $(n + \theta)$ rev. for any integer n. If the measure of an angle (undirected) has principal value θ rev., it may also be expressed as $(n + \theta)$ rev. or as $(n - \theta)$ rev. for any integer n.

The above comparison indicates several of the common properties of linear measurements and angular measurements. It also indicates the basic differences arising from the fact that any integral number of revolutions is equivalent, as far as the resulting geometric figure is concerned, to the identity transformation.

EXERCISES

1. Prove that the principal value of the measure of any angle may be expressed as θ right angles where $0 \leq \theta \leq 2$.

2. Prove that two angles with measures θ and ϕ, respectively, are equal if and only if $\theta = (\phi + n)$ rev. or $\theta = (n - \phi)$ rev.

3. Define complementary angles, supplementary angles, acute angles, obtuse angles.

4. Prove that two angles are equal if and only if they correspond under a euclidean transformation.

5. Prove that if two angles correspond under a similarity transformation, they are equal.

6. Prove that any two right angles are equal.

7. Prove that any line m contains two distinct rays having a common initial point, say O, and that every point of m that is distinct from O is on exactly one of the rays.

8. Consider two distinct parallel lines m and m' and a *transversal* t (i.e., a line that is not parallel to m). Define corresponding angles, alternate interior angles, and alternate exterior angles.

9. Prove that when parallel lines are cut by a transversal, any two corresponding angles are equal, alternate interior angles are equal, and alternate exterior angles are equal.

10. Prove that if the sides of one angle are parallel to the sides of another angle, then the angles are equal or supplementary.

11. Prove that two angles are equal if they are
   (a) supplementary to the same or equal angles,
   (b) complementary to the same or equal angles.

12. Given a point O on a plane, use the measures of angles with vertex O to obtain a distance relationship on the ideal line.

13. Is the distance function obtained in Exercise 12 a euclidean distance function? Explain.

14. Prove that in euclidean geometry any angle ACB where AB is a diameter of a circle and C is a third point on that circle (i.e., any angle inscribed in a semicircle) is a right angle.

15. Use the property obtained in Exercise 13 and give another definition of a circle.

6–8 Common figures. We have developed euclidean plane geometry from the undefined terms (point and line), the undefined relations of incidence and separation, and the postulates of incidence, existence, projectivity, harmonic sets, separation, and continuity (based upon the isomorphism between the set of real numbers and the set of real points on a line). This synthetic development has been proved to be equivalent to an algebraic (or analytic) development in terms of triples of real numbers, the properties of numbers, and the properties of algebra. The euclidean plane has been defined by considering only real points on a projective plane with one line deleted. The group of euclidean transformations has been defined so that the incidence of points and lines, parallelism, perpendicularity, shape, area, distances, and the principal values of the measures of angles are invariant. These invariants will now be used as we consider properties of several common figures.

All figures of euclidean geometry may be considered in terms of their dependence upon the invariants of euclidean geometry. For example, a general conic may be defined in projective geometry and therefore depends only upon the incidence, existence, continuity, sets of points, and sets of lines that are invariant in projective geometry. The ideal line of affine geometry was needed to define an ellipse, which therefore depends also upon parallelism. The euclidean distance function or some equivalent such as the rotations of euclidean geometry is needed to define a circle, which therefore depends also upon the invariance of perpendicularity.   In general, we may identify the invariants upon which a given figure depends by identifying the geometries in which the figures may be defined.   Since any figure that is defined in a geometry is defined in all special cases of that geometry, we shall simply identify the most general geometry that we have discussed and in which each figure is defined.   Briefly then, a figure in euclidean geometry is also a figure in affine geometry if it can be defined without using area and perpendicularity or an equivalent assumption such as the equality of distances on intersecting lines. It is a figure in projective geometry if it can be defined without using parallelism.   These concepts are used in Exercise 1.

We may also identify the most general geometry that we have studied and in which various concepts may be defined (Exercise 2) and the most general geometries in which various theorems may be proved (Exercise 3).   The determination of these geometries as indicated in the following set of exercises is highly recommended as a means of recognizing the fundamental concepts upon which these figures, definitions, and theorems of euclidean geometry are based. A mastery of these exercises and our development of euclidean geometry should enable any reader to identify the fundamental concepts underlying any figure, definition, or theorem in a high-school or college geometry text.   Such a mastery should also provide an effective foundation for further work in geometry or in any field requiring the use of geometric principles.

EXERCISES

1. Identify the most general geometry (euclidean, affine, projective) in which each of the following figures is defined: (a) triangle, (b) right triangle, (c) isosceles triangle, (d) quadrilateral, (e) parallelogram, (f) square, (g) rhombus, (h) conic, (i) ellipse, (j) circle, (k) parabola, (l) hyperbola, (m) equilateral hyperbola, (n) regular polygon, (o) equilateral triangle, (p) trapezoid.

2. As in Exercise 1, identify the most general geometry in which each of the following may be defined: (a) angle bisector, (b) mid-point of a line segment, (c) triangle inscribed in a circle, (d) tangent to a conic, (e) median of a triangle, (f) altitude of a triangle, (g) regular hexagon.

3. As in Exercise 1, identify the most general geometry in which each of the following theorems may be proved:

    (a) the altitudes of a triangle are concurrent,

    (b) the medians of a triangle are concurrent,

    (c) the angle bisectors of a triangle are concurrent,

    (d) the line segment joining the mid-points of two sides of a triangle is parallel to the third side and equal to half of it,

    (e) the diagonals of a parallelogram bisect each other,

    (f) there exists a unique line parallel to a given line and through a given point,

    (g) lines parallel to the same line are parallel to each other,

    (h) the sum of the angles of a triangle is equal to two right angles,

    (i) a line intersecting one of two parallel lines intersects the other also.

4. Select one of the theorems stated in Exercise 3 that may be proved in affine geometry and prove it

    (a) in euclidean geometry,

    (b) in analytic affine geometry, and

    (c) in synthetic affine geometry.

6–9 Survey. We have now completed our development of euclidean plane geometry from projective geometry. In the present chapter we started with affine geometry, defined the euclidean plane to be the real affine plane, selected an elliptic involution on the ideal line as the absolute involution, used this involution to define perpendicular lines and orthogonal line reflections, and defined all finite products of orthogonal line reflections to be euclidean transformations. Since intuitively the euclidean transformations were rigid motions when the plane was considered as a subset of space, we have used the words "rigid motion" and "euclidean transformation" interchangeably.

The set of euclidean transformations was found to be equivalent to the set of similarity transformations that were also equiareal transformations. The set of euclidean transformations is also equivalent to the set of affine transformations that preserved the euclidean distance function. Accordingly, incidence, parallelism, perpendicularity, shape, area, distance, and angles were found to be invariants of euclidean geometry. Finally, these invariants were used to identify the geometric properties underlying many common figures, definitions, and theorems of euclidean geometry. In this sense we have seen that euclidean geometry may be considered as a study of properties which are invariant under the group of euclidean transformations. We have also seen (Chapters 1 to 6) how euclidean geometry may be considered as a postulational system from either a synthetic or an algebraic point of view.

We have developed euclidean geometry by postulating projective geometry and then placing restrictions upon the transformations (i.e., considering special cases of the geometry) until we obtained euclidean geometry. The following chart indicates the basic sequence of groups of transformations that we have used to obtain the group of euclidean transformations.

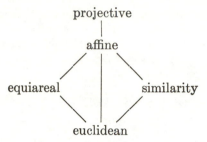

The group of affine transformations may be obtained by requiring that the ideal line be invariant as a line; the group of similarity transformations by requiring that the absolute involution be invariant; the group of equiareal transformations by requiring that the matrices of the transformations have determinant of numerical value 1; and the group of euclidean transformations by requiring that distances be invariant. We have obtained the group of euclidean transformations by placing successive restrictions upon the transformations of projective geometry. We have obtained euclidean geometry by considering a sequence of special cases of projective geometry (i.e., by specializing the geometry).

We may also study the properties of euclidean geometry by making a sequence of generalizations of euclidean geometry. For example, if we remove the single requirement that area be invariant, we necessarily lose some other properties as well and obtain the group of similarity transformations. Also if we remove the single requirement that perpendicularity (angles) be invariant, we obtain equiareal transformations; if we remove the two requirements for the invariance of area and angles, we obtain affine geometry; if we also remove the requirement that parallelism be invariant, we obtain projective transformations. In this sense we may study the significance of certain invariant properties of euclidean geometry by studying geometries in which these properties are not invariant.

We may start with projective geometry and specialize the geometry until we obtain euclidean geometry or we may start with euclidean geometry and generalize the geometry until we obtain projective geometry. Chapters 2 to 6 of this text form a unit based upon the specialization of projective geometry to obtain euclidean geometry. The remaining chapters are included to increase the reader's understanding of euclidean plane geometry through a brief historical account of the development of euclidean geometry (Chapter 7), a comparison of euclidean geometry with the two noneuclidean geometries (Chapter 8), and an introduction to topology — a geometry having projective geometry as a special case (Chapter 9).

Chapter 7 provides some background for Chapter 8. Chapter 9 is completely independent. Each of these three chapters is concerned with a topic requiring a large book for even a reasonably comprehensive treatment. Accordingly, each of the remaining chapters should be considered as a sketch of a topic that will increase the reader's understanding of euclidean geometry but which cannot be treated thoroughly in the space available.

REVIEW EXERCISES

1. Consider a general projective transformation on a plane and outline the restrictions used to obtain euclidean transformations
    (a) by considering the restrictions in terms of invariants,
    (b) by considering the successive restrictions upon the elements of the matrix of the general projective transformation.
2. Give a set of postulates for euclidean plane geometry.
3. Define angle bisector.
4. Prove that in any triangle
    (a) the altitudes are concurrent,
    (b) the medians are concurrent,
    (c) the perpendicular bisectors of the sides are concurrent, and
    (d) the angle bisectors are concurrent.
5. Define the angles of a triangle and prove that in euclidean geometry the sum of the angles of any triangle is two right angles.
6. Prove that two triangles are congruent if
    (a) corresponding sides are equal,
    (b) two sides and the included angle of one are equal to two sides and the included angle of the other,

(c) two angles and the included side of one are equal to two angles and the included side of the other.

7. Prove that if two sides of a triangle are equal, the angles opposite those sides are equal, and conversely.

8. Prove that if two triangles correspond under a similarity transformation, then their corresponding angles are equal and their corresponding sides are proportional.

9. For any circle define: central angle, inscribed angle, major arc, minor arc, and chord.

10. Prove that in the same or equal circles, equal central angles determine equal chords.

11. Prove the converse of Exercise 10.

12. If A, B, C are points of a circle with center O, prove that $\angle AOB = 2\angle ACB$.

CHAPTER 7

THE EVOLUTION OF GEOMETRY

What is geometry?   Do we mean the same thing by "geometry" today that Euclid meant over two thousand years ago?   Do we mean the same thing that was meant fifty years ago?   In this chapter we shall endeavor to observe the change, the growth, and the evolution of the concept of geometry.   We shall find that man's concept of geometry is constantly changing as new principles and new applications of principles are discovered.   Thus geometry is not a static subject but rather a growing body of knowledge with ever widening applications and with an inherent beauty in its systematic organization and structure.   We shall consider primarily the evolution of euclidean geometry and the origins of several other geometries through the applications of new principles to certain fundamental concepts of euclidean geometry.   In this sense euclidean geometry is the "father" of nearly all other geometries.

7–1 Early measurements.   We use the word "geometry" (derived from the Greek word for "earth measure") to represent the present state of development of a body of knowledge that, about four thousand years ago, consisted primarily of a few practical procedures for measuring areas of fields.   Originally geometry was an empirical science, that is, a science based upon experience and observations. General theories, postulates, and proofs had not been formalized. Thus our geometry has evolved from a few formulas and procedures arising from practical needs.   Today we usually think of geometry as a deductive science based upon undefined terms, postulates, and the logical deduction of theorems.   From this point of view different systems of postulates often give rise to geometries with different properties and different applications.   Most mathematicians look upon all practical applications of geometric properties as by-products rather than as an integral part of the subject.

We do not know the complete history of early geometry.   We do know from pictorial tablets that about 4000 B.C. the Babylonians found the area of a rectangular field as the product of its length and its width.   We also see evidences of early engineering accomplishments in Babylon and in Egypt that must have required the use of geometric concepts.   Indeed, it appears reasonable to assume that every culture has developed some geometric concepts at its own rate and in its own era.

The pyramids offer striking evidence of early achievements in Egypt.   The great pyramid at Giza was built about 2900 B.C.   Its sandstone blocks are fitted together and have an average weight of about two and a half tons.   This pyramid also contains chambers having large granite roof members about 200 feet above ground level. These roof members weigh about fifty tons each and probably were brought from a quarry over 600 miles away.   Even though we have no written records of the geometric knowledge of these early Egyptians, the engineering feats involved in the construction of the pyramids and the erection of huge stone obelisks certainly required the use of many concepts of areas and volumes.

By about 2000 B.C. the Babylonians were using the areas of fields for taxation purposes.   Between 2000 and 1600 B.C. the Babylonians used specific cases of and probably knew the usual general formulas for the area of a rectangle, area of a right triangle, area of a trapezoid with one side perpendicular to the parallel sides, volume of a rectangular parallelepiped, and the volume of a right prism with a trapezoidal or circular base.   They also appear to have known that the altitude from the vertex of an isosceles triangle bisects the base, that corresponding sides of similar right triangles are proportional, and that any angle inscribed in a semicircle is a right angle.   The early Babylonians used $\pi = 3$ to find the circumference and area of a circle. They also found general formulas for the sides of Pythagorean triangles (i.e., right triangles with sides of integral lengths such as the 3, 4, 5 right triangle) [1; 8–9].

Meanwhile, the Egyptians were developing many of these same geometric rules.   Two manuscripts written in Egypt about 1850 B.C. and 1650 B.C. supply the following information.   The Egyptians at this time knew the general formula for the area of any triangle. They developed a procedure for computing the area of a circle of diameter d as $[d - (1/9)d]^2$.   This is equivalent to using π as $256/81$ $= 3.1605. \ldots$ The expression $[d - (1/9)d]^2$ was not written as $(8d/9)^2$ at that time because only unit fractions (i.e., only fractions with numerator 1) were used.   As in Babylon, the volume of a right circular cylinder (such as a granary) was obtained as the product of the area of the base and the height.   The Egyptians also considered the cotangents of the angles formed by the faces and bases of pyramids and appear to have known that the volume of a frustum of a pyramid with height h and square bases of sides a and b is given by the formula $(1/3)h(a^2 + ab + b^2)$ [1; 14–15].

The geometric rules that were developed prior to 1500 B.C. in Babylon and Egypt are very impressive.   Undoubtedly there were incorrect as well as correct formulas in common use in both cultures. One writer [2; 5] asserts that in both cultures the area of any quadrilateral with opposite sides a and a', b and b' was assumed to be $(a + a')(b + b')/4$.   This formula gives a correct answer only in the case of rectangles.   However, all such rules must be recognized as empirical results based solely upon experience and observations. They must be considered on their own individual merits rather than as evidences of a large body of theoretical knowledge of geometry. It is probable that both Babylonian and Egyptian concepts of geometry remained at a utilitarian and empirical level until about 600 B.C. when the influence of Greek thinkers began to have an effect, especially in Egypt.

7–2   Early Greek influence.   About 600 B.C. the Greek culture, with its emphasis upon reason, truth, beauty, and knowledge for its own sake, was becoming an important factor in the ancient world. The evolution of geometry from a set of empirical rules to a body of knowledge with importance in its own right and to a deductive system was very slow.   The early Greeks made the first recognizable progress in the study of geometry as a science independent of its practical applications.   Indeed, the Greeks provided the major impetus for the development of geometry prior to the intellectual reawakening in Europe in the 16th and 17th centuries.   Thus we find that

(i) the empirical geometry of the Babylonians and Egyptians was further developed as a subject in its own right by the early Greeks (600 to 300 B.C.) (Section 7–2);

(ii) the Greeks used geometric representations and constructions when they operated with numbers (Sections 7–2, 7–3, and 7–4);

(iii) the cultural environment that gave birth to the systematic reasoning and logic of Aristotle in the 4th century B.C. also provided a basis for the systematic organization of geometry by Euclid (Section 7–3);

222   THE EVOLUTION OF GEOMETRY   [CHAP. 7

(iv) since the work of Euclid included a summary and organization of the geometry of his time, the geometric developments through the next few centuries may be considered with reference to Euclid's outstanding work (Section 7–4); and

(v) from the 4th to the 15th centuries A.D. the most significant progress in geometry was made through Greek commentaries on Euclid in the 4th and 5th centuries, by Arabian, Hindu, and Persian mathematicians, and through the awakening in Europe of an interest in astronomy and trigonometry starting in the 13th century (Sections 7–4 and 7–5).

From the 16th century to the present, the development of geometry has been rapid and we shall have difficulty gaining a proper perspective of its growth. However, at all times the development of geometry has reflected the general intellectual climate of the times and has been closely related to progress in such related fields as arithmetic, astronomy, trigonometry, algebra, calculus, and engineering.

The early Greek traders and scholars encountered both the Egyptian and the Babylonian mathematical achievements. However, the Greeks who took an unusual interest in mathematics appear to have studied in Egypt. The first individual to be given credit for definite mathematical discoveries was Thales (640 to 546 B.C.). He had a fine reputation as a mathematician, astronomer, statesman, engineer, businessman, and philosopher. Indeed, he was considered to be one of the Seven Wise Men of early times. His breadth of interests and accomplishments emphasizes the narrow view that we are taking as we consider only the evolution of geometry. In general, it is highly desirable that progress of any sort be considered with respect to the complete culture in which it occurs. We shall consider only geometry, in an effort to condense the discussion so that readers may visualize the complete evolution of euclidean geometry to date. Some distortions and many omissions are inevitable in this process.

Thales studied in Egypt and demonstrated his genius in many ways. His most spectacular accomplishment in Egypt may have been the indirect measurement of the height of a pyramid. According to one legend he used a vertical stick, observed the instant at which the length of the stick's shadow was equal to the length of the stick, indicated the extent of the shadow of the pyramid at that instant, and measured the shadow of the pyramid. Such an experiment would be easily understood by a high-school geometry class and by

most eighth graders today.   However, the Egyptians had been primarily concerned with measurements of areas and volumes.   There
did not exist a geometry of lines or of similar triangles.   Thales
started the trend of geometry away from a study of direct measurements of areas and volumes to a study of properties of lines and
figures formed by lines.   In this sense he is often said to have created
the geometry of lines — a subject with a potential importance in its
own right and a subject that, although originating in man's needs
and experiences, attained an abstract character in a relatively few
years.

The following results are usually attributed to Thales [1; 17]:

(i) a circle is bisected by any diameter;
(ii) the base angles of an isosceles triangle are equal;
(iii) vertical angles are equal;
(iv) if two angles and one side of a triangle are equal respectively
to two angles and the corresponding side of another triangle, the
triangles are equal in all respects; and
(v) any angle inscribed in a semicircle is a right angle.

All these results are included later in Euclid's *Elements*.   Also, as
we have seen, the last result had been in part recognized over a
thousand years before in Babylonia.   Such rediscovery and often
simultaneous independent discoveries are very common in mathematics.   They imply that whenever a culture is ready for a discovery,
some person of exceptional insight, and often several such people,
will make the discovery.   Thales had great insight.   He extended
the scope of the geometry of the Egyptians to obtain properties of
triangles and used these properties in such practical tasks as the
measurement of the height of a pyramid and the measurement of the
distance of a ship from shore.   Thus he broadened the concept of
geometry and laid a foundation for the development of geometry as a
science.

Intellectual progress among the early Greeks was greatly influenced
by the formation of schools centered around outstanding scholars and
leaders.   For example, the pupils of Thales and later their pupils
constituted the Ionian school that flourished for about a hundred
years.   The next important school was the Pythagorean school
(about 500 to 350 B.C.).   Pythagoras had considerable ability, traveled widely, studied mathematics in Egypt, studied mysticism in

Persia, had difficulty finding a place to settle, and finally formed a secret society at Crotona in southeastern Italy. The Pythagorean school was a brotherhood involving a strict way of life and in this respect was different from other Greek schools. The members of the school pooled their results and are usually given credit for significant results regarding

    (i) the properties of parallel lines and their relation to the angle sum of a triangle;

    (ii) the transformations of polygons into polygons of the same area and the use of these methods to find real roots of linear and quadratic equations;

    (iii) the properties of similar figures related to proportions with rational elements;

    (iv) at least three of the regular solids; and

    (v) the incommensurability of the diagonal and side of a square (i.e., the fact that their ratio cannot be represented as a rational number).

The association of the following theorem with the Pythagoreans may be in recognition of the first proof of this relation:

PYTHAGOREAN THEOREM: *In any triangle with sides of lengths a, b, and c and hypotenuse c, we have* $c^2 = a^2 + b^2$.

We know that at least by 1600 B.C. the Babylonians had a table of integers satisfying this relation, and there is reason to believe that they knew a general rule for such numbers. Many writers assume that the Babylonians and Egyptians used the relation $c^2 = a^2 + b^2$ to construct right angles. If such is the case, the Pythagorean Theorem is the converse of this well-known construction for right angles. In other words, the name of the theorem may give recognition to the Pythagoreans for taking the empirical rule:

If a, b, and c are the lengths of the sides of a triangle and if $c^2 = a^2 + b^2$, then the sides of lengths a and b form a right angle,

formulating the converse statement:

If a, b, and c are the lengths of the sides of a triangle and if the sides of lengths a and b form a right angle, then $c^2 = a^2 + b^2$, and giving a proof for the converse statement. This conjecture is not well established. However, it is known that the Pythagoreans considered this a very important theorem. Indeed, this theorem is probably the basis for their discovery of the irrationality of $\sqrt{2}$, since, if $a = b = 1$, then $c = \sqrt{2}$ and c cannot be written as an integer or as a quotient of the integers. The number $c = \sqrt{2}$ is irrational and the corresponding line segment is said to be *incommensurable* with respect to the segments of length 1.

The problems associated with irrational numbers, such as $\sqrt{2}$, arising from incommensurable lengths provided one basis for the Greek emphasis on constructions with straightedge and compasses, since the constructions are the same for all line segments, whether commensurable or incommensurable with respect to a given unit line segment. The Greeks used geometric constructions not only to add, subtract, multiply, and divide, but also to find square roots and to perform all the operations that we associate with finding the real roots of linear and quadratic equations. The use of line segments to represent numbers and the use of areas to represent products of numbers enabled the Greeks to avoid many problems of number notation and questions of rationality. It also provided a basis for the reluctance of early mathematicians to accept negative numbers as numbers.

The Pythagoreans grouped geometry, arithmetic, music, and astronomy together. The word "mathematics" is derived from a Greek word that, at the time of the Pythagoreans, meant "a subject of instruction" and for several centuries referred to the above four subjects. In general, the Pythagoreans continued the trend toward the recognition of geometry as a subject in its own right. Probably Pythagoras, and certainly the group of Pythagoreans, considered a study of geometry as a part of a liberal education, made several geometric discoveries, and through the Sophist school in Athens had a profound influence upon the Greek concepts of geometry.

The Sophists were paid teachers in the young democracy at Athens. They were motivated by a spirit of understanding rather than of utility, introduced the ideas of the Pythagoreans at Athens, and, among other things, considered the following construction problems:

(i) the squaring of a circle (i.e., the construction of a square with area equal to that of a given circle);

(ii) the duplication of a cube (i.e., the construction of a cube with volume double that of a given cube); and

(iii) the trisection of any given angle.

These problems became classics as far as constructions with straight-edge and compasses were concerned [10; 228–249]. They were solved within a century using other methods, but it was over two thousand years before geometry, the theory of equations, the theory of numbers, and the use of algebra in solving geometric problems had developed sufficiently to enable anyone to prove that all three of the classical construction problems were impossible with only straightedge and compasses. The importance of these three classical construction problems is due in a large part to the many discoveries made by mathematicians who were trying in vain to solve these problems.

Another mathematical concept that was destined to cause difficulty for many centuries was considered at about this same time. As long as geometry was composed of empirical results, all results and all processes were finite. When geometry began to be studied on its own, there were many problems related to processes that could be continued *ad infinitum*, problems related to infinitesimal distances, and problems related to sets containing infinitely many elements. For example, the visualization of a circle as coinciding with the limiting figure obtained by considering regular polygons of 4, 8, 16, ... sides inscribed in the circle involves a concept of a nonending process; the process of dividing a line segment or distance AB to obtain $\frac{1}{2}$, $\frac{1}{4}$, $\frac{1}{8}$, ... of the given distance is a nonending process in a theoretical sense and involves infinitesimal distances; the set of points on a line segment is an infinite set. These concepts are related to the continuity of line segments and to the concept of a limit, which provides the basis for the calculus. The early Greeks and indeed nearly all scholars before the 17th century A.D. recognized the problems associated with these concepts and seldom used them. This was in part due to the paradoxes of Zeno, who emphasized the confusions arising from a partial understanding of these concepts. One of his most famous arguments was roughly as follows: A fast runner can never catch a tortoise because the runner must first reach the point where the tortoise started and the tortoise will have meanwhile moved to another spot. When the runner reaches this spot, the tortoise will again have moved to another spot, and so forth *ad infinitum*. The runner gets closer and closer to the tortoise but never catches him. This paradox was best clarified by Aristotle (384 to 322 B.C.), who, in addition to his famous work systematizing deductive logic, also gave a discussion of continuity that was not surpassed for many centuries.

We have emphasized the growing tendency among the early Greek scholars to recognize geometry as a part of a liberal education. This was also emphasized by the philosopher Plato (about 400 B.C.), who reportedly placed the inscription "Let no one who is not acquainted with geometry enter here" over the entrance of his school near Athens. This emphasis upon geometry as an essential part of the training of the mind has influenced many scholars, both ancient and modern.

Plato and his school made a notable contribution to the reasoning used in obtaining proofs in geometry. They devised a method for discovering a proof by starting with the statement to be proved and working back (analyzing the desired result) to obtain the given data. When all the steps used are reversible this method constitutes a proof by *analysis*. We contrast it with the deductive method of proof, *synthesis*, which is used when we start with the given data and deduce a sequence of statements leading to the statement to be proved. The Greeks usually followed their analysis with a synthetic proof.

The outstanding mathematical genius of this time was Eudoxus, a pupil of Plato. Eudoxus made significant contributions to the knowledge of areas and volumes, to the theory of proportions, and to the study of incommensurables as discussed in the fifth book of Euclid's *Elements*. He is credited with the development of the "method of exhaustion," which was used to obtain theorems regarding ratios of areas. Since rational numbers and their positive square roots may be constructed by classical methods, the ratio of the area of a unit square to the area of a circle of unit diameter is closely related to the classical construction of a square with area equal to that of a given circle. This ratio is greater than 1, since the circle may be inscribed in the square. After the chords through the points of tangency have been drawn, the circle circumscribes a square of area 1/2 and thus the original ratio is less than 2. In these two steps we have shown that the ratio of the area of the square to the area of the circle is between 1 and 2 and therefore on an interval of length 1. Given any positive length of interval, the method of exhaustion could be used to determine the ratio of the areas on an interval of at most the given length. In other words, the error indicated by the difference between the determined ratio and the actual ratio $4/\pi$ could be made less than any preassigned positive number.

The method of exhaustion was a forerunner of integral calculus. It also provides a method for obtaining formulas for areas. For example, given a circle and any positive element of area, it is possible to inscribe a regular polygon in the circle such that the difference between the area of the circle and the area of the polygon is less than the given element of area. This can also be done for circumscribed polygons. Many of our high-school texts now use such methods to rationalize formulas for areas, perimeters, and volumes.

Menaechmus, a pupil of Eudoxus, discovered the parabola, ellipse, and hyperbola as plane sections of cones. The study of these conics then led to the construction of the mean proportional between two line segments and to nonclassical methods for solving the three classical construction problems. However, since the parabola, non-circular ellipse, and hyperbola cannot be constructed using only straightedge and compasses, these nonclassical solutions were not accepted as solutions of the classical construction problems proposed by the Sophists.

We have now seen that the early Greeks studied the geometry of the Egyptians, gave status to geometry as a subject and as a part of liberal education, discovered many new geometric results, considered constructions by straightedge and compasses, and endeavored to apply the principles of their philosophy to geometry. The concept of geometry as a deductive system received considerable impetus from Aristotle. Even though Aristotle was not primarily a geometer, he improved some of the definitions of geometry and probably found in geometry the inspiration for his great work as a systematizer of logic. Thus the early Greeks set the stage for the work of Euclid.

7–3 Euclid. In Section 7–1 we considered the growth of empirical geometric rules in Babylon and Egypt. In Section 7–2 we observed the influence of the early Greeks and the shifting of the centers of mathematical achievement to the Ionian Islands (Thales), to southern Italy (Pythagoras), and then to Athens (the Sophists, Plato, Aristotle). About 300 B.C. we find the center of mathematical activity shifting back to Egypt and, in particular, to the university at Alexandria. Euclid was a professor of mathematics at Alexandria. Unfortunately, very little else is known about his personal life. Since most of the scholars of his time were trained at Athens and since his work reflects that of the early Greeks, it is probable that Euclid studied at Athens. He may have studied in Plato's school. Even though such details of his personal life must remain as conjectures, we may be more definite regarding his works.   Euclid probably wrote at least ten scholarly treatises covering a wide range of scientific knowledge and methods.   Three of the treatises were on applied mathematics, including optics and music.   His most famous work is his *Elements*.   This treatise contains thirteen books, in which he presents an elegant organization of

> plane geometry (Books I to IV),
> the theory of proportions (Books V and VI),
> the theory of numbers (Books VII to IX),
> the theory of incommensurables (Book X), and
> solid geometry (Books XI to XIII).

The books on geometry included nearly all the concepts considered in a 20th century high-school geometry course.   They also included geometric proofs of algebraic identities and geometric solutions of quadratic equations.   In general, Euclid's *Elements* contained a summary of the mathematical knowledge of his time, an organization of this knowledge into logical units, and the filling in of gaps in the logical organization by new mathematical discoveries.   The genius of Euclid was manifested in his organization of the concepts.   He may also have discovered many of the concepts, but this aspect of his work is very hard to establish.

Euclid's *Elements* is the earliest comprehensive Greek mathematical text that has been copied over and over and in this way passed down through the ages.   Its great influence upon the development of geometry and upon our western culture is illustrated by the fact that since its first printing in 1492 there have been over a thousand editions of this treatise.*   Euclid's outstanding contribution and the basis for the broad influence of the *Elements* is its logical structure.   Even though the Greek love of reason and truth, the philosophy of Plato, and the systematizing of Aristotle provided an ideal environment for Euclid's organization of mathematics, Euclid deserves a great deal of credit for his monumental work.   The first book of the *Elements* starts with a statement of the definitions and assumptions (postulates and common notions) that are used in the book.   Each succeeding

---

* In view of the unavoidable differences that arise when a manuscript is passed down for many centuries, we shall adopt Heath's treatise [8] as the basis for our discussion of the organization and content of the *Elements*.

book opens with a statement of the definitions required in that book. The proofs of the propositions (or theorems) are synthetic proofs without any indication of the analysis (in the sense used by Plato) that led to the proofs.

The excellent logical structure of the exposition of the proofs has influenced all scientific thinking. This logical structure is essentially as follows:

> a statement of the proposition,
> a statement of the given data (usually with a diagram),
> an indication of the use that is to be made of the data,
> a construction of any additional lines or figures,
> a synthetic proof, and
> a conclusion stating what has been done.

We cannot be sure whether this logical structure is due primarily to Euclid or to his training. In either case it is a logical consequence of the trend in early Greek philosophy and it has had a very great influence.

Lest we idealize the *Elements*, it should be mentioned that there are logical fallacies and unstated assumptions. For example, Euclid appeared to try to define everything. The first eight definitions of Book I were [8; 153]

> (i) A *point* is that which has no part.
> (ii) A *line* is breadthless length.
> (iii) The extremities of a line are points.
> (iv) A *straight line* is a line that lies evenly with the points on itself.
> (v) A *surface* is that which has length and breadth only.
> (vi) The extremities of a surface are lines.
> (vii) A *plane surface* is a surface which lies evenly with the straight lines on itself.
> (viii) A *plane angle* is the inclination to one another of two lines in a plane which meet one another and do not lie in a straight line.

We have difficulty accepting some of these definitions. We also recognize that it is not possible to define every term that we use (Section 1–1). In this sense we must think of some of Euclid's definitions as intuitive descriptions of the terms.

With regard to unstated or *tacit assumptions*, Euclid assumes without any explicit statement or explanation that

    (i) points and lines exist;

    (ii) not all points are on the same line;

    (iii) not all points are on the same plane;

    (iv) two distinct lines have at most one point in common;

    (v) a straight line that contains a vertex B and an interior point of a triangle ABC must also contain a point of the line segment AC;

    (vi) things which are equal may be made to coincide;

    (vii) all sets of objects are finite;

    (viii) a line segment joining the center of a circle to a point outside the circle must contain a point of the circle (continuity); and

    (ix) the existence of order relations on a line.

However, the definitions and tacit assumptions that have been discovered in over two thousand years of study should not seriously detract from the significance of Euclid's *Elements*.

Let us now consider the *Elements* from the point of view of the geometry of the 4th century B.C. The Pythagoreans and the Sophists had previously recognized the study of geometry as a part of a liberal education. This point of view predominated in Athens at least through the time of Plato. It appears probable that the Sophists had emphasized plane geometry from the point of view of straightedge-and-compasses constructions. This emphasis was also continued by Plato and his followers. Euclid's *Elements* reflects this emphasis upon geometry as a body of knowledge, the emphasis upon straightedge-and-compasses constructions, and the systematizing of Plato and Aristotle. Some commentators upon Euclid have placed great importance upon the constructions. A few have stated that the postulates are designed to restrict the geometry to the geometry of constructions by straightedge and compasses. However, Heath [8; 124] states that there is no foundation in the *Elements* for this concept of euclidean geometry. It appears much more likely that the constructions were used in the sense of demonstrating the existence of the points and lines under consideration. From this point of view we should expect Euclid to emphasize the constructions of all points, lines, and figures needed in his propositions. Such is the case.

Straightedge-and-compasses constructions involve the following assumptions:

> (i) *any two points may be joined by a line segment;*
> (ii) *any line segment may be extended to form a line;* and
> (iii) *a circle may be drawn with any given center and distance.*

These three assumptions, usually ascribed to Plato, were the first three of Euclid's five postulates. The remaining two were

> (iv) *any two right angles are equal;* and
> (v) *if a line m intersects two lines p, q such that the sum of the interior angles on the same side of m is less than two right angles, then the lines p and q intersect on the side of m on which the sum of the interior angles is less than two right angles.*

These two postulates and especially the last one have presented special problems to mathematicians. Since Euclid avoided using the fifth postulate until it was definitely needed in the proof of Proposition 29 (Book I), it appears reasonable to assume that he felt a bit uneasy about it. We shall consider the fifth postulate in more detail in Section 7–10. For the present we note that Euclid's five postulates include the assumptions underlying constructions with straightedge and compasses.

This emphasis upon constructions is also evident in the first three propositions of Book I.

PROPOSITION 1. *An equilateral triangle may be constructed on any given line segment.*

PROPOSITION 2. *Given any point, a line segment may be constructed having that point as an extremity and equal to a given line segment.*

PROPOSITION 3. *Given two unequal line segments, a segment equal to the smaller may be cut off from the larger.*

These constructions suffice for Propositions 4 through 6 regarding triangles. In general, constructions are developed as they are needed throughout the *Elements*. The lines and figures needed are based upon straightedge-and-compasses constructions, and these constructions served to establish their existence. Modern mathematicians feel that explicit postulates of existence and continuity are necessary to provide a basis for the existence of the points needed in euclidean plane geometry.

Proposition 2 emphasizes a technicality of constructions with straightedge and compasses. Formally (i.e., from a literal translation of Euclid's first three postulates), it is considered improper to pick up the compasses in order to transport a line segment or distance (radius) from one point to another. We indicate this by saying that the early Greeks used "collapsible" compasses. Proposition 2 states that it is possible, in effect, to transport distances, as most readers probably assumed they could do anyway. In other words, according to the postulates, compasses may be used only to draw a circle having a given center and through a given point. Proposition 2 states that even under these conditions, it is possible to use the length of a given line segment as a radius of a circle even though the center of the circle is not at an end point of the given line segment. Most high-school text books take this as an assumption and allow their students to transport distances by picking up their compasses.

The above five postulates and the following five common notions constituted the stated assumptions upon which the proofs in the *Elements* are based. The postulates involve geometric properties. The common notions were sometimes called "self-evident truths" and were assumed to be applicable to all sciences. The *common notions* of Euclid are [8; 155]:

(i) *things which are equal to the same thing are equal to each other;*

(ii) *if equals are added to equals, the sums are equal;*

(iii) *if equals are subtracted from equals, the remainders are equal;*

(iv) *things which coincide are equal;*

(v) *the whole is greater than the part.*

The first three common notions are properties of the equivalence relation "equal" [10; 7]. The remaining two common notions may appear acceptable to a casual reader, but they deserve detailed examination, since they have been the source of considerable confusion.

The fourth common notion raises a philosophical question as to whether two distinct things may coincide. It also has produced mathematical questions in that the converse statement,

things which are equal may be made to coincide,

has been used. For example, Euclid used this converse statement in the proof of Proposition 4 of Book I of the *Elements*. The assumption of the converse of the fourth common notion is therefore one of Euclid's tacit or unstated assumptions. In the geometry of rigid motions (Chapter 6), these difficulties may be easily removed by replacing the fourth common notion by the statement

congruent figures are equal and conversely.

The fifth common notion also requires special attention, since it holds only for finite sets of objects. For example, the following one-to-one correspondence between the positive integers n and the positive even integers $2n$

$$1 \quad 2 \quad 3 \quad 4 \quad \ldots \quad n \quad \ldots$$
$$2 \quad 4 \quad 6 \quad 8 \quad \ldots \quad 2n \quad \ldots$$

indicates that the set of positive even integers contains as many elements as the set of positive integers. Indeed, this property of the whole being equivalent to a part is a characteristic property of infinite sets [10; 35]. Thus Euclid tacitly assumed that the fifth common notion would be applied only to finite sets of elements. Under this assumption the fifth common notion is generally accepted.

The distinction between postulates and common notions reflects the influence of Aristotle. Heath [8; 119] reports that Aristotle recognized the need for unproved principles (so that the demonstration would not be endless) and indicated that some of these principles are common to all sciences while others are peculiar to the particular science under consideration. The distinction between common notions and postulates persisted until the recent emphasis upon axiomatic thinking that started near the end of the 19th century. Modern mathematicians use many sets of postulates and do not distinguish between self-evident truths (common notions) and postulates. Indeed, the philosophy underlying the concept of self-evident truths has been replaced, at least in the sciences, by the consideration of consequences of specified sets of postulates.

We have seen that Book I of the *Elements* was an outgrowth of the mathematical work of the early Greeks. The same is true of the other twelve books. In general, Euclid organized the mathematical knowledge of his time. We do not know how many of the details he filled in himself. Most of our knowledge of Euclid's work is based upon the commentaries of his successors. We shall consider these commentaries as we consider the evolution of geometry during the first few centuries after Euclid.

7-4 Early euclidean geometry. Euclid's *Elements* constitutes a logical organization of the geometry, algebra, and number theory of his time. This organization was based upon geometric concepts that provided a basis for numbers (line segments), for products of pairs of numbers (areas of rectangles), for products of triples of numbers (volumes), and even for the "existence" of numbers satisfying specified conditions (construction of line segments). The entire concept of mathematics was and still is constantly changing. The Greeks gradually included the subjects of logistica (the art of calculation), geodesy (surveying), mechanics, and optics under the heading of mathematics. Trigonometry was a part of astronomy. Early Greek notations for numbers were very crude, and indeed the concept of a number was only beginning to develop. The difficulty of finding notations for irrational numbers may have led to their representation by line segments. Negative numbers could not be represented by undirected line segments and were not accepted.

As knowledge increased, it became desirable to restrict the concept of mathematics and the concept of geometry. For example, optics, surveying, music, astronomy, and mechanics gradually became recognized as separate bodies of knowledge. Algebra, analysis, trigonometry, and the theory of numbers became separate branches of mathematics. We shall be primarily concerned with concepts that are commonly recognized as geometric, even though this specialization is highly artificial. After striving for specialized developments, mathematicians are today becoming increasingly aware of the interdependence of the various branches of mathematics. For example, a modern research problem may arise in classical algebra, have significance in geometry, and be solved using the theories of analysis. In this sense our emphasis upon strictly geometric concepts is against the modern trend, but it appears to be necessary to reduce the mass of details to a comprehensible unit.

Since Euclid's *Elements* may be considered as a logical organization of all the mathematical concepts of his time, we may consider the mathematical achievements of the next thousand years in relation to the concepts included in the *Elements*. Most of the mathematicians of this period were associated with the school at Alexandria or at least studied there. Their works involved concepts related to irrational numbers, number theory (Diophantos), astronomy and trigonometry (Ptolemy and Heron), conics (Apollonius), and commentaries upon

Euclid's *Elements*. Archimedes (3rd century B.C.) was the outstanding genius during this time; Pappus (3rd century A.D.) was the
last creative Greek mathematician of note; Proclus (5th century A.D.)
wrote an excellent commentary upon the *Elements*. The interval of
a thousand years is taken as a unit since Alexandria was razed by the
Arabians in the 7th century A.D. Thus the next thousand years after
Euclid's time constitutes the completion of an era. The first half
of this era was characterized by a continued development of mathematical concepts; the second half has decreasing mathematical significance. Our discussion of this era is entitled "early euclidean
geometry" to emphasize its dependence upon the assumptions underlying Euclid's *Elements*.

We next sketch the development of astronomy and trigonometry
up to their recognition as independent bodies of knowledge. During
the 3rd century B.C., Aristarchus asserted that the earth and the
other planets (only five others were known at that time) revolved
about the sun. This assertion antedates Copernicus by about seven-
teen centuries. Also during the 3rd century B.C. Eratosthenes, the
librarian at Alexandria, calculated the circumference of the earth.
About 140 B.C. Hipparchus developed a systematic theory regarding
the sun and the moon, calculated their distances and sizes, catalogued
the places and sizes of 850 stars [1; 25], introduced the concept of a
degree as 1/360 of a circle into Greece (previously used in Babylon),
prepared a table of the lengths of chords of circles, and used similar
triangles to provide a better foundation for the body of knowledge
that we call trigonometry. The concepts, definitions, and even some
formulas of spherical trigonometry were developed by Menelaus
(probably 1st century A.D.), who modeled his treatise on spheres after
Euclid's *Elements*. In the 2nd century A.D. another scholar from
Alexandria, Claudius Ptolemy, developed a new theory of the motion
of planets, catalogued the positions of 1028 stars [1; 26], made both
terrestrial and celestial maps, used stereographic projections (Section 7–7), prepared a table of the lengths of chords of circles for central
angles of $0.5°$, $1°$, $1.5°$, . . ., $180°$, presented formulas equivalent to
those for the sine and cosine of the sum and difference of two angles,
and included a few formulas for spherical trigonometry. The table
of the lengths of chords was a forerunner of a table of sines, since
$\sin \theta$ is half the chord of a central angle 2θ in a unit circle. The
change from the whole chord to the half chord was an important

Hindu contribution.   The work on projections was continued in the 6th century by Philoponus, who also wrote a treatise on the astrolabe (an early form of sextant used to observe the positions of stars and planets).   The work of the Greek and the Hindu mathematicians was organized further and extended by the Arabians in the 9th through the 13th centuries.   The work of Nasr-ed-Din on astronomy and trigonometry in the 13th century was noteworthy as the first thorough treatment of plane and spherical trigonometry as an independent body of knowledge (i.e., not as a part of astronomy).   This work included treatment of all six trigonometric functions as well as methods for solving spherical triangles.

In the 15th century a Persian astronomer Ulugh Beg prepared a table of sines and cosines for angles expressed to the nearest minute and with an accuracy equivalent to at least eight decimal places. About this same time a German mathematician (Johann Müller, known as Regiomontanus) wrote the first European book on plane and spherical trigonometry.   During the 16th century more detailed trigonometric tables were prepared, the trigonometric functions were defined in terms of the sides of right triangles (instead of in terms of circles), and trigonometry gained recognition both in terms of its uses in astronomy and as an independent body of knowledge.   Also, astronomy rapidly gained stature based upon the work of Copernicus, Kepler, and Galileo.   By the middle of the 17th century both astronomy and trigonometry were recognized as independent bodies of knowledge.

We have just sketched the development of trigonometry as a study of a few ratios, an aid to astronomical calculations, and an independent body of knowledge.   If this sketch were brought up to date, we would find in trigonometry as in other branches of mathematics a trend of successive generalizations, abstractions (removal of visual and physical connotations), and arithmetizations (increasing reliance upon properties of numbers).   Thus even though the trigonometric functions were based upon studies of chords and half chords of circles, we now have general definitions of the trigonometric functions in terms of x, y, and r, in terms of line values, and in terms of infinite series.

We now restrict our concept of euclidean geometry so as to exclude both astronomy and trigonometry.   We shall also exclude the theory of numbers except for its direct applications to our narrow concept of geometry. We shall be primarily concerned with points, lines, curves, and surfaces; lengths, areas, and volumes; and the association of numbers with these concepts.

Archimedes has been called the greatest mathematical genius up to at least the 17th century. He lived from about 287 B.C. until 212 B.C., studied at Alexandria, and made noteworthy contributions to the knowledge of hydrostatics (the study of pressures, equilibrium, etc., in liquids), mechanics (especially in connection with centers of gravity), operations with large numbers, areas of curves and curved surfaces, volumes of solids bounded by curved surfaces (especially volumes obtained by revolving conics about an axis), and approximations for irrational numbers (especially π and irrational square roots). The breadth of concepts included in this list indicates Archimedes' versatility. His methods for finding areas and volumes were equivalent to the methods of integration that were developed nineteen centuries later. He is even credited with destroying Roman ships by using mirrors to reflect the sun's rays upon them. Many students of geometry remember the axiom that now bears his name. This axiom, probably based on the work of Eudoxus, may be stated in the language of arithmetic as follows:

AXIOM OF ARCHIMEDES: *If a and b are positive numbers, there exists an integer n such that na > b.*

Euclid and other earlier writers had used various forms of this statement. Archimedes assumed that areas have this property [2; 39] (i.e., if A and B represent areas, there exists an integer n such that $nA > B$). Archimedes was particularly pleased with his discovery that the volume of a sphere was equal to two-thirds of the volume of the circumscribed cylinder. This result illustrates his extension of geometric concepts, especially those involving measurements.

The early Greek interest in conics and the work of Menaechmus have already been mentioned (Section 7-2). This work provided the basis for an extensive treatise on conics by Apollonius in the 3rd century B.C. Apollonius studied at Alexandria. His treatise on conics includes the knowledge of his time and many results of his own. The terms "parabola," "ellipse," and "hyperbola" are due to Apollonius. These figures are described in terms of a diameter and a tangent at an extremity of the diameter. In one sense he used coordinates and transformations of coordinates. He considered normal lines, conjugate diameters, and foci of central conics. He recognized that the sum of the distances of a point from the two foci is constant for the points on an ellipse and that the difference of these distances is constant for the points on a hyperbola. In view of this work on loci and the work in some of his other treatises, Apollonius is credited with providing the best early example of the study of the positions of points.

In the 2nd and 1st centuries B.C. the most famous developments were made by Hipparchus in astronomy (mentioned above) and by Heron, who is often credited with the formula $\sqrt{s(s-a)(s-b)(s-c)}$ for the area of a triangle with sides a, b, c, where $2s = a + b + c$. Heron was a prolific writer with a primary interest in the practical uses of mathematics and mechanics. He obtained the above formula in a purely geometric form (sometimes attributed to Archimedes).

In the 2nd century A.D. Ptolemy, whom we have already mentioned as a great astronomer, attempted to prove Euclid's parallel postulate as a theorem, and wrote about optics, music, and geography. The latter treatise contained several maps and thereby illustrates another application of geometry (cartography, Section 7-7).

The last of the creative Greek mathematicians was Pappus (3rd century A.D.), also of Alexandria. Pappus' fame rests upon the originality that he used in collecting and discussing the entire scope of Greek geometry. He gave a construction for a conic through five points and found that the volume of a solid of revolution is equal to the product of the area revolved and the circumference described by the center of gravity of the area. His treatise on geometry also includes several special curves, including spirals. In general, the Greeks considered many special curves and used them to solve problems. Such dependence upon geometric processes was made necessary by the absence of algebraic notations and methods.

The remaining mathematicians of note prior to the destruction of Alexandria in the 7th century were primarily commentators. In the 4th century Theon wrote about the works of Euclid and Ptolemy. In the 5th century the philosopher Proclus wrote about Euclid's *Elements*. These commentaries marked the end of this era. By the 7th century the geometry of measurements of the Babylonians and Egyptians had been molded according to the Greek love of knowledge and reason and was in need of new influences. This is not intended to minimize the Greek influence but rather to indicate that it had run its course.

In geometry, as in astronomy, the only contributions of note between the 6th and 13th centuries were by Hindu, Arabian, and Persian mathematicians. The rationality of lengths and of areas was considered in the 7th century. Number notation based upon zero and place value was developed. In the last part of the 11th century Omar Khayyám, author of the *Rubáiyát*, wrote a treatise on algebra, determined some roots of cubics as intersections of conics (discarded negative roots), and devised a calendar that was more accurate than the one we use [1; 30]. Also Nasr-ed-Din, whom we have already mentioned as an astronomer, and probably Omar Khayyám, considered the parallel postulate of Euclid. Thus the influence of Euclid continued for many centuries and indeed continues today.

7–5 The awakening in Europe. We have observed that the center of mathematical achievement has shifted from place to place with the location of the dominant culture of the time. Thus the measurements of the Babylonians and Egyptians characterized one era, the abstractions of the Greeks characterized another, and the preservation and expansion of knowledge by the practical Arabians characterized a third era. After the fall of the Roman Empire, there was very little creative work in mathematics or any other subject in Europe for several centuries. Thus we find another example of the dependence of mathematical progress upon a culture in which there is sufficient specialization to allow at least a few individuals to withdraw from the manual labor of providing the food and shelter needed by their families and to devote their energies to abstract theories. All sciences and arts depend upon favorable cultural environments and flourish more or less simultaneously.

About 500 A.D. the Roman writer Boethius attained fame by collecting some of the works of Euclid and other Greek mathematicians into elementary texts on geometry and arithmetic that were widely used in monasteries. There were evidences of mathematical life in England in the 8th century and in France in the 10th century. But in general mathematical progress was very slow.

In mathematics, as in trade and art, we find the first signs of reawakening in Italy. By the 12th century Gherardo of Cremona was industriously translating Euclid's *Elements* and many other mathematical works from Arabic into Latin. As in the early stages of

Greek influence, the first outstanding achievements were made by men who traveled widely. Thales was a businessman who visited both Babylon and Egypt before making his place in Greek mathematics. At the end of the 12th century we have Leonardo (or Fibonacci, i.e., son of Bonaccio), who traveled extensively before returning to Pisa, Italy, where he published several treatises presenting the knowledge that he had gained. His presentation indicated his originality and understanding. He recommended the adoption of Hindu-Arabic numerals and emphasized practical problems in both arithmetic and geometry. Basically, his works provided an accessible source for a vast amount of information regarding previous achievements in arithmetic, algebra, geometry, and trigonometry. In this way the knowledge gathered by Leonardo during his travels and his original presentations of that knowledge stimulated a renewed interest in mathematics in Europe.

The founding of European universities in the 13th century had a profound influence on mathematics. These groups of scholars provided the stimulus to each other that is necessary for great achievements. Also in the 13th century we find a renewed interest in the concept of infinity and the continuity of the set of points on a line. Thomas Aquinas (Italy) adopted Aristotle's point of view with some modifications. Roger Bacon (England) argued against the concept of infinity. Zeno's paradoxes received further attention. During this period the nature and "truth" of certain geometric concepts attained prominence in philosophy and, to some extent, in religious theories.

The invention of movable type and the resulting increased availability of books during the 15th century and thereafter has had a great influence upon all areas of knowledge. Information could be widely disseminated and scholars became less dependent upon a few "schools" led by outstanding mathematicians. Trigonometric tables were printed for the use of astronomers. The reawakening of intellectual activity spread rapidly throughout Europe. At first it was necessary to obtain the achievements of the Greeks and other mathematicians, which would provide a basis for the contributions of the scholars of the new culture. Thus we find an intellectual smoldering in Europe for a few centuries. The smoldering in Italy broke into noteworthy achievements in the solution of equations (Ferro, Tartaglia, Cardan, Ferrari) in the early part of the 16th century while the rest of Europe was still gathering its intellectual strength.   There were other occasional bursts of early activity such as that of Regiomontanus (Germany) in trigonometry and Copernicus (Poland) in astronomy.   Then Vieta (France) introduced letters for numbers, applied algebraic concepts (as they were understood in the 16th century) to geometry, and prepared a foundation for analytic trigonometry.   Soon there burst forth an avalanche of activity that appears to be still expanding without any definite signs of spending its force. Such intense activity presents serious difficulties when one attempts to assess the evolution of geometric concepts.   We shall attempt to overcome this difficulty by isolating three particular topics or areas of geometry for treatment aside from our chronological treatment of geometry.   The three topics are constructions (Section 7–6), descriptive geometry (Section 7–7), and the parallel postulate (Section 7–10). The isolation of these topics will give us a clearer picture of them and will facilitate our chronological development of other topics of geometry.

7–6 Constructions.   We have already observed that the Pythagoreans used constructions with straightedge and collapsible compasses to perform operations with numbers considered as lengths of line segments.   Euclid used such constructions also to establish the existence of points and figures.   In both cases the straightedge could be used only to draw straight lines and the compasses could be used only to draw circles or arcs of circles.   As mentioned in Section 7–3 the compasses could not formally be picked up to transport distances from one line to another, but it was proved (Proposition 2, Book I of the *Elements*) that this operation could be performed without picking up the compasses.

Undoubtedly the above restrictions upon the classical constructions were based upon the searching of the early Greeks for fundamental concepts of numbers and geometry.   Even though some modern readers may consider the restrictions unnecessary, they were indeed based upon concepts that we now recognize as fundamental.   In any case, whether we accept the classical constructions as necessary or simply as rules of a game that has been popular for many centuries, they have played an important role in the development of geometric and other mathematical concepts.   The attempts of the early Greeks to solve the three classical construction problems (Section 7–2) led to the discovery of many mathematical figures and properties. Throughout the ages these constructions have aided our understanding of elementary mathematical concepts. The following summary of the efforts of mathematicians to analyze the concepts underlying the classical constructions illustrates the manner in which mathematicians have tested concepts and arrived at our present stage of mathematical development and understanding.

In the 10th century the Arabian mathematician Abu'l Wefâ considered constructions with straightedge and compasses with a fixed opening. In the 19th century the French mathematician Poncelet conjectured and the German mathematician Steiner proved not only that Wefâ's constructions included all classical constructions but that the fixed compasses would be needed only once (i.e., all points that could be determined using a straightedge and compasses could be determined using a straightedge and one circle given on the plane with its center). These ideas were extended in the early part of the 20th century by the Italian mathematician Francesco Severi (1879–   ), who proved that only the center and an arbitrarily small arc of a given circle are needed.

We have just sketched the efforts of mathematicians to minimize the use of the circle in classical constructions. Other mathematicians have endeavored to minimize the use of the straightedge. While recognizing that a circle cannot be drawn with a straightedge and a straight line cannot be drawn with compasses, these mathematicians also recognize that the points obtained in the classical constructions arise as intersections

    (i) of two lines,
    (ii) of two circles, or
    (iii) of a line and a circle.

Accordingly, they endeavor to construct such points minimizing the use of the compasses or the straightedge or both (i.e., minimizing the number of steps in the construction). In the 17th century Georg Mohr (Denmark), and independently in the 18th century Lorenzo Mascheroni (Italy), discovered that all points that can be determined using straightedge and compasses in classical constructions can be determined using compasses alone. Thus the straightedge is not needed at all when the determination of two points on a line is accepted as the determination of the line.

The classical constructions may be completely analyzed algebraically in terms of addition, subtraction, multiplication, division, and the extraction of square roots. These operations and only these operations may be performed using the classical constructions [10; 228–238]. Thus when Wantzel (19th century, France) proved that certain cubic equations could not be solved using these operations, he proved that the classical trisection problems and the duplication of a cube were impossible. Similarly, when Lindemann (19th century, Germany) proved that π did not satisfy any polynomial equation with rational coefficients (i.e., was transcendental), he proved that the classical problem of squaring a circle was impossible.

Finally, we note that if the classical restrictions are relaxed to allow marks on the ruler, the use of both sides (width) of the ruler, or the use of rulers hinged together, the scope of possible constructions is radically changed. Such modifications have been considered for both practical and theoretical reasons. The hinged or pivoted bars provide the basis for many machines, including the transfer of the straight-line motion of a locomotive's piston to the circular motion of its wheels. Systems of bars (or linkages as they are called) may also be used to find all real roots of algebraic equations. However, for complicated equations the systems of bars also become very complicated. The theoretical study of linkages enjoyed a brief prominence at the end of the 19th century while the basic principles were being discovered. Since that time, linkages have been considered primarily from the point of view of practical applications or as visual aids in the teaching of mathematics.

In the above consideration of constructions, we have observed the analyzing of earlier concepts, the use of both algebraic and geometric techniques, the further restriction of earlier assumptions, the relaxing of earlier restrictions, and the international nature of mathematical discoveries. These aspects of the development of constructions permeate the development of all geometric, and indeed all scientific, theories.

7–7 Descriptive geometry. Throughout recorded history artists young and old have endeavored to sketch or represent space figures on plane surfaces. These plane surfaces may be on rocks, clay tablets, strips of papyrus, an artist's canvas, or a sandy beach, or in many other forms. In each case the problem is to sketch a person, object, or scene so that other people will recognize its similarity with the original.   In some cases the recognition is desired only for amusement; in other cases it is to enable someone to construct or to find an object or place.   Whatever the purpose of the drawings, the geometric principles underlying the recognition or desired similarity between space figures and plane drawings are the same.   These geometric principles were developed very slowly.   Apparently the mathematicians weren't interested and the artists did not have sufficient training in geometry.

Leonbattista Alberti (Italy, 1446*) was one of the first to recognize the problem of the artist as a geometrical problem.   He visualized a "pyramid of rays" running from the eye of the observer to the object.   The plane of the picture (you may think of this as a sheet of clear glass) was to be held vertically between the observer and the object.   Then the problem was to represent each point of the object by the point at which its ray entered the plane of the picture.   Alberti set up formal rules involving the horizon line and methods for constructing on the plane of the picture the image of a network of squares on the ground.   Then the picture of any figure on the ground could be obtained using the points corresponding to the intersections of the figure with the sides of the squares.   This technique was improved by Piero (in the 1460's) and by many of the Italian artists.   The theory was extended by Ubaldo (Italy, 1600), Gravesande (Holland, 1700), and Taylor (England, 1715).   We can now see the forerunner of coordinate systems in this method of locating points with reference to a network of squares and the forerunner of projective geometry in the representation of points (figures) by considering their point section (from the observer's eye) and then the plane section of these projectors (rays).   These concepts from the 15th century were developed in the 17th century (Section 7–8) by Desargues, Pascal, and La Hire, who obtained several basic theorems which, in the 19th century (Section 7–11), became part of projective geometry.

Dürer (Germany, 1591) considered projections of figures onto two perpendicular planes and onto three mutually perpendicular planes. However, it was nearly two centuries before Gaspard Monge (Sec-

---

* Single dates refer to the time of the major discovery or publication of the individual; pairs of dates refer to the individual's life span (sometimes estimated).

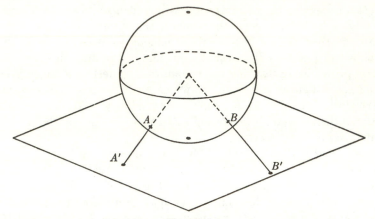

FIG. 7–1

tion 7–9), who is often called the father of descriptive geometry, organized these ideas into an effective body of knowledge.   Monge's work was initially developed for the French Army and was considered of such military importance that secrecy was maintained for many years.   By the beginning of the 19th century, descriptive geometry was gaining recognition as a particular part of geometry.

The description of objects is closely related to the description of the earth's surface.   The making of maps has gradually developed into an application of geometry called *cartography*.   The first maps were crude sketches.   Then, after the shape of the world was recognized as spherical, the problems associated with the representation of the surface of a sphere on a plane were studied in detail.

If the plane (map) is considered as a tangent plane of the sphere and the points on the sphere projected from the sphere's center onto the plane (gnomonic projection, Fig. 7–1), then all the great circles on the sphere (i.e., all intersections of the sphere and planes through the center of the sphere) appear as straight lines.   However, angles, areas, and distances are distorted by this projection.   The significance of this type of map is based upon the appearance of great circle (shortest) routes as straight lines.

If the plane is taken as a diametral plane (such as the equatorial plane) and the points are projected from a pole of that plane (stereographic projection, Fig. 7–2), then angles on the map correspond to angles on the sphere, but distances and areas are distorted.

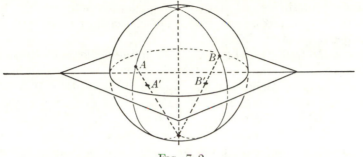

FIG. 7–2

There are many other types of maps. Some are based upon projections; others are not. All important types are based upon the preservation of some properties (distance, areas, angles, . . .) and the recognition that it is not possible to preserve all of the properties of the surface of a sphere in making a plane map. In general, the development of cartography has paralleled that in geometry. Recently tremendous strides have been taken with the use of airplanes and several synchronized cameras. Such developments emphasize the importance of technological as well as theoretical advances. Both cartography and descriptive geometry provide examples of the application of geometric properties and the development of separate bodies of knowledge, or disciplines.

7–8 Seventeenth century. We now return to our chronological development of geometry. The concepts and methods considered in Sections 7–6 and 7–7 will be noted when they provide a basis for other developments, but they no longer need to be considered as a part of the special aspects of geometry considered in the following sections.

At the beginning of the 17th century, Johann Kepler (Germany, 1571–1630) formulated his famous laws regarding the motions of planets. He also made important discoveries regarding conics, suggested that parallel lines should be considered as meeting at infinity, and developed methods for neglecting infinitesimals in his calculations. He considered a circle as composed of an infinite number of triangles having their common vertex at the center, a sphere as composed of an infinite number of pyramids. He also considered the volumes obtained by revolving conics about lines. Such considerations of infinitesimals and motions provided a basis for the development of the calculus and an indication of the shift away from the static geometry of the Greeks.

Galileo (Italy, 1564–1642) considered falling bodies and recognized that, when air resistance is neglected, the path of a projectile is a parabola. One of Galileo's pupils, Cavalieri (Italy, 1598–1647), developed a theory of "indivisibles." He assumed that a line could be generated by a moving point, a plane by a moving line, a solid by a moving area. *Cavalieri's Principle* asserts that two solids have the same volume if they have the same height and the same cross-sectional areas at equal heights. These concepts were forerunners of the integral calculus.

We now find the first systematic use of the algebraic notation of Vieta (Section 7–5) and the coordinate concepts developed in descriptive geometry (Section 7–7). René Descartes (France, 1596–1650) applied algebraic notation to the analysis of conics by Apollonius (Section 7–4), visualized all algebraic expressions as numbers instead of geometric objects, and found equations representing several curves (considered as loci). His interpretation of such symbols as x^2 and x^3 as numbers and therefore as lengths of line segments was very important. Previously, linear terms such as x or $2y$ had been considered as line segments, quadratic terms such as x^2 or xy as areas, and cubic terms such as x^3 or x^2y as volumes. The old interpretations were restrictive in the sense that only like quantities could be added. For example, it was permissible to add x^2 and xy (areas), but it was not previously permissible to add x^2 and x (i.e., an area and a line segment). Descartes' interpretation of all algebraic expressions as numbers and therefore as line segments made it possible to consider sums such as $x^2 + x$. This new point of view provided a basis for the representation of curves by equations. At first these representations were based upon proportions and the Greek geometry of conics.

The study of geometric figures as loci corresponding to equations introduced a new era and an entirely new point of view into the study of geometry. Since we shall find enthusiastic supporters of both the new and the old points of view, we shall endeavor to distinguish between the two as follows: The study of figures in terms of their algebraic representation by equations will be called *analytic geometry;* the study of figures directly, without use of their algebraic representations, will be called *synthetic geometry.* Since many geometers will use both algebraic and synthetic methods, the above distinction is, at best, a relative one. Modern geometers consider the two geometries as two equivalent points of view of the same body of knowledge.

Pierre de Fermat (France, 1601–1665) used an algebraic approach in his treatment of straight lines, conic sections, and areas. He also found maximum and minimum points on curves by considering the effect of slight changes in the variables. In this respect he had some of the basic ideas of the differential calculus but did not see any relationship between the differential and the integral calculus.

Simultaneously with the beginnings of analytic geometry there were important developments in synthetic geometry. Girard Desargues (France, 1593–1661) and Blaise Pascal (France, 1623–1662) visualized the conic sections (circle, ellipse, parabola, hyperbola) as projections of circles, discovered other properties of conics, and prepared the foundations for synthetic projective geometry. Their work was continued by La Hire (France, 1640–1718) but was not further developed until the 19th century.

Several of the great thinkers of the 16th century were searching for general methods for understanding the universe. The rigor of Euclid's *Elements* and the apparent correspondence between the geometry based upon the *Elements* and the geometry of the physical world led many people to believe that the fundamental assumptions of euclidean geometry were laws of the universe. This concept of geometry is not accepted in the 20th century.

One of the great universal scientists was Christian Huygens (Holland, 1629–1695), whose geometric discoveries were primarily in synthetic geometry. He used regular polygons to calculate π correct to nine decimal places and developed several properties of special curves.

Isaac Barrow (England, 1630–1677) is renowned as Newton's teacher and a man who, at the early age of 39, resigned his professorship to his talented pupil, Newton. Barrow developed a new method for obtaining areas of curves and published several mathematical texts. His most important work was probably the simplification of Fermat's work on the tangent lines of curves by the introduction of several infinitesimals instead of one. These concepts were extended by Newton in his development of the calculus.

The last half of the 17th century is marked by the independent discovery of the relationship between differentiation (often visualized as rate of change) and integration (often visualized as summation), by Isaac Newton (England, 1643–1727) and Gottfried Leibniz (Germany, 1646–1716). It now appears that Newton made the first discovery (1665–1666 as compared to Leibniz' 1673–1676), but Leibniz published his results first (1684–1686 as compared to Newton's 1704–1736) [13; 149]. Newton's discovery was based upon a study of points in motion; Leibniz' upon a study of properties of static figures. Although the early methods of both Leibniz and Newton appear vague compared with modern standards of rigor, they discovered a new technique which gave rise to a new type of geometry, differential geometry, and greatly influenced all branches of mathematics. Our present notation is largely that of Leibniz.

Newton made contributions to the development of both synthetic and analytic geometry. He classified nearly all cubic curves (i.e., curves corresponding to cubic equations). In his famous *Principia* he gave many applications to synthetic geometry of his work with the calculus. The application of new ideas and techniques to the geometry of the Greeks typifies the progress in geometry during the 17th century. There was the application of algebra initiated by Descartes and Fermat, the applications of projections initiated by Desargues and Pascal, and the applications of the calculus initiated by Newton and Leibniz. These new approaches were the beginnings of three major phases of geometry — analytic geometry, synthetic projective geometry, and differential geometry.

7–9 Eighteenth century. The geometry of the 18th century was based upon applications of the 17th-century concepts of algebra and the calculus. The analytic plane geometry of Descartes and Fermat was extended to three dimensions. The known properties of conics were restated, using algebraic notations. Maclaurin (Scotland, 1698–1746) extended Newton's work on cubic curves and discovered many special curves, including the cissoid, cardioid, and lemniscate. Leonard Euler (Switzerland, 1707–1783) used the techniques of analytic geometry and introduced transformations of coordinates in three-dimensional space. He wrote the first text on the calculus of variations and used the techniques of differential geometry (in a sense, he made the first real progress in that subject) to study curvature. Given any triangle, Euler proved that the center of the circumscribed circle (circumcenter), the center of gravity (centroid), and the intersections of the altitudes (orthocenter) were collinear. Finally, he is often considered the first topologist because of his work on traversable networks (Section 9–6).

During the last half of the 18th century, we find the German mathematician Johann Lambert (1728–1777) considering the geometry on the surface of a sphere, developing a new type of map projection, obtaining properties of conics from his study of astronomy, and publishing a book on perspective. We also find evidence of the forthcoming crescendo of mathematical activity that characterizes the 19th and 20th centuries. For example, we have Joseph Louis Lagrange (France–Italy, 1736–1813), Gaspard Monge (France, 1746–1818), Pierre Simon Laplace (France, 1749–1827), Adrien Marie Legendre (France, 1752–1833), Karl Friedrich Gauss (Germany, 1777–1855), and many others.

Lagrange provided an algebraic basis for the calculus. Monge has already been mentioned (Section 7–7) as the founder of descriptive geometry. He also made contributions involving curvature and differential geometry. Laplace published a very important five-volume treatise on celestial mechanics. Legendre did most of his work in the theory of numbers, theory of functions, and the calculus. He also wrote an elementary geometry rearranging and modifying Euclid's *Elements*. Legendre's geometry text was used for many years and has had a noticeable influence upon high-school texts — even in the United States. Gauss is said to have decided to become a mathematician after discovering a method for constructing a regular polygon of 17 sides at the age of 19. He made other contributions to the theory of constructions, curved surfaces, differential geometry, noneuclidean geometry (Section 7–10), and topology (Chapter 9).

This intensity of mathematical activity renders our task of identifying the evolution of geometry very difficult. Also, there are problems associated with the increasing specialization of terminology and the detailed or abstract character of many of the contributions. We shall be forced to compromise in each of these regards. In general, as the mathematical level of the geometrical contributions advances beyond that expected of most readers of the present text, our treatment of the evolution of geometry will, of necessity, become more expository.

7–10 Euclid's fifth postulate. Euclid apparently felt uneasy about his fifth postulate. For example, in Book I of his *Elements* he avoided using the fifth postulate in his proofs of Proposition 27:

If a line intersects two lines such that the alternate interior angles are equal, the two lines are parallel.

and Proposition 28:

If a line intersects two lines such that an exterior angle is equal to an opposite interior angle on the same side of the line, or such that the sum of the interior angles on the same side of the line is equal to two right angles, the two lines are parallel.

He first used the fifth postulate in his proof of Proposition 29, which is the converse of the above two propositions:

If a line intersects two parallel lines, it makes the alternate interior angles equal, each exterior angle equal to the opposite interior angle on the same side of the line, and the sum of two interior angles on the same side of the line equal to two right angles.

Other early geometers shared this uneasiness about the fifth postulate. Many of them tried to prove this postulate, but gradually it was found that other assumptions were necessary for any proof of the fifth postulate; i.e., the fifth postulate is independent of Euclid's other postulates.

We shall find that the attempts to prove the fifth postulate were generally of one of the following types:

(i)   direct proof from Euclid's other postulates,

(ii)   replacement of the fifth postulate by a more "self-evident" postulate (either explicitly or tacitly and unknowingly) and a proof of the fifth postulate as a consequence of the new assumption and Euclid's other postulates, or

(iii)   indirect proof by showing that the fifth postulate cannot fail to hold.

All these approaches were doomed to failure as "proofs" of the fifth postulate.   They did, however, encourage the study of several possible geometries using primarily synthetic methods, algebraic concepts, curvature and differential geometry, distance relationships, and groups of transformations.

As a result of these studies, mathematicians have considered the axiomatic basis for euclidean geometry in detail and now recognize the consistency of the noneuclidean geometries.

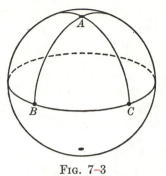

Fig. 7–3

The fifth postulate is the converse of Proposition 17:

The sum of any two angles of a triangle is less than two right angles.

This probably encouraged early geometers to seek a proof of the fifth postulate. Indeed, one of the subtleties of euclidean geometry lies in the fact that Proposition 17 is based upon Euclid's tacit assumption that a straight line containing a vertex B and an interior point of a triangle ABC must also contain a point of the line segment AC. This assumption implies that the sum of the angles of a triangle is less than or equal to two right angles and excludes the geometry on the surface of a sphere as a euclidean geometry. Note that angles B and C are both right angles in triangle ABC of Fig. 7–3, where lines are taken as arcs of great circles (i.e., circles obtained as the intersections of the sphere and planes through the center of the sphere). The first 15 propositions in Book I of the *Elements* are independent of this tacit assumption of the infinitude of lines.

For many centuries all mathematicians continued to use this tacit assumption regarding lines and triangles. Euclid's assumption may be visualized in many ways. It is equivalent to the assumption that a line is not re-entrant in the sense that a circle is re-entrant. Intuitively, this implies that an astronomer cannot see the back of his head by looking into a telescope. Similarly, if a line is in an east-west position, the above assumption implies that one cannot, as a few geographers assumed, go to the "Far East" by traveling westward. The assumption that lines are not re-entrant is equivalent to the assumption that a point on a line separates the line into two segments as, for example, into an eastern and a western segment in the above example. It is also equivalent to the assumption that given any three distinct points on a line, exactly one of the points is between the other two. Furthermore, when considered with Euclid's first four postulates and his other tacit assumptions, the foregoing assumption regarding lines implies the existence of at least one line parallel to a given line and through a given point. Since Euclid first used his fifth postulate in the proof of Proposition 29, the early geometers tried to prove that Euclid's first four postulates, his common notions, Propositions 1 to 28, and the tacitly assumed existence of at least one line through P parallel to m implied the existence of a unique line through a given point P parallel to a given line m.

Posidonius (about 100 B.C.) defined parallel lines essentially as equidistant coplanar lines. This definition is based upon the following assumptions:

(i) the locus of points equidistant from a given line and on a given plane forms at least one line, and

(ii) two coplanar lines that do not meet are equidistant.

Thus the change in definition suggested by Posidonius introduces new assumptions rather than solving our difficulties in proving the fifth postulate.

Ptolemy (2nd century A.D.), a Greek geometer not to be confused with the Egyptian rulers of the same name, endeavored to rationalize the fifth postulate as follows: Any two lines cut by a transversal are no more parallel on one side of the transversal than on the other. Accordingly, if two parallel lines are cut by a transversal such that the sum of the interior angles on one side of the transversal is greater than two right angles, then the sum of the interior angles on the other side must be greater than two right angles also, contrary to the fact that the sum of all four interior angles must add up to four right angles. This reasoning is based upon the assumptions that

(i) there exists at least one line parallel to a given line through a given point, and

(ii) two lines are as parallel on one side of a transversal as on the other side.

The second assumption is now recognized as a postulate asserting that there is at most one line parallel to a given line and through a given point.

Proclus (5th century) attempted to prove the fifth postulate in the form

If a straight line intersects one of two parallel lines, it will intersect the other also.

Proclus also appears to have been acquainted with another form of the fifth postulate:

PLAYFAIR'S AXIOM: *Given any line m and any point P, there is a unique line through P and parallel to m.*

In the 13th century Nasr-ed-Din (Section 7–4) attempted to prove the fifth postulate in the form

The angle sum of a triangle is always equal to two right angles.

Later, John Wallis (England, 1616–1703) attempted to prove the fifth postulate in the form

Given any triangle, we may construct a similar triangle of any size whatever.

Early in the 18th century, Girolamo Saccheri (Italy, 1667–1733) attempted to prove the fifth postulate by the method of *reductio ad absurdum*. That is, he endeavored to consider all possible cases of a figure and to exclude all but the case based upon the fifth postulate. Saccheri's work involves the properties of a quadrilateral such as $ABCD$ in Fig. 7–4 having right angles at A and B, $AD = BC$, $DH = HC$, and $HK \perp AB$. He proved that in general $\angle C = \angle D$ and discussed three hypotheses according as $\angle C$ is a right angle, an acute angle, or an obtuse angle. Furthermore, he proved that if any one of the three hypotheses holds for a single quadrilateral, then it holds for all quadrilaterals. His three cases are respectively equivalent to the cases in which the sum of the angles of a triangle is equal to, less than, or greater than two right angles. Saccheri

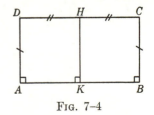

FIG. 7–4

disposed of the last two hypotheses by assuming that two lines cannot merge into a single line at infinity and by accepting Euclid's tacit assumption that lines are not re-entrant. We now recognize these assumptions as postulates asserting the existence of at most one and at least one line parallel to a given line and through a given point.

About the middle of the 18th century some mathematicians began to feel that there might exist geometries in which Euclid's fifth postulate did not hold. For example, Lambert proved that the area of a triangle was a function of the excess or deficiency of its angle sum relative to two right angles and suggested that Saccheri's acute-angle hypothesis might hold for an imaginary sphere (i.e., a sphere with an imaginary radius). However, other mathematicians were still trying to prove the fifth postulate.

In the early 19th century, Gauss apparently visualized the possibility of geometries in which the fifth postulate did not hold but did not publish his results. He probably observed that Euclid's fifth postulate is quite different from his first four postulates. The first four postulates are concerned with finite segments of lines and the possibility of extending a finite segment to form a line. The fifth postulate asserts a property of lines in their full extent. Then, since all measurements and constructions are finite (even the most distant star visible in the most powerful telescope is only a finite distance from the telescope), the fifth postulate is based upon a faith or conviction that if two lines are cut by a transversal such that the sum of the interior angles on one side of the transversal is less than two right angles, then the lines will intersect if they are extended sufficiently far. When the sum of the interior angles differs very slightly from two right angles, this assumption appears reasonable but cannot be definitely established.

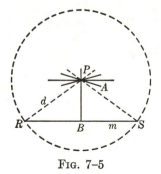

Given a line m and a point P that is not on m, there are infinitely many lines through P that do not intersect m within any given distance d of the point P. When d is the radius of the circle in Fig. 7-5, the existence of infinitely many lines (diameters of

Fig. 7-5

the circle) through P and not intersecting m is expected. Why shouldn't there also be infinitely many lines through P and not intersecting m when d is a mile, three miles (approximately the distance to the horizon when a person stands on level ground), the distance to the moon, or the distance to the most distant visible star? This question could not be dogmatically answered 150 years ago. It is now known that one may obtain equally consistent geometries by assuming that either

(i) there are infinitely many lines through P that do not intersect m, or

(ii) there is exactly one line through P that does not intersect m.

The above point of view can be very helpful in understanding Euclid's fifth postulate, its significance, and the form in which it was stated. If the line PB is perpendicular to m in Fig. 7–5, then Euclid's fifth postulate asserts that any line PA such that angle BPA is less than a right angle must intersect the line m if the lines PA and m are extended sufficiently far.

Bolyai Janos (Hungary, 1802–1860) and Nicolai Lobatschewski (Russia, 1793–1856) independently developed theorems in a geometry based essentially upon Saccheri's acute-angle hypothesis. They both obtained their results by considering a figure similar to Fig. 7–5. Later this geometry was called hyperbolic geometry in recognition of the two distinct lines (PR and PS in Fig. 7–5) parallel to a given line and through any given point that is not on the given line. We shall study some of the properties of this geometry in Chapter 8.

Also in the 19th century, Bernard Riemann (Germany, 1826–1866) developed a geometry based essentially upon Saccheri's obtuse-angle hypothesis. Riemann considered space as a set or manifold of undefined objects called points, where each point was determined by the values of n independent measures, or coordinates. He assumed that there existed a distance function and that the square of the differential of the distance function was homogeneous and of the second degree in the differentials of the coordinates. For example, in euclidean plane geometry the distance function

$$s^2 = (x_1 - x_2)^2 + (y_1 - y_2)^2$$

may be expressed in the form

$$ds^2 = dx^2 + dy^2$$

in terms of the differentials of the distance and the coordinates. Then just as we consider transformations (rotations and translations) that do not change the euclidean distance function, Riemann considered transformations that do not change his generalized distance function. Thus, as in euclidean geometry, Riemann assumed that the measurements of figures and objects do not depend upon their position. Under this assumption the figures on any surface must be freely movable, as on a plane (on which figures may slide or rotate) or the surface of a sphere. Technically, this implies that the curvature must be the same at all points of the surface. For example, in euclidean geometry the curvature at any point on a plane is zero; the curvature at any point on the surface of a sphere of radius r is $1/r^2$. Thus in euclidean geometry a plane and a sphere are each surfaces of constant curvature.

The distance functions or metrics were developed further by Arthur Cayley (England, 1821–1895) and Felix Klein (Germany, 1849–1925). The new theories were extended to three dimensions and then to n dimensions. Felix Klein considered any geometry to be a study of the properties that are invariant (not changed) under a group of transformations and gave a classification of geometries using invariance as a criteria. Sophus Lie (Norway, 1847–1899) considered groups of transformations leaving distance functions invariant and proved that if figures are to be freely movable (slide in any direction and rotate about any point), there are exactly four possible types of geometry in three-dimensional space. These are euclidean geometry, hyperbolic geometry, spherical geometry, and elliptic geometry. Hyperbolic plane geometry is associated with surfaces of constant negative curvature. Spherical and elliptic plane geometries are both considered on surfaces of constant positive curvature. Elliptic geometry may be visualized as the geometry of points on a sphere, with each pair of diametrically opposite points identified as a single point (Section 8–4). Elliptic and hyperbolic geometries are often called the *noneuclidean geometries* (Chapter 8).

We have sketched the stages in the discovery of the significance of Euclid's fifth postulate and the development of the noneuclidean geometries. The use of algebra (especially the theory of groups) and differentials in the development of these geometries provides another illustration of the interdependence of the various branches of mathematics. This interdependence is also seen in the search for proofs of the consistency of geometries. It can now be proved that euclidean geometry is consistent if the real number system is consistent. (This dependence of geometry upon properties of numbers is an example of the arithmetization of modern geometry.) The elliptic and hyperbolic geometries are also consistent if the real number system is consistent. Accordingly, even though several prominent philosophers have based their beliefs in natural logical systems upon the existence of a natural geometry — euclidean geometry, we must now recognize that euclidean geometry may not be the inherent geometry of our universe. It is probable that the development of philosophy as well as mathematics would have been noticeably changed if this had been recognized several centuries ago.

7-11 Nineteenth and twentieth centuries. The enormous productivity of the mathematicians of the 19th and 20th centuries makes it very difficult for us to view their work in proper perspective. Are we steadily pushing back the horizons of mathematical knowledge? Are we gaining an ever-firmer grasp of this body of knowledge, its underlying principles, and its implications? Such appears to be the case. The early Greeks made tremendous advances, considering the notation and mathematical tools at their disposal. The new techniques made possible by the development of algebra and the calculus in the 17th century were followed in the 18th century by a period of restating old ideas and finding new ones using the new techniques. After this exploratory work in the 18th century the mathematicians of the 19th century were able to use the new techniques more systematically, to gain considerable insight into the fundamental concepts of geometry, and to start organizing all geometric concepts in terms of these fundamental concepts. This process has continued in the 20th century with an emphasis upon arithmetization, more rigorous deduction of geometric properties from recognized assumptions, and generalizations of previously obtained results. In a sense the work of the 20th-century mathematicians appears as a tremendous surge for ever-increasing generalizations and abstractions with an underlying search for a more rigorous axiomatic foundation for the present body of mathematical knowledge.

In geometry the influence of intellectual curiosity has been evident since the time of the early Greeks. Many phases of geometry have been primarily intellectual achievements, and practical applications were not found for them until they were well developed. In other cases utility came earlier in the development. Recently technological advances have provided a wide range of activity for the mathematically curious as well as those who are trying to accomplish specific tasks. The electronic and mechanical computers provide a striking example of the influence of technological advances. Within the past decade these machines have gained recognition for the ways in which they may remove most of the drudgery from mathematical computations and make possible the use of very long computations in both applied and theoretical branches of mathematics. It seems reasonable to predict that within another decade these machines will revolutionize many of our approaches to mathematical problems. This new technique appears destined to go through the same stages that we have observed in the use of the techniques of algebra and the calculus. At present we are in the period of the discovery of the technique and initial explorations. Problems of rigor are arising. We may expect a rash of exploitations of the new technique comparable to the exploitations of the calculus in the 18th century. As in the case of algebraic and synthetic methods in geometry, there will undoubtedly be strong proponents of old and new techniques. Also, as in the case of algebraic and synthetic methods in geometry, we may look forward to the eventual absorption of the new technique into accepted mathematical procedures with a resulting unification of viewpoints and increase in our knowledge of mathematics.

The calculation of π offers an impressive example of an application of modern mathematical machines. A chronological sketch of the history of calculations of π may be found in [6; 90–96]. We have previously mentioned approximations of π by the early Babylonians, Greeks, and others. After the reawakening in Europe, each century produced one or more individuals who endeavored to express π as a fraction, an infinite series, or a decimal. In the 19th century William Shanks (England, 1812–1882) spent fifteen years calculating π to 707 decimal places. An error in the 528th place was found in 1946, new results were published in 1947, and by 1948 the first 808 decimal places were known. In 1949 the effectiveness of the electronic computers was dramatically illustrated by the calculation on the ENIAC of 2035 decimal places of π. The speed with which such machine calculations may be made is illustrated by the calculation of decimal places of e on the ILLIAC at a rate of over a thousand decimal places per hour.

Mathematicians are using machines more and more. At present many geometric problems, even problems in synthetic projective geometry, are being prepared for machine computation. Not all the results are as dramatic as the calculation of the first part of the decimal expression for π. However, an increasing influence and use of techniques based upon machine calculations must be expected.

Let us now return to the early 19th century. At the end of the 18th century, Monge was active in France. As we mentioned in Section 7-9, he organized descriptive geometry and applied the new techniques of algebra and the calculus to curves and surfaces. One of his pupils, Victor Poncelet (France, 1788–1867), wrote the first text on projective geometry, considered ideal points (intersections of parallel lines), used synthetic methods to extend the theory of constructions (Section 7-6), and developed the concept of duality. Planar duality had also been considered earlier by Charles Julien Brianchon (France, 1785–1864) and Joseph Diaz Gergonne (France, 1771–1859). Julius Plücker (Germany, 1801–1868) is given credit for extending this concept to three-dimensional and higher-dimensional spaces.

Jacob Steiner (Switzerland, 1796–1863), a pupil of Pestalozzi, was a specialist in synthetic geometry. He contributed to the theory of constructions (Section 7-6) and unified the classical methods of synthetic projective geometry. Steiner, Poncelet, and Karl von Staudt (Germany, 1798–1867) highlighted the organization of synthetic geometry in the 19th century. Von Staudt considered a geometry of positions independent of all measurements and proved that geometry did not need the techniques of algebra and the calculus.

Meanwhile, as should be expected, there was a corresponding development of analytic geometry. Gauss' ideas on curvature were extended by Riemann (Section 7-10) and eventually provided a basis for relativity theory. Riemann also used n sheets (in a sense, planes) to render n-valued functions single-valued on these so-called Riemann surfaces. This concept had implications in both the theory of functions and geometry. Plücker classified cubic curves, listed quartic curves, introduced a new type of coordinates, and studied properties of curves. His student, Felix Klein, has already been mentioned for his work on noneuclidean metrics and his classification of geometries (Section 7–10). Cayley (Section 7–10) and Hermann Grassman (Germany, 1809–1877) developed new coordinate systems, considered distance relations (metrics), and extended their results to n-dimensional geometries. Ludwig Schläfli (Switzerland, 1814–1895) discovered additional significant properties of n-dimensional geometries. August Moebius (Germany, 1790–1868) introduced new coordinates and discovered a new type of surface, a one-sided surface (Section 9–9). As indicated by the introduction of several types of coordinates, by Riemann's concept of a point as an undefined entity determined by its coordinates, and by the introduction of n-dimensional spaces, the concept of a point was undergoing a change in the middle of the 19th century. The complete abandonment of all visual intuitive concepts of a point was gradually accepted by theoretical mathematicians. The new concepts of a point were used in the noneuclidean geometries (Section 7–10) and as a basis for abstract geometry. Thus near the end of the 19th century we find geometry extending its break with the physical world. This break was completed with the axiomatic developments of the 20th century.

The concept of a line was changed along with the concept of a point. However, the concept of a line had another stage in its development. The existence of a mechanical procedure or device for constructing a line which would be "straight" at least in theory had posed a problem for centuries. Peaucellier (France, 1832–1913) solved this problem using a linkage (Section 7–6) and constructing the inverse of a circle with respect to a point on it. This theoretical development was soon followed by an abstract concept of lines analogous to that for points.

In addition to the developments mentioned above there were notable geometric contributions by Lazare Carnot (France, 1753–1832) in synthetic projective geometry, Michel Chasles (France, 1793–1880) in the history of geometry and in synthetic projective geometry, William R. Hamilton (Ireland, 1806–1865) in the application of differential geometry to optics, George Salmon (Ireland, 1819–1904) in cubic curves and elementary textbooks, Elwin Bruno Christoffel (Germany, 1829–1900) in differential geometry, Alfred Clebsch (Germany, 1833–1872) in analytic geometry, Georg Zeuthen (Denmark, 1839–1920) in synthetic projective geometry, Émile Lemoine (France, 1840–1912) in the simplification of constructions, Jean

Gaston Darboux (France, 1842–1917) in the study of surfaces and the use of imaginary elements, Georg Cantor (Germany, 1845–1918) in synthetic geometry and the study of continuity, Henri Brocard (France, 1845–1922) in the geometry of the triangle, Henri Poincaré (France, 1854–1912) through his popular works on the foundations of geometry, and many others.   Among the others are the members of a very active center of geometric activity in Italy and several mathematicians who were particularly interested in the foundations of geometry.   The discussion of both of these groups has been purposely postponed for special treatment as we now conclude our discussion of the evolution of geometry.

About the end of the 19th century the Italian geometers enjoyed a period of leadership, based upon the work of Luigi Cremona (1830–1903) in synthetic geometry, Francesco Brioschi (1824–1897) in the theory of invariants, Luigi Bianchi (1856–1928) in differential geometry, Eugenio Beltrami (1835–1900) in differential geometry and hyperspaces, Giuseppe Peano (1858–1922) in postulational systems, Francesco Severi (1879–   ) in analytic geometry and algebraic geometry (a recent theory at an advanced level), and others. In recent years this leadership has probably shifted to the United States with the influx of scientific immigrants and the development of abstract algebraic geometry under Solomon Lefschetz (1884–   ), Oscar Zariski (1899–   ), Andre Weil (1906–   ), Claude Chevalley (1909–   ), and others.   The shift in leadership appears to be due to the introduction of the arithmetic ideas of Richard Dedekind (Germany, 1831–1916) and Heinrich Weber (Germany, 1842–1913) and to the modern algebraic concepts of group, ring, field, and ideal.   Since the trend in algebraic geometry is indicative of trends in other branches of geometry, the following introductory remarks of Oscar Zariski in a paper presented at the 1950 International Congress of Mathematicians are included to give the viewpoint of one of the leaders in this area [20; 77].

The past 25 years have witnessed a remarkable change in the field of algebraic geometry, a change due to the impact of the ideas and methods of modern algebra.   What has happened is that this old and venerable sector of pure geometry underwent (and is still undergoing) a process of arithmetization.   This new trend has caused consternation in some quarters.   It was criticized either as a desertion of geometry or as a subordination of discovery to rigor.   I submit that this criticism is unjustified and arises from some misunderstanding of the object of modern algebraic geometry. This object is not to banish geometry or geometric intuition, but to equip the geometer with the sharpest possible tools and effective controls. It is true that the lack of rigor in algebraic geometry has created a state of affairs that could not be tolerated indefinitely. Effective controls over the free flight of geometric imagination were badly needed, and a complete overhauling and arithmetization of the foundations of algebraic geometry was the only possible solution. This preliminary foundational task of modern algebraic geometry can now be regarded as accomplished in all its essentials.

But there was, and still is, something else more important to be accomplished. It is a fact that the synthetic geometric methods of classical algebraic geometry, operating from a narrow and meager algebraic basis and faced by the extreme complexity of the problems of the theory of higher varieties, were gradually losing their power and in the end became victims to the law of diminishing returns, as witnessed by the relative standstill to which algebraic geometry came in the beginning of this century. I am speaking now not of the foundations but of the superstructure which rests on these foundations. It is here that there was a distinct need of sharper and more powerful tools. Modern algebra, with its precise formalism and abstract concepts, provided these tools.

An arithmetic approach to the geometric theories which we were fortunate to inherit from the Italian school could not be undertaken without a simultaneous process of generalization; for an arithmetic theory of algebraic varieties cannot but be a theory over arbitrary ground fields, and not merely over the field of complex numbers. For this reason, the modern developments in algebraic geometry are characterized by great generality. They mark the transition from classical algebraic geometry, rooted in the complex domain, to what we may now properly designate as *abstract algebraic geometry*, where the emphasis is on abstract ground fields.

We have seen in our previous discussion and in the above quotation that the trend of geometry in the first half of the 20th century has been toward generalization, arithmetization, and the axiomatic foundations of geometry. Moritz Pasch (Germany, 1843–1931) visualized geometry as a deductive science based upon a set of postulates. Giuseppe Peano and Mario Pieri (Italy, 1860–1913), David Hilbert (Germany, 1862–1943), E. V. Huntington (United States, 1874–1952), Oswald Veblen (United States, 1880–   ), George David Birkhoff (United States, 1884–1944), John Wesley Young (United States, 1879–1932), Robert Lee Moore (United States, 1882–   ), and others have formalized this concept of geometry. Hilbert and Veblen have been especially active. Most readers should appreciate

Hilbert's concept of mathematics as a game played according to certain rules with meaningless marks on paper.

7-12 Survey. We have considered the empirical geometry of early cultures, the rational approach of the Greeks, the influence of Euclid's postulates, and the development of special branches of geometry, such as constructions, descriptive geometry, synthetic geometry, analytic geometry, and differential geometry. Our concept of geometry has spread like a fan from very simple concepts of earth measure to a myriad of topics, concepts, and techniques too numerous to be completely comprehended by any one person. We still have not answered and cannot answer the question "What is geometry?" We have seen that the word "geometry" meant very different things to the early Egyptians, the Greeks, and the European mathematicians. Likewise, our geometry today is different from that of the 19th century and even somewhat different from that of a decade ago. Thus the meaning of the word "geometry" must be considered relative to a certain culture and indeed to a certain time in a culture. Even under these restrictions a precise definition is impossible, since at any given time "geometry" means different things in different countries and to different mathematicians. We must be satisfied with a general impression of the concepts that have been included in the past and the influence of our culture upon those concepts. Especially as geometric novices we must accept indications of trends without extensive details. Such has been the point of view of the treatment in this chapter. We have neglected many advanced geometrical concepts and also several concepts that could easily have been included. Probably the most noticeable omission is topology — a very general and important branch of geometry. Chapter 9 is devoted entirely to topology.

In spite of the inevitable omissions of many individuals, difficult concepts, and details in the present chapter it is hoped that the following sequence of developments in the evolution of geometry has been clear to all readers:

(i) the use of practical rules for measurements of areas and volumes by the Babylonians and Egyptians;

(ii) the study of lines and figures formed by lines (Thales);

(iii) the development of geometric constructions and of geometry as a body of knowledge (early Greek);

(iv) the recognition of geometry as a part of a liberal education;

(v) the influence of Greek reasoning and logic leading up to Euclid's organization of geometry;

(vi) the further development of euclidean geometry by the Greeks;

(vii) the decline of the Greek civilization and the rise of the Arabian civilization;

(viii) the reawakening in Europe with the invention of the printing press, the easier dissemination of ideas, and gradually the formulation of new concepts;

(ix) the early notions of projective geometry, analytic geometry, and differential geometry in the 16th and 17th centuries;

(x) the exploitation, at times reckless, of the new techniques in the 18th century;

(xi) the recognition of several geometries (especially euclidean, elliptic, and hyperbolic) as independent and equally valid branches of mathematics;

(xii) the abandonment of the study of geometry as a study of relations in the physical universe;

(xiii) the continued application of techniques of algebra and the calculus, the organization of concepts, and a tendency to more rigorous treatments of geometry in the 19th century;

(xiv) the emphasis upon generalization, arithmetization, and the axiomatic foundations of geometry in the first half of the 20th century; and finally

(xv) the current development of machine calculators which, it is conjectured, will give rise to new techniques and have considerable influence upon the evolution of geometry.

The evolution of geometry sketched above should not be interpreted as implying that all developments of synthetic and analytic geometry will soon be outmoded. Recent theories in the foundations of geometry suggest the interpretation of synthetic and analytic geometries as equivalent representations of geometry considered as a postulational system. Thus the synthetic and algebraic methods continue to be useful in our search for knowledge and yet are brought together in our concept of a geometry as a deductive system. Historically speaking, we are witnessing at any one time in history a fleeting instant of an ever-changing body of knowledge called geometry. By geometry we therefore mean a certain body of knowledge that has been discovered in the past together with concepts that are in the process of being developed and absorbed at the present. We must expect geometry to change constantly under the influence of new theories and new technological advances.

The consideration of projective geometry in the present text as a deductive system (Chapters 2 and 3), as a union of synthetic and algebraic concepts (Chapter 4), and as a study of properties that are invariant under a group of transformations illustrates the points of view that are prevalent in the middle of the 20th century. Euclidean geometry, as we now use the term, has been noticeably restricted relative to the wide range of topics considered in Euclid's *Elements*. We have considered euclidean geometry as the study of properties that are invariant under the group of rigid motions (translations and rotations). Euclidean geometry may also be considered as a postulational system and from either an algebraic or a synthetic point of view. These considerations provide an excellent background for alert students who wish to study the geometries of the present or to be aware of the continued evolution of geometry in the future.*

---

* In any historical survey there is a temptation to give an extensive bibliography. Only the references used in the presentation are given here. The author believes that the reader can gain an understanding only by perusing several sources, recognizing their contrary statements, and attempting to distil from a mass of detail the elements of truth. The reference [6] is unique in that it contains many exercises correlated with the discussion of historical developments.

CHAPTER 8

NONEUCLIDEAN GEOMETRY

For many centuries euclidean geometry was considered to be *the* geometry of the space in which we live. During the last century mathematicians, scientists, and philosophers have become increasingly aware that there exist other geometries and that it is possible that one of the noneuclidean geometries (elliptic or hyperbolic) may provide a better approximation than euclidean geometry to the geometry of the space in which we live (Section 7–10). In the present chapter we shall consider a few properties of the non-euclidean geometries and their relation to projective and euclidean geometries.

8–1 The absolute polarity. In euclidean geometry the pairs of an absolute involution on the ideal line are invariant and the points on the ideal line are called ideal points. In the two noneuclidean geometries an absolute polarity (Section 4–9) is invariant and the points on the conic (set of self-conjugate points) determined by the absolute polarity are called ideal points. Thus in this chapter we shall be concerned with an ideal conic instead of an ideal line. More precisely, we shall be concerned with a polarity (4–34) with self-conjugate points satisfying an equation (4–35). We shall consider sets of projective transformations leaving the polarity, and therefore its set of self-conjugate points, invariant. We shall call the corresponding geometry hyperbolic if there exists at least one self-conjugate point with real coordinates, elliptic if there are no such points. The remainder of the present section will include an explanation of the last two sentences and a brief discussion of the two geometries. As in the case of the absolute involution (6–1), we shall use changes of coordinates to express the absolute polarity in a convenient form (8–4).

The definition of a polarity in Section 4–9 was based upon correlations. A correlation on a plane is a nonsingular linear homogeneous transformation (4–31) of points (x_1, x_2, x_3) into lines $[u_1, u_2, u_3]$. Under a correlation any point P corresponds to a line p and any line q corresponds to a point Q. When these correspondences are such

268

that Q is on p if and only if P is on q, the correlation is called a polarity. In general any polarity has equations of the form (4–34)

$$u_1 = a_{11}x_1 + a_{12}x_2 + a_{13}x_3,$$
$$u_2 = a_{12}x_1 + a_{22}x_2 + a_{23}x_3,$$
$$u_3 = a_{13}x_1 + a_{23}x_2 + a_{33}x_3,$$

$$|\,a_{ij}\,| \neq 0$$
$$(i,j=1,2,3).$$

As in Section 4–9, the line $[u_1,u_2,u_3]$ is called the polar of the point (x_1,x_2,x_3) and the point is called the pole of the line with respect to the given polarity. If we select our coordinate system such that the polar of $(1,0,0)$ is $[1,0,0]$ and the polar of $(0,1,0)$ is $[0,1,0]$, then from (4–34)

$$[1,0,0] = [a_{11},a_{12},a_{13}]$$

and

$$[0,1,0] = [a_{12},a_{22},a_{23}],$$

whence $a_{12} = a_{13} = a_{23} = 0$ and the polarity may be represented in the form

$$u_1 = a_{11}x_1,$$
$$u_2 = a_{22}x_2,$$
$$u_3 = a_{33}x_3,$$

$$(a_{11}a_{22}a_{33} \neq 0)$$

in the new coordinate system.

We now note that

(i) since the coefficients a_{11}, a_{22}, a_{33} are three real numbers different from zero, at least two of the three coefficients must have the same sign;

(ii) since $[u_1,u_2,u_3] = [ku_1,ku_2,ku_3]$ whenever $k \neq 0$, the three coefficients may be multiplied by any real number $k \neq 0$ without changing the polarity; and

(iii) since any two coordinates may be interchanged by changing the coordinate system, as for example by the projective transformation

$$x_1' = x_2, \quad x_2' = x_1, \quad x_3' = x_3,$$

we may consider the coefficients in any order that we wish.

Accordingly, we may assume that a_{11} and a_{22} are positive and write the polarity in the form

(8–1)

$$u_1 = a^2x_1,$$
$$u_2 = b^2x_2,$$
$$u_3 = c^2ex_3,$$

$$(a^2b^2c^2 \neq 0,\ e^2 = 1).$$

Then the condition (4–35) that a point be self-conjugate under the polarity (8–1) has the form

$$(8\text{–}2) \qquad a^2x_1^2 + b^2x_2^2 + c^2ex_3^2 = 0, \qquad (e^2 = 1).$$

If there exist real points (i.e., points having triples of real numbers as coordinates) satisfying (8–2), the polarity is called a hyperbolic polarity and the set of self-conjugate points is called a point conic (Section 4–9). If there do not exist real points satisfying (8–2), the polarity (8–1) is called an elliptic polarity. In either case we may identify the polarity (8–1) by means of the left member of the equation (8–2). Also, in either case, since $a^2b^2c^2 \neq 0$, we may use a change of coordinates

$$x_1' = ax_1, \quad x_2' = bx_2, \quad x_3' = cx_3$$

and write the equation (8–2) in the form

$$(8\text{–}3) \qquad x_1^2 + x_2^2 + ex_3^2 = 0, \qquad (e^2 = 1).$$

We have seen that by a suitable choice of coordinates any polarity (4–34) may be expressed in the form

$$(8\text{–}4) \qquad \begin{aligned} u_1 &= x_1, \\ u_2 &= x_2, \\ u_3 &= ex_3, \end{aligned} \qquad (e^2 = 1)$$

and identified by means of the left member of the equation (8–3). We now select (8–4) as the *absolute polarity* and consider the set of projective transformations that leave the condition (8–3) invariant, i.e., the set of projective transformations that leave the absolute polarity invariant.

The polarity (8–4) is defined (Section 4–9) to be elliptic if $e = +1$ (since in this case (8–3) is not satisfied by any real point) and hyperbolic if $e = -1$. The geometry associated with the set of all projective transformations leaving (8–3) invariant is called *elliptic geometry* if $e = +1$, *hyperbolic geometry* if $e = -1$. Intuitively, we may think of the set of points (complex coordinates) satisfying the equation

$$(8\text{–}5) \qquad x_1^2 + x_2^2 + x_3^2 = 0$$

as an imaginary conic and visualize elliptic geometry as the geometry of the real projective plane that leaves the imaginary conic (8–5) invariant. Similarly, we may consider the point conic with equation

$$(8\text{–}6) \qquad\qquad x_1^2 + x_2^2 - x_3^2 = 0$$

and visualize hyperbolic geometry as the geometry of the real projective plane that leaves the conic (8–6) invariant. We shall call the conics (8–5) and (8–6) the *absolute* or *ideal conics* of their respective geometries. Associated with each ideal conic is an absolute polarity (8–4) that is invariant under a projective transformation if and only if the ideal conic is invariant.

We may now gain an even broader point of view by comparing the condition (8–3) with the expression

$$(8\text{–}7) \qquad\qquad a(x_1^2 + x_2^2) + bx_3^2,$$

where a and b are not both zero. Under the assumption that all ordinary points are to be represented in the form $(x_1, x_2, 1)$, the plane geometry associated with the set of projective transformations leaving (8–7) invariant

when $a = 0$	is affine geometry,
when $a = 0$ and $b = 0$	is euclidean geometry (Exercise 3),
when $a = b \neq 0$	is elliptic geometry,
when $a = -b \neq 0$	is hyperbolic geometry.

Finally, we note that the sign rather than the numerical value of e is the basic factor when (8–3) is used to distinguish elliptic from hyperbolic polarities or geometries. Accordingly, under the identification of e with b/a and the assumption that $e = 0$ when $b = 0$ and $e = \infty$ when $a = 0$, the above classification for (8–7) may be restated and extended for (8–3) as follows: The subgeometry of projective geometry that leaves the left member of (8–3) invariant

when $e = \infty$	is affine geometry,
when $e = \infty$ and $e = 0$	is euclidean geometry,
when e is positive	is elliptic geometry,
when e is negative	is hyperbolic geometry.

In this sense euclidean geometry ($e = 0$ or ∞) is a limiting case of elliptic geometry ($e > 0$) and also of hyperbolic geometry ($e < 0$).

EXERCISES

1. Write down a set of equations (4–31) for
    (a) a correlation that is not a polarity,
    (b) a general polarity,   (c) an elliptic polarity,
    (d) a hyperbolic polarity.

2. Write down the condition that a point be self-conjugate under the polarity given in answer to
    (a) Exercise 1(c),   (b) Exercise 1(d).

3. Prove that when all ordinary points are represented in the form $(x_1,x_2,1)$, the plane geometry associated with the set of projective transformations leaving (8–7) invariant when $a = 0$ and also when $b = 0$ is euclidean geometry.

8–2 Points and lines. There exist noneuclidean geometries in three-dimensional and higher-dimensional spaces. However, as in our discussion of euclidean geometry in Chapter 6, we shall be primarily concerned with noneuclidean plane geometries. Thus we shall be primarily concerned with points, lines, and transformations of points and lines.

Any geometry may be defined in terms of a set of elements (possibly undefined) and a set of transformations. We shall consider the elements of any plane geometry to be "points" and "lines" with the understanding that these elements may be interpreted or represented in various ways. For example, the points and lines of euclidean plane geometry may be visualized as points and lines on a plane in our physical universe, as pairs of real numbers (x,y) and sets of such points satisfying a linear equation $ax + by + c = 0$, or in other ways. When euclidean plane geometry is considered as a subgeometry of projective geometry, we may define the elements of euclidean plane geometry as the points and lines that are on a real projective plane but are not on an ideal line $x_3 = 0$. The points and lines of the noneuclidean geometries may be similarly defined.

The points and lines of elliptic geometry are precisely the points and lines on a real projective plane. The ordinary points and lines of hyperbolic plane geometry are the points (x_1,x_2,x_3) on a real projective plane that satisfy the relation

$$x_1^2 + x_2^2 - x_3^2 < 0,$$

that is, the interior points of the absolute conic (8–6); the ordinary lines are segments of the projective lines containing these points.

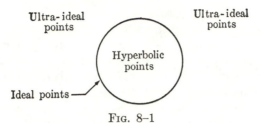

Fig. 8-1

These elements of hyperbolic geometry may be visualized more easily after we make the following definitions.

Given any hyperbolic plane geometry with its ideal conic (8–6), we define the points (x_1, x_2, x_3) on the real projective plane to be

$$\begin{aligned}
\text{ordinary points} &\quad \text{if } x_1^2 + x_2^2 - x_3^2 < 0, \\
\text{ideal points} &\quad \text{if } x_1^2 + x_2^2 - x_3^2 = 0, \\
\text{ultra-ideal points} &\quad \text{if } x_1^2 + x_2^2 - x_3^2 > 0.
\end{aligned}$$

This definition is equivalent (see Fig. 8–1) to defining each point on the real projective plane to be an ordinary point, ideal point, or ultra-ideal point according as it is an interior point of the conic, a point on the conic, or an exterior point of the conic (Exercise 4). The above classifications of points may be summarized as follows: The points on a real projective plane are ordinary or ideal with respect to euclidean geometry; are all ordinary points with respect to elliptic geometry; are ordinary, ideal, or ultra-ideal points with respect to hyperbolic geometry. We next use this classification of points to classify the lines of the respective geometries.

On a real projective plane any two distinct lines have a unique point in common. A projective line contains a euclidean line if it is distinct from the ideal line; is always a line of elliptic geometry; and contains a line of hyperbolic geometry if it contains an interior point of the ideal conic. In euclidean geometry any two distinct lines intersect if the corresponding projective lines have an ordinary point in common, and they are parallel if the projective lines have an ideal point in common. In elliptic geometry any two distinct lines intersect, since all points on the projective plane are ordinary points in elliptic geometry. In hyperbolic geometry any two distinct lines are said to be *intersecting* if the corresponding projective lines have an ordinary point in common, *parallel* if the projective lines have an ideal point in common (i.e., intersect at a point on the ideal conic), *nonintersecting* if the projective lines have an ultra-ideal point in common. Thus any two distinct lines are intersecting or parallel in euclidean geometry; are intersecting in elliptic geometry; and are intersecting, parallel, or nonintersecting in hyperbolic geometry.

We have now defined and compared the elements (points and lines) of euclidean, elliptic, and hyperbolic geometries when these geometries are considered as subgeometries of projective geometry. As indicated in Chapter 7, this systematic view of the three geometries is an outgrowth of many centuries of study and searching for the fundamental concepts of geometry. For over a thousand years mathematicians have fumbled and grasped for ideas that would help them to understand euclidean geometry. Indeed, it is entirely conceivable that a thousand years from now our present efforts will appear as disorganized graspings for ideas. However, compared with past efforts and concepts, our ideas are becoming organized. The above presentation of the euclidean and noneuclidean geometries, their elements and their transformations, as special cases of projective geometry represents one of the advances of the late 19th and early 20th centuries. It is based upon an abstract concept of lines and points. These concepts will require that we be able to interpret points and lines in various ways as we consider properties and representations of the various geometries. Since hyperbolic geometry evolved first historically and probably contains concepts that may be visualized more easily than those of elliptic geometry, we shall next consider properties of hyperbolic plane geometry.

EXERCISES

1. Prove that given any line m and any point P in euclidean plane geometry, there exists a unique line through P that is parallel to m.

2. Restate Exercise 1 and prove that for elliptic geometry there does not exist a line through P that is parallel to m.

3. Restate Exercise 1 and prove that in hyperbolic geometry for any point P that is not on m there are two distinct lines through P that are parallel to m.

4. Prove that the interpretation of interior points as ordinary points and of exterior points as ultra-ideal points is consistent with the definitions of interior and exterior points in Section 4–9.

8–3 Hyperbolic geometry. We have defined hyperbolic plane geometry as the subgeometry of real projective geometry that leaves the ideal conic (8–6) invariant. We have also seen (Section 7–10) that the discovery of hyperbolic geometry was based upon efforts to prove Euclid's fifth postulate from the other four postulates and the assumption that lines are not re-entrant. In the present section we shall use our definition of hyperbolic geometry as a subgeometry of projective geometry to obtain several properties of hyperbolic geometry.

The properties of hyperbolic geometry that depend upon incidence relations may be easily visualized by considering the ideal conic

$$x_1^2 + x_2^2 - x_3^2 = 0$$

as the unit circle

(8–8) $$x^2 + y^2 - 1 = 0$$

on a euclidean plane. Then the ordinary points of the hyperbolic plane are the interior points of the unit circle, and a euclidean line $ax + by + c = 0$ contains an ordinary hyperbolic line if and only if $a^2 + b^2 - c^2 > 0$ (Exercise 1). As in Section 8–2, we define two hyperbolic lines to be intersecting, parallel, or nonintersecting, according as the corresponding projective lines intersect at an interior point of the conic, a point of the conic, or an exterior point of the conic. These definitions and representations enable us to prove that

(i) Any two distinct hyperbolic points determine a unique hyperbolic line (Exercise 3).

(ii) Any two distinct hyperbolic lines determine at most one hyperbolic point (Exercise 4).

(iii) Every hyperbolic line has two distinct associated ideal points, which Hilbert called its two *ends* (Exercise 5).

(iv) Given any hyperbolic line m and any hyperbolic point P that is not on m, there are exactly two distinct hyperbolic lines through P parallel to m (Exercises 6 and 7).

(v) There exist pairs of nonintersecting hyperbolic lines (Exercise 8).

(vi) Two hyperbolic lines that are parallel to the same hyperbolic line in the same sense (i.e., having a common end) are parallel (Exercise 9).

The above properties are the basic properties of hyperbolic geometry that depend upon incidence relations. There are also many properties of hyperbolic geometry that depend upon distance, angles, and congruence. These properties may also be defined in terms of the projective transformations of the hyperbolic plane onto itself. However, we shall not attempt a detailed treatment here. Rather we shall consider a second model of hyperbolic geometry that will facilitate our visualization of a few more properties of hyperbolic geometry.

Henri Poincaré at one time visualized physical space as a sphere of radius R such that the absolute temperature t at any point at a distance r from the center of the sphere is given by

$$t = c(R^2 - r^2),$$

where c is a constant of proportionality. Under the assumption (not completely accepted at present) that physical bodies decrease in volume with decreasing temperature and vanish altogether at the bounding surface of the above sphere where the temperature is absolute zero, the shortest path between any two given points may be shown in differential geometry to be along an arc of a circle that intersects the sphere at right angles. We may obtain a model for hyperbolic plane geometry by considering the points on a plane through the center of Poincaré's sphere representing physical space. Any plane through the center of this sphere intersects the sphere in a circle that we shall call the *fixed circle* on the plane. The geometry of circles that intersect the fixed circle at right angles, i.e., the *geometry of circles orthogonal to a fixed circle*, is a hyperbolic geometry. Thus in Poincaré's model the plane intersects the sphere in a fixed circle, points are interior points of the fixed circle, and lines are arcs of circles that intersect the fixed circle at right angles; i.e., the tangents to the circles are perpendicular at their point of intersection.

Poincaré's model of hyperbolic plane geometry has the same incidence relations as the geometry of segments of straight lines interior to a conic that we considered above (Exercise 12). Accordingly, we shall find it convenient to use the same terminology in discussing the elements of the two geometries, i.e., to consider the two models simply as different representations of the points and lines of the geometry. In the geometry of segments of lines interior to a conic, we were able to use the properties of lines on the projective and euclidean planes corresponding to the plane of the conic. This model (geometry) has a special place in our discussion because of the form of our definition of hyperbolic geometry. In the geometry of circles orthogonal to a fixed circle, we may use properties of circles on the euclidean plane of the fixed circle. These properties will be needed whenever properties of the geometry of circles orthogonal to a fixed circle are considered. In particular they will be needed in Exercise 10. Since properties of circles are often not as well known as properties of lines, we shall consider only a few of these properties. Further properties may be found in texts such as [5].

Given a fixed circle with center O, a circle with center O' intersects the fixed circle orthogonally (i.e., at right angles) at R and S if and only if OR is perpendicular to $O'R$ and OS is perpendicular to $O'S$. Given any three noncollinear points, the center of the circle through the given points may be determined as the intersection of the perpendicular bisectors of any two of the segments determined by the given points. When three given points are collinear, the line through them may be considered as a circle of infinite radius. Under this convention there exists a circle through any three points and we may construct a circle orthogonal to a given circle through any two given points [19; 18]. Let a fixed circle with center O and two points A, B be given where at least one of the points, say A, is an interior point of the fixed circle. Draw OA and construct RS perpendicular to OA at A where R and S are points of the fixed circle (Fig. 8–2). Let the tangents of the fixed circle at R and S intersect at A'. The circle determined by the three points A, A', and B is orthogonal to the fixed circle (Exercise 10). When the terminology of the geometry of circles orthogonal to a fixed circle is used, the above construction provides a construction for a hyperbolic line through any two points. It also provides a construction for a hyperbolic line parallel to a given hyperbolic line and through a point that is not on the given line (Exercise 11). If the centers of the circles in euclidean geometry are called the *centers* of the corresponding hyperbolic lines, we may prove (for example, by algebraic methods) that the locus of centers of the hyperbolic lines through a given point $A \neq 0$ is a euclidean line perpendicular to the line OA where O is the center of the fixed circle (Exercise 13). This result may be used to give a second

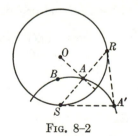

FIG. 8–2

construction for a hyperbolic line through any two hyperbolic points (Exercise 14). We may also prove that the locus of centers of hyperbolic lines perpendicular to a given hyperbolic line RS where R and S are points of the fixed circle is the set of ultra-ideal points on the euclidean line RS (Exercise 15). This result may be used to construct a hyperbolic line perpendicular to any given hyperbolic line and through any given hyperbolic point (Exercise 16).

The hyperbolic length $\overline{AB}^h$ of any segment AB on a hyperbolic line with ideal points R and S may be expressed in the form

$$(8\text{-}9) \qquad \ln \frac{(AR)(BS)}{(BR)(AS)} = \overline{AB}^h,$$

where ln stands for the natural logarithm and the quantities AR, AS, BR, and BS are the lengths of the chords of the corresponding hyperbolic lines. As in the case of euclidean geometry, we could base our hyperbolic geometry upon the assumption of this hyperbolic distance function. The properties of this function may be compared with those of the euclidean distance function (Section 6–5) and found to be as follows (Exercise 18):

$$\overline{AB}^h = 0 \text{ if and only if } A = B,$$
$$\overline{AB}^h = -\overline{BA}^h,$$
$$\overline{AB}^h + \overline{BC}^h = \overline{AC}^h.$$

The hyperbolic distance function (8–9), the model provided by the circles orthogonal to a fixed circle, and the constructions considered in Exercises 10 through 17 may be used to visualize without formal proof the following properties of hyperbolic geometry. In many cases the proofs may be readily obtained from the constructions. In all cases figures illustrating the properties may be obtained either freehand or from the constructions.

(vii) Two parallel hyperbolic lines converge continuously in the direction of parallelism and diverge in the opposite direction.

(viii) There is a unique hyperbolic line perpendicular to any given hyperbolic line and through any given hyperbolic point (Exercises 19 and 20).

(ix) Two hyperbolic lines that are perpendicular to the same hyperbolic line are nonintersecting (Exercise 21).

(x) The angle sum of a hyperbolic triangle is less than 180°.

(xi) The area of a hyperbolic triangle is proportional to the deficiency of its angle sum from 180°.

(xii) Hyperbolic triangles with the same angle sum are equal in area.

(xiii) All similar hyperbolic triangles are congruent.

(xiv) There is an upper bound for the areas of hyperbolic triangles.

(xv) There is a natural unit of angle, a natural unit of length, and a natural unit of area.

(xvi) Any two nonintersecting hyperbolic lines have a unique common perpendicular.

When hyperbolic triangles are allowed to have ideal points (points of the ideal conic or fixed circle) as vertices, we also have:

(xvii) There exists a hyperbolic triangle with angle sum 0° (this is a hyperbolic triangle of maximum area) (Exercise 24).

(xviii) Given any hyperbolic line m and any hyperbolic point P that is not a point of m, there exists a hyperbolic line segment AP perpendicular to m where A is on m, and there also exists a hyperbolic *angle of parallelism APR* of the segment AP where R is an ideal point on m (Exercise 25).

(xix) Two hyperbolic segments having the same angle of parallelism are equal.

(xx) The area of any hyperbolic triangle with maximum area provides an upper bound for the areas of hyperbolic triangles and a basis for a natural unit of area.

(xxi) The distance from the center of the fixed circle to a side of a hyperbolic triangle of maximum area may be used to obtain a natural unit of length.

The above list of properties could be extended in several ways. However, we shall leave detailed treatments for texts devoted entirely to noneuclidean geometry. Thus in the next section we shall turn our attention to the properties of elliptic geometry. In general, our interest in the noneuclidean geometries is enhanced by the possibility that the natural geometry of the space in which we live may be a noneuclidean geometry. The possibility that the natural geometry of the space in which we live is hyperbolic geometry appears almost reasonable when the fixed circle or absolute conic is taken as many euclidean miles in diameter instead of a few inches as we have found convenient in our figures.

EXERCISES

1. Prove that a euclidean line, $ax + by + c = 0$, intersects the circle (8–8) in distinct points if and only if $a^2 + b^2 - c^2 > 0$.

2. Sketch an absolute conic and draw pairs of hyperbolic lines that are (a) intersecting, (b) parallel, (c) nonintersecting.

3. Prove that any two distinct hyperbolic points determine a unique hyperbolic line.

4. Prove that any two distinct hyperbolic lines determine at most one hyperbolic point.

5. Prove that every hyperbolic line has two associated ideal points (ends).

6. Prove that given any hyperbolic line m and any hyperbolic point P that is not on m, there are exactly two distinct hyperbolic lines through P parallel to m.

7. Draw a figure for Exercise 6.

8. Show that for any given hyperbolic line m there exist hyperbolic lines that do not intersect m and are not parallel to m.

9. Prove that two hyperbolic lines that are parallel to the same hyperbolic line in the same sense (i.e., all three lines have a common end) are parallel.

10. Prove that the circle determined by the points A, A', B in Fig. 8–2 is orthogonal to the fixed circle.

11. Given a fixed circle, a hyperbolic line m in the geometry of circles orthogonal to the fixed circle, and two hyperbolic points P, Q that are not on m, construct (a) the hyperbolic line PQ, (b) a hyperbolic line through P and parallel to m.

12. Repeat Exercises 1 through 9 for the geometry of circles orthogonal to a fixed circle.

13. Prove that in the geometry of circles orthogonal to a fixed circle with center O, the locus of the centers of the hyperbolic lines through a given hyperbolic point $A \neq O$ is a euclidean line perpendicular to the euclidean line AO.

14. Give a construction for a hyperbolic line through any two hyperbolic points using the result in Exercise 13.

15. Prove that the locus of centers of hyperbolic lines perpendicular to a given hyperbolic line RS, where R and S are points on the fixed circle, is the set of ultra-ideal points on the euclidean line RS [5; 182].

16. The mid-point of a hyperbolic segment AB on a hyperbolic line RS, where R and S are on the fixed circle, is the intersection with AB of the hyperbolic line with center at the intersection of the euclidean lines AB

and RS [18; 210–211].   Draw a figure showing this construction.

17. Prove that the hyperbolic distance function (8–9) has the properties (a) $\overline{AB}^h = 0$ if and only if $A = B$, (b) $\overline{AB}^h = -\overline{BA}^h$, (c) $\overline{AB}^h + \overline{BC}^h = \overline{AC}^h$.

18. Give a construction for a hyperbolic line perpendicular to any given hyperbolic line and through any given hyperbolic point.

19. Prove that there is a unique hyperbolic line perpendicular to any given hyperbolic line and through any given hyperbolic point.

20. Sketch or construct a figure showing two hyperbolic lines perpendicular to the same hyperbolic line.

21. Sketch or construct a hyperbolic triangle with a positive angle sum.

22. Sketch several figures showing common perpendiculars between nonintersecting hyperbolic lines.

23. Sketch a hyperbolic triangle of maximum area.

24. Sketch or construct a figure showing that the angle of parallelism of a hyperbolic segment AP approaches 90° as $\overline{AP}^h$ approaches zero and approaches 0° as $\overline{AP}^h$ increases without bound.

8–4 Elliptic and spherical geometries.   We have seen in Section 7–10 that when figures are freely movable, there can be only four possible types of geometry in three-dimensional space: euclidean geometry (Chapter 6), hyperbolic geometry (Section 8–3), elliptic geometry, and spherical geometry. Elliptic and spherical plane geometries are both associated with spaces of constant positive curvature.   Elliptic geometry is a subgeometry of projective geometry; spherical geometry is not.   We shall consider a few properties of spherical plane geometry and then use these properties and a model of an elliptic plane to obtain properties for elliptic plane geometry.

Spherical plane geometry has been defined as a geometry of points and great circles on a sphere.   We shall refer to the great circles as lines of the spherical geometry.   Then among the properties of spherical geometry we have the following:

(i) Any two lines intersect in two diametrically opposite (*antipodal*) points.

(ii) Any two points that are not antipodal points determine a unique line.

(iii) There is a natural unit of angle measurement (based upon a revolution), a natural unit of length (based upon the circumference of a great circle), and a natural unit of area (based upon the area of the sphere).

(iv) Each line has two antipodal points as its poles, the common intersections of the set of lines perpendicular to the given line.

(v) Each point has a unique polar line, the line on the plane through the center of the sphere and perpendicular to the diameter through the given point.

If we consider only triangles with sides that are minor arcs of great circles, we also have the following properties:

(vi) The angle sum of a triangle is greater than 180° and less than 540°.

(vii) The area of a triangle is proportional to the excess of its angle sum over 180°.

(viii) Two triangles with the same angle sum are equal in area.

(ix) There is an upper bound for the area of triangles.

(x) The product of two line reflections (orthogonal) may be considered as a rotation about either of the points of intersection of their axes.

(xi) Two triangles are congruent if and only if they correspond under a finite product of line reflections.

(xii) Two triangles with corresponding angles equal are congruent (i.e., all similar triangles are congruent — see the assumption of Wallis in Section 7–10).

The above list of properties could be extended, but it will suffice for our present purposes. Proofs of these properties may be found in comprehensive texts on solid geometry. Since in spherical geometry any two lines intersect in a pair of diametrically opposite points (i.e., antipodal points), spherical geometry is not a subgeometry of projective geometry. Accordingly, we next consider a method for modifying spherical geometry to obtain a model for a subgeometry of projective geometry. Our method will be equivalent to the identification of pairs of antipodal points as points of the new geometry.

Consider any diameter of the sphere as a "point" and any diametral plane of the sphere as a "line." Then any two points determine a unique line, any two lines determine a unique point, and, in general, the postulates of projective geometry are satisfied. The geometry of points (considered as diameters) and lines (considered as diametral planes) is a subgeometry of projective geometry and indeed is an elliptic geometry.   In view of the restrictions upon the sides of triangles in the second list of properties of spherical geometry, each of the above properties of spherical geometry may be restated as properties of the elliptic geometry of diameters and diametral planes (Exercise 1).

The elliptic geometry of diameters and diametral planes may also be visualized as a geometry of the lines and planes on a point.   We shall use this interpretation to gain some insight into the relationship between euclidean and elliptic geometries.   Consider the origin $(0,0,0,1)$ in three-space as the center of a sphere and think of the geometry of diameters and diametral planes of the sphere as a geometry of the lines and planes on the origin.   Each line through the origin intersects the ideal plane $x_4 = 0$ in a point.   Each plane through the origin intersects the ideal plane in a line.   Thus any elliptic geometry of the lines and planes on the origin imposes an elliptic geometry of the points and lines on the ideal plane, and conversely.   We next observe that any euclidean geometry of three-space imposes an elliptic geometry on the ideal plane.

Euclidean plane geometry may be defined (Exercise 3, Section 8–1) as the special case of projective plane geometry in which the line $x_3 = 0$ is invariant and the expression $x^2 + y^2$ [i.e., the expression $x_1^2 + x_2^2$ when all ordinary points are represented in the form $(x_1,x_2,1)$] is invariant.   In three-space, euclidean geometry may be defined as the special case of projective geometry in which the plane $x_4 = 0$ is invariant and the expression $x_1^2 + x_2^2 + x_3^2$ is invariant when all ordinary points are represented in the form $(x_1,x_2,x_3,1)$.   Then, in particular, any euclidean geometry of three-space implies the invariance of (8–5) on the plane $x_4 = 0$.   In other words, euclidean geometry of three-space imposes an elliptic geometry on the ideal plane $x_4 = 0$.   In this sense, a thorough understanding of euclidean geometry in three-space requires an understanding of elliptic plane geometry.

We have now considered a few properties of hyperbolic, spherical, and elliptic geometries in terms of properties of certain models of the geometries.   The distinction between a geometry and one of its models is very important.   However, the euclidean, hyperbolic, spherical, and elliptic geometries each have categorical sets of postulates, and therefore any two representations or models are equivalent (isomorphic).   Accordingly, any property that holds for one model must hold for all other models.   We shall conclude this chapter with a comparison of the properties of euclidean, hyperbolic, and elliptic plane geometries, i.e., a comparison of euclidean plane geometry with the noneuclidean plane geometries.

EXERCISES

1. Restate and prove the twelve properties of spherical geometry listed above for the elliptic geometry of diameters and diametral planes.

2. Complete the following table of corresponding elements of the elliptic geometries on a plane and on a point, where the correspondence is based upon incidence (one could obtain a different correspondence based upon duality).

on a plane	on a point
point	line
line	_____
angle	_____
triangle	trihedral angle
unit length	unit plane angle
unit angle	_____
unit area	_____
line reflection	_____
_____	rotation about a line

8–5. Comparisons.   We have now considered properties of euclidean, hyperbolic, and elliptic geometries in terms of their definitions and some of their models.   We next consider the comparisons and contrasts of the properties of the three geometries.   The following properties hold in all three geometries.

(i) Any two distinct points determine a unique line.

(ii) There is a unique line perpendicular to any given line and through any given point on the line.

(iii) All transformations may be expressed as products of line reflections.

(iv) The base angles of an isosceles triangle are equal.

(v) Vertical angles are equal.

(vi) Space has constant curvature.

Table 8–1 may be used to compare the three geometries.

There also exist other fundamental comparisons that we have not considered. For example, trigonometric formulas in elliptic geometry are identical with the formulas in spherical trigonometry [4; 232–238]. When the radius of the sphere is increased without bound, the limiting forms of the trigonometric formulas in elliptic geometry are the trigonometric formulas in euclidean geometry. When the radius k of the sphere is replaced by ik, the trigonometric formulas in elliptic geometry become the trigonometric formulas in hyperbolic geometry. Other comparisons may be found in detailed treatments of noneuclidean geometry, such as [4].

Euclidean, elliptic, and hyperbolic geometries appear from the above comparisons to have very different properties. However, in the neighborhood of any point, the three geometries approximately coincide. The geometry on a sphere in the neighborhood of a point P is essentially the same as the geometry in the neighborhood of the point P on the plane that is tangent to the sphere at P. The geometry in the neighborhood of a point in the geometry of circles orthogonal to a fixed circle is essentially the same as the euclidean geometry in the neighborhood of a point considered as an interior point of an absolute conic. The term "neighborhood of a point" appears to include here at least the portion of physical space in which man can make direct measurements. Thus physical space is locally euclidean; i.e., physical space is a euclidean space in the neighborhood of any point. Astronomical space may have small positive curvature (elliptic geometry), zero curvature (euclidean geometry) or small negative curvature (hyperbolic geometry). According to Einstein's General Theory of Relativity, astronomical space has small positive curvature wherever there is matter. We shall never be able to prove that empty space is euclidean. We may, as at present, prove that if it has a curvature different from zero, then that curvature is less than the smallest errors in the instruments used in our measurements.

We have observed that euclidean, elliptic, and hyperbolic geometries are equally consistent and in a sense equally likely as the geometry of empty space. Euclidean geometry is at least a close approximation to the geometry of the portion of physical space in which man is able to make measurements. There is some evidence that astronomical space may have an elliptic geometry wherever there is matter. Also, there is no measurable reason for discarding the

TABLE 8–1

	Euclidean	Elliptic	Hyperbolic	
Any two lines intersect in	at most one	one	at most one	point.
All lines are	intersecting or parallel	intersecting	intersecting, parallel, or noninter- secting.	
Every line is associated with	one	no	two	ideal points.
Given any line m and any point P that is not on m, there exist	one	no	two	lines through P and parallel to m.
All points asso- ciated with a line are	real or ideal	real	real, ideal, or ultra-ideal.	
Every line	is	is not	is	separated into two segments by a point.
Parallel lines	are equidis- tant	do not exist	converge in one direc- tion, diverge in the other.	
If a line inter- sects one of two parallel lines, it	must	_____	may not	intersect the other.
The valid Saccheri hy- pothesis is the	right-angle	obtuse-angle	acute-angle	hypothesis.
Two lines per- pendicular to the same line are	parallel	intersecting	noninter- secting.	
The angle sum of a triangle is	equal to	greater than	less than	two right angles.
The area of a triangle is	independent	proportional to the excess	proportional to the defi- ciency	of its angle sum.

TABLE 8–1 (*Cont'd*)

	Euclidean	Elliptic	Hyperbolic
Two triangles with corresponding angles equal are	similar	congruent	congruent.
The areas of triangles are	unbounded	bounded	bounded.
There are natural units of	angle	angle, length, and area	angle, length, and area.
Curvature (Section 7–10) of space is constant and	zero	positive	negative.

possibility that empty space has a hyperbolic geometry. In view of this uncertainty as to the nature of the geometry of the space in which we live, we are free to select the geometry that is easiest to work with and the most productive in the sense of having useful properties (theorems). It is not sufficient to select projective geometry with its duality and inherent simplicity, since in this case we would still have the three possible subgeometries: euclidean, elliptic, and hyperbolic geometries. Accordingly, we shall undoubtedly continue to use euclidean geometry with the intuitive concepts of rigid motions. Since the ideal plane associated with euclidean three-space has an elliptic geometry, we need to understand both euclidean and elliptic geometries. Finally, as an aid to man's never-ending search for knowledge of the space in which he lives, all possibilities (euclidean, elliptic, and hyperbolic) should be understood. In this text an attempt has been made to introduce the fundamental concepts upon which all geometries are based and which are necessary if we are to understand the relation of euclidean geometry (high-school geometry) to other geometries and to the space in which we live.

CHAPTER 9

TOPOLOGY

Topology is often called "analysis situs," the analysis of position. It is concerned with the relative positions of elements. In this chapter we shall introduce several intuitive aspects of topology and consider some of the properties that are invariant under topological transformations. Throughout the chapter we shall strive only for an intuitive understanding of the fundamental concepts of topology and shall leave detailed discussions of topological concepts and spaces for specialized courses in topology. Our primary goal is an appreciation for topology as a geometry from which projective geometry may be obtained as a special case.

9–1 Topology. The word geometry probably brings to most readers of this text visualizations of points, lines, congruent triangles, and other recollections from a secondary-school course in euclidean plane geometry. However, we have seen that there exist more general meanings for the word geometry. The elements need not be ordinary points, lines, and planes. They may remain undefined or they may even be students in a class.

Consider a classroom with twenty-five students and twenty-five chairs. The students may change seats in many ways without moving the chairs and such that after each of the changes every student has a chair and every chair is occupied by a student, i.e., such that there is always a one-to-one correspondence between the set of students and the set of chairs. We shall use this concept to obtain a correspondence of the set of chairs with itself. Let each of the students be seated in one of the chairs. The resulting one-to-one correspondence between the chairs C_j and the students S_j may be indicated by the subscripts. Suppose that each student S_j sits in chair C_j for $j = 1, 2, \cdots, 25$. Next, we allow the students to leave their present seats and seat themselves (one in each chair) in any manner that they wish. Let us designate the new seat of student S_j by C_j'. The correspondence of seats C_j with seats C_j' is a one-to-one correspondence of the set of chairs with itself. Such a correspondence is often called a *biunique correspondence*, since each

C_j corresponds to a unique C_j' and each C_j' is associated with a unique C_j. We shall think of this correspondence as giving rise to a transformation $C_j \rightarrow C_j'$ ($j = 1, 2, \cdots, 25$) of the set of chairs into itself. Finally, in order to illustrate continuity, the basic property of topological transformations, we shall consider the chairs around any given chair as neighbors of that chair. Suppose that the neighbors of C_j are C_r, C_s, and C_t. Then if the neighbors of C_j' are C_r', C_s', and C_t', the transformation takes the neighbors of C_j into the neighbors of C_j', or, as we usually say, the neighborhood of C_j corresponds to the neighborhood of C_j'. A transformation that takes the neighborhood of C_j into the neighborhood of C_j' for all elements C_j is called a *continuous transformation*. A continuous transformation of C_j into C_j' such that the inverse transformation of C_j' into C_j is also a continuous transformation is called a *bicontinuous transformation*. A transformation that is biunique and bicontinuous is a *topological transformation*.

The above definitions of biunique correspondences, bicontinuous transformations, and topological transformations may be applied to very general sets of points, subject to certain conditions upon the space in which the points are considered. For example, the following definition of a continuous transformation assumes that there is a distance function in the space under consideration. More general definitions and applications may be found in texts on topology. Consider a correspondence of P to P' between a set S of elements P and a set S' of elements P'. The correspondence is biunique if each element of S corresponds to exactly one element of S', and, conversely, each element of S' is the correspondent of exactly one element of S. This is the basic property of all one-to-one correspondences. Continuous transformations may, as before, be defined in terms of points and neighborhoods of points. On a line, any segment $(p-b, p+c)$ where b and c are positive numbers may be visualized as a neighborhood of the point P with coordinate p. On a plane, the interior points of any circle of positive radius and having P as center may be taken as a neighborhood of P. In space, the interior points of any sphere of positive radius and having P as center form a neighborhood of P. A biunique transformation of a set S of points P into a set S' of points P' is a continuous transformation of P into P' if, corresponding to every neighborhood N' of P', there exists a neighborhood N of P such that all points of N correspond to points of N'.   It is a continuous transformation of S into S' if it is continuous for all points P of S.   As before, the transformation of S into S' is bicontinuous if it is continuous and the inverse transformation of S' into S is also continuous.   The transformation of S into S' is a topological transformation if it is biunique and bicontinuous.

Let us now consider a few examples of transformations.   Consider the points C of a circle of radius three units with center at the origin in ordinary euclidean plane geometry and the points S of the square having its sides on the lines $x = 1$, $y = 1$, $x = -1$ and $y = -1$.   Let the correspondence between the points C and the points S be obtained by associating with each point C the point S on the radius OC where O is the origin.   This correspondence is biunique and bicontinuous. The transformation of the circle into the square is a topological transformation.   The circle and the square are *topologically equivalent figures* or *homeomorphic figures* (Section 9–2), since there exists a topological transformation of the circle into the square.   Intuitively, the circle can be shrunk onto the square without any cutting, tearing, or folding.   In general, subject to a few unnecessary limitations due to the nature of the space in which we live, any continuous transformation may be performed without any cutting or tearing apart of the figures and without folding together any points that did not originally coincide.   From this point of view, we may visualize topological transformations on the plane by thinking of a plane as a rubber sheet that can be stretched or shrunk as desired but cannot be torn, cut, or folded together.   Then any regular polygon may be shrunk onto a circle or any other simple closed curve (Section 9–2). A unit square is topologically equivalent to a circle of radius 500 miles.   The equator of the earth is topologically equivalent to the coastline of Australia or even Iceland.

The above examples indicate that size and shape are not invariant in topology.   Indeed, topology is a geometry in which size and shape have no meaning.   It is interesting and possibly surprising to some readers to find that there exists a geometry in which "an inch is as good as a mile."   Perhaps it is even more surprising that such a geometry has importance in many advanced mathematical theories. For example, the solutions of algebraic equations and the evaluation of integrals may often be studied as topological problems.   Although the details of these applications are beyond the scope of the present discussion, they are cited here to emphasize that topology has importance not only as a generalization of euclidean geometry but also in relation to other advanced mathematical theories.

When topology is considered as a generalization of euclidean geometry, we find that the group of topological transformations has the group of projective transformations as a subgroup and therefore has the group of euclidean transformations as a subgroup. Projective geometry evolved as a geometry based upon the incidence and existence of points and lines. There are also other basic properties, such as separation, but, especially on a real projective plane, the concept of a projectivity as a collineation (i.e., a correspondence of lines with lines) is of primary importance. From this point of view topology is a generalization of projective geometry in which incidence relations are invariant but lines no longer need to correspond to lines. Intuitively, lines may correspond to simple arcs of curves. The invariance of incidence relations provides a basis for the concept of topology as "analysis of position" and also for the concept of topology as a geometry based primarily upon continuity (invariance of neighborhoods). Projective geometry may be considered as a special case of topology in which lines correspond to lines; euclidean geometry may be considered as a special case of topology in which lines correspond to lines, parallel lines correspond to parallel lines, and the euclidean distance function is invariant.

EXERCISES

1. Give five common plane curves that are topologically equivalent to a unit circle.

2. The correspondence between the points on the xy-plane and the points on the x-axis obtained by associating with each point (x,y) the point $(x,0)$ is not biunique. Explain the above statement and give a second example of a correspondence that is not biunique.

3. Give an example of a transformation that is not a topological transformation.

4. Consider the set S of points on the circumference of a unit circle and give a correspondence of the set S into itself that is
   (a) not biunique,
   (b) biunique but not continuous,
   (c) biunique and continuous.

5. Repeat Exercise 4 for the points in the first quadrant of an ordinary euclidean plane.

FIG. 9–1

9–2 Homeomorphic figures. We now adopt the terminology "homeomorphic figures" in place of the longer term "topologically equivalent figures." Throughout this section we shall be primarily concerned with examples of homeomorphic figures. Later we shall be concerned with the common properties of all figures that are homeomorphic to a given figure. In this regard it is important to observe that a figure B is homeomorphic to a figure A if and only if the figure A is homeomorphic to the figure B (Exercise 1). Also, two figures that are homeomorphic to the same figure are homeomorphic to each other (Exercise 2).

We have already observed that a regular polygon is homeomorphic to a circle. In general, any plane curve that is homeomorphic to a circle is a *simple closed curve.* Several simple closed curves are shown in Fig. 9–1. We may visualize the wide range of possible simple closed curves by joining the ends of a piece of string, placing the string on a flat surface (possibly a table top) in the form of a circle, and then deforming the figure formed by the string subject to the conditions that the string cannot be cut, broken, or made to coincide at any of its distinct points (no tangencies and no crossings). This approach to simple closed curves illustrates the continuous transformation of the circle into any other simple closed curve as well as the variety of possibilities. For example, three points of the circle of string may be selected and pulled away from the center such that a triangle is formed, and the continuous transition of the figure from a circle to a triangle may be clearly observed. Similarly, polygons of n sides may be obtained by selecting n points for any positive integer n greater than two. Furthermore, the new figures do not need to be polygons and may indeed take many surrealistic shapes.

From a slightly different point of view, a plane figure is a simple closed curve if one may start at any point of the figure, traverse the complete figure passing through each point exactly once, and thereby

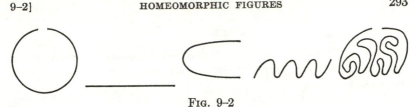

FIG. 9–2

return to the starting point. A parabola is not a closed curve in the euclidean plane. A straight line is not a closed curve in the euclidean plane. A figure eight is not a simple curve since it crosses itself or, in the above terminology, one must pass through one of the points twice in traversing the curve.

A sphere is homeomorphic to a cube, to any regular polyhedron, and to many other figures. We define any polyhedron that is homeomorphic to a sphere to be a *simple polyhedron*. Thus a brick is a simple polyhedron, but a builder's block with one or more holes in it is not a simple polyhedron.

A circle with a point removed, i.e., a *punctured circle*, is homeomorphic to a straight line and many other figures called *simple arcs* (Fig. 9–2). The correspondence between a punctured circle and a straight line may be indicated by drawing the line tangent to the circle at the point diametrically opposite the removed point P and making a point R of the punctured circle correspond to a point R' on the line if and only if R and R' are collinear with P where P is not considered as a point of the punctured circle (Fig. 9–3). Similarly, a sphere with a point removed, i.e., a *punctured sphere*, is homeomorphic to a plane (Exercise 5). Several other examples of homeomorphic figures will be considered in the following exercises. In the next section we shall consider a common property of all curves homeomorphic to a circle, i.e., of all simple closed curves.

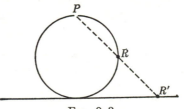

FIG. 9–3

EXERCISES

1. Use the definition of a topological transformation and prove that a figure B is homeomorphic to a figure A if and only if the figure A is homeomorphic to the figure B.

2. Prove that two figures that are homeomorphic to the same figure are homeomorphic to each other.

3. Draw three plane curves satisfying each of the following conditions:
   (a) not closed,
   (b) closed and not simple,
   (c) closed and simple.

4. Describe a homeomorphism between a doughnut and the builder's block with one hole in it.

5. Draw a figure indicating a homeomorphism between a punctured sphere and a plane.

6. Describe a homeomorphism between a doughnut and a teacup with one handle.

7. Draw three curves that are homeomorphic to
   (a) a circle,   (b) a straight line,   (c) a figure eight.

8. Describe three surfaces that are homeomorphic to
   (a) a cube,
   (b) a punctured sphere,
   (c) a teacup with one handle,
   (d) a sugar bowl with two handles.

9. Sketch five homeomorphic closed curves that cross themselves:
   (a) not at all,   (b) once,   (c) twice,   (d) three times.

10. Describe five surfaces that are homeomorphic to a sphere.

11. Describe five surfaces that are homeomorphic to the surface of a doughnut.

9–3 Jordan Curve Theorem. Topology is concerned with the common properties of all figures that are homeomorphic to a given figure. In particular, it is concerned with the common properties of all figures that are homeomorphic to a circle, i.e., all simple closed curves. The following theorem states one of the most important of these properties.

THEOREM 9–1. JORDAN CURVE THEOREM. *Any simple closed curve in a plane divides that plane into exactly two regions.*

Essentially, the Jordan Curve Theorem states that every simple closed curve on a euclidean plane has an inside and an outside; i.e.,

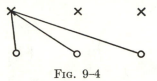

Fig. 9–4

it divides the plane into two regions. The curve is a common boundary of the two regions and is not considered a part of either region. The original proof of this theorem was neither short nor simple. A proof of the special case of the theorem involving polygons may be found in [3; 267–269]; more general proofs may be found in specialized texts on topology. We shall not consider a formal proof of this theorem but rather shall seek an intuitive understanding of the theorem through a discussion of a few of its applications. These applications are based upon the fact that on the plane one cannot cross from a point of one of the two regions to a point of the other region without crossing the curve. The Jordan Curve Theorem is a very powerful theorem and yet a very simple theorem. It is independent of the size and shape of the curve. It is a topological theorem.

How can such a simple theorem have any significance? It may be used to prove the validity of Euclid's assumption that any line segment joining the center of a circle to a point outside the circle must contain a point of the circle. It provides a basis for Venn diagrams in any two-valued logic. For example, true statements may correspond to points inside the curve, false statements to points outside the curve. This simple theorem regarding the existence of an inside and an outside of any simple closed curve may also be used to answer questions raised by a problem on which most of you have probably spent many hours.

Consider three houses in a row and three other objects in a second row. The problem is to join each house to each of three objects by an arc in such a way that no two arcs cross. Several years ago I tried this problem when three houses and three utilities were involved. Tucker and Bailey [14; 23] considered three houses, a haystack, a well, and a dovecote. Whatever the objects are called, the problem is equivalent to joining each cross to each circle in Fig. 9–4. It is easy to designate paths from one cross, say the one on the left, to each of the three circles. One can also designate the

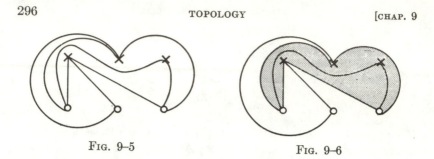

FIG. 9–5   FIG. 9–6

three paths from the second cross as in Fig. 9–5. Then one can designate two of the paths from the third cross as in Fig. 9–5, but it is not possible on a euclidean plane to draw the path from the third cross to the remaining circle. This assertion is based upon the fact that the simple closed curve indicated in Fig. 9–6 divides the plane into two regions; the third cross is inside the curve (shaded area), the remaining circle is outside, and the two cannot be joined without crossing the curve.

Another problem that is based upon the properties of a simple closed curve may be found in [9; 274–276]. A Persian Caliph reportedly endeavored to select the best suitor for his daughter by posing two problems. The problems appear very similar at first sight. The suitors were asked to draw arcs joining the corresponding numbers in each of two figures (Figs. 9–7 and 9–8) in such a way that the arcs do not intersect with each other or the given figures. In the case of Fig. 9–7 the problem is trivial. However, we can be confident that the Caliph's daughter died an old maid if her father insisted upon a solution of the second problem. The first and second pairs of numbers may be joined, but the third pair cannot be joined. This assertion is also based upon the properties of a simple closed curve as indicated in Fig. 9–9. One element of the third pair is inside the curve (shaded area) and the other is outside. By the Jordan Curve Theorem the two points cannot be joined in the plane without crossing the curve.

The Jordan Curve Theorem also holds when the curve is drawn upon a sphere instead of a plane. In this case it does not matter which region is called the inside of the curve and which region is called the outside, since the points of the two regions are homeomorphic. When the simple closed curve is on a euclidean plane, it is customary to call the bounded region the inside and the unbounded

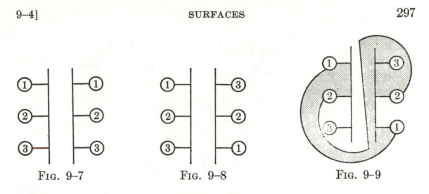

FIG. 9–7   FIG. 9–8   FIG. 9–9

region the outside. This distinction is intuitively evident even though we have not defined the concept of a bounded region.

We have observed that any simple closed curve divides a euclidean plane into two regions. Furthermore, if AB is any arc on the plane where A is in one region and B is in the other region relative to a given simple closed curve, then the arc AB contains a point of the simple closed curve. This property of simple closed curves provides one basis for our definition of genus in the next section.

EXERCISES

1. Solve the Caliph's problem in Fig. 9–7.

2. Use the Jordan Curve Theorem and prove that no matter how the lines are drawn subject to the conditions of the problem, it is not possible to solve the Caliph's problem in Fig. 9–8.

3. As in Exercise 2, prove that given any six distinct coplanar points A, B, C, D, E, F, there cannot exist nonintersecting arcs $AD, AE, AF, BD, BE, BF, CD, CE,$ and CF on the plane of the given points.

9–4 Surfaces. We now shift our attention from curves to surfaces such as planes, surfaces of spheres, surfaces of doughnuts, surfaces of pretzels, and regions on these surfaces. Although we shall not attempt to classify all surfaces, a few elementary definitions will be helpful as we consider some of the topological invariants of surfaces.

A surface such that any two of its points may be joined by an arc lying completely on the surface is called a *connected surface*. Thus any fixed point P of a connected surface may be joined to any point Q of the surface by an arc lying on that surface. A surface that is not connected is said to be *separated*. The following are examples of separated surfaces: the total surface of two distinct parallel planes,

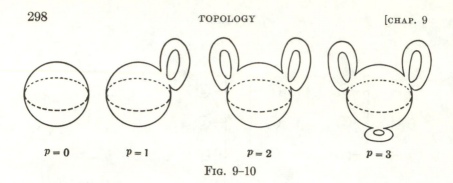

$p = 0$ $p = 1$ $p = 2$ $p = 3$

FIG. 9–10

the total surface of two nonintersecting spheres, the total surface of the treads of the four tires on an automobile.

A simple closed curve and its interior (or bounded) region is called a *closed bounded surface*, or *disk*. The interior region without the simple closed curve is called an *open bounded surface*. It is open in the sense that there exists a boundary, but the points of the boundary are not included in the surface under consideration (namely, the interior region). Intuitively, a surface that includes all the points on its boundary is *closed;* a surface for which there exists at least one boundary point that is not considered as a point of the surface is *not closed*. A formal definition of a closed surface may be made in terms of sequences of points. A separated surface is closed if and only if each of its "component parts" is closed. Unless otherwise specified we shall hereafter use the word surface to mean connected surface.

A surface that does not have an edge or boundary, such as a sphere, is called a *closed surface without a boundary*. A euclidean plane is a surface without a boundary in euclidean space but is not a closed surface, since it is homeomorphic to a punctured sphere and therefore to a disk without its boundary. The surface of a doughnut (technically a *torus*), a pretzel, or of any solid in euclidean space is a closed surface without a boundary. Each such surface is homeomorphic to the surface of a sphere with handles (Fig. 9–10).

The *genus* of a surface without a boundary is defined as the largest number of nonintersecting simple closed curves that can be drawn on the surface without separating it. The Jordan Curve Theorem thus states that the genus of a plane is zero. Similarly, a sphere has genus zero; a doughnut has genus 1; and a sphere with p handles has genus p (Fig. 9–10), since there exist simple closed curves around

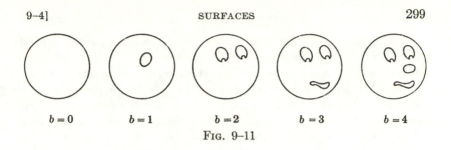

$b = 0$ $b = 1$ $b = 2$ $b = 3$ $b = 4$

Fig. 9–11

each handle that do not separate the surfaces, and, after these simple closed curves have been deleted, the surfaces are connected. In general, any ordinary closed surface without a boundary is homeomorphic to a sphere with p handles for some nonnegative integer p and is said to have genus p. The significance of the genus is indicated by the fact that for the ordinary closed surfaces without boundaries any two surfaces with the same genus are homeomorphic. Also, when the solution of algebraic equations and the evaluation of integrals are considered as topological problems, the genus of a related surface may be used to interpret each problem.

Let us next consider a few closed surfaces with boundaries. Any closed connected surface with a boundary is homeomorphic to a disk with b holes for some nonnegative integer b and is said to have *Betti number* b (Fig. 9–11). Technically, the Betti numbers of closed surfaces with boundaries may be defined in terms of crosscuts. A *crosscut* is a simple arc joining two distinct points on the boundary of a surface. The Betti number of any closed surface with a boundary may be defined as the largest number of nonintersecting crosscuts that can be drawn without separating the surface. Then, as in Fig. 9–11, a disk has Betti number 0, a ring-shaped region bounded by two concentric circles has Betti number 1, and, in general, a disk with b holes in it has Betti number b (Fig. 9–12).

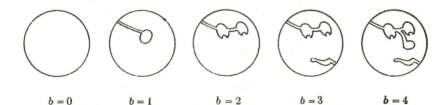

$b = 0$ $b = 1$ $b = 2$ $b = 3$ $b = 4$

Fig. 9–12

The genus p and the Betti number b of a surface are topological invariants; that is, any two homeomorphic surfaces have the same genus and the same Betti number whenever these numbers are defined. We have defined the genus of a closed connected surface without a boundary and defined the Betti number of a closed connected surface with a boundary. We shall not attempt complete definitions. The above definitions may be extended somewhat by considering simple closed curves formed by two crosscuts and two segments of the boundary. Under the extended definitions, we have $p = b/2$ when b is even.

The genus and the Betti number b of a surface are related to a third topological invariant — the connectivity h of the surface. A disk is said to be *simply connected* and to have connectivity $h = 1$. A ring is said to be *doubly connected* and to have connectivity $h = 2$. A disk with $m - 1$ holes is said to be *m-tuply connected* and to have connectivity $h = m$. Thus, in general, if a closed (connected) surface with a boundary has Betti number b, it has connectivity $h = b + 1$. When b is even, we also have $b = 2p$. For example, a disk with 6 holes in it has Betti number $b = 6$, connectivity $h = 7$, and genus $p = 3$.

The significance of the genus has already been mentioned. The concept of Betti number has been generalized for many types of surfaces and for spaces of arbitrary dimensions. Both concepts are topological invariants and are important in advanced mathematical theories. We shall leave all further details such as the relationship between the genus and the Betti number of a curve or the relationship between the Betti number of a surface and the Betti number of its boundary for texts in topology.

EXERCISES

1. Describe three surfaces that are
   - (a) connected,
   - (b) separated,
   - (c) closed,
   - (d) not closed,
   - (e) bounded,
   - (f) not bounded,
   - (g) connected and closed,
   - (h) separated and closed,
   - (i) connected and not closed,
   - (j) separated and not closed,
   - (k) connected and bounded,
   - (l) separated and bounded,
   - (m) connected and not bounded,
   - (n) separated and not bounded.
2. Describe three closed surfaces without boundaries having genus
   - (a) 0,
   - (b) 1.

3. Draw three closed surfaces with boundaries having genus
    (a) 0,   (b) 1,   (c) 2,   (d) 3.

4. Define the Betti number of a curve as the largest number of cuts (use interior points of arcs rather than vertices) that can be made in the curve without separating it and draw three curves having Betti number
    (a) 0,   (b) 1,   (c) 2,   (d) 3,   (e) 5.

9–5 Euler's Formula.   The original formula of Euler is concerned with the numbers of vertices, edges, and faces of a simple polyhedron in euclidean space.   Generalizations of the formula, as well as the original, are very important in the study of topology, since the Euler characteristic obtained by the formula is a topological invariant.   A polyhedron is a simple polyhedron if it is homeomorphic to a sphere (Section 9–2).   Euler's formula states that for any simple polyhedron with V vertices, E edges, and F faces we have the relation

$$V - E + F = 2.$$

An elementary intuitive proof of the formula may be given [3; 236–240], although we shall not do so.

An important application of the formula may be found in the derivation of the possible regular polyhedra in euclidean space [3; 240].   A polyhedron is a *regular polyhedron* if its faces are congruent regular polygons and the polyhedron is convex (i.e., no line segment joining points on the surface of the polyhedron contains an exterior point of the polyhedron).   Thus a cube is a regular polyhedron. In general, suppose a regular polyhedron has F faces where each face is a regular n-sided polygon.   Suppose also that r of the E edges of the polyhedron meet at each vertex.   Then since each face has n edges and each edge is the boundary of two faces, $nF = 2E$. Similarly, since each vertex is on r edges and each edge joins two vertices, $rV = 2E$.   Then all regular polyhedra correspond to integral solutions of Euler's Formula, which may now be written in the form

$$2E/r - E + 2E/n = 2.$$

In other words, the integral values of n and r satisfying this relation for integral values of E give rise to regular polyhedra, and all regular polyhedra may be obtained in this way.

The above formula may also be written in the form

$$1/r + 1/n = 1/2 + 1/E.$$

Then, since every polygon has at least three edges, we shall seek
integral solutions of this equation where $n \geq 3$. Similarly, since
any vertex of a polyhedron is on at least three faces, we have $r \geq 3$.
When $n = 3$, we have $1/r - 1/6 = 1/E$, whence r is less than 6,
since E is positive. For $n = 3$ and $r = 3$, 4, and 5, we find $E = 6$, 12,
and 30 and $F = 4$, 8, and 20, corresponding to the regular polyhedrons
of 4, 8, and 20 faces, i.e., the tetrahedron, octahedron, and icosahedron
respectively. For $n > 3$ we must have $r = 3$, since if n and r were
both greater than 3, then E would not be positive. Thus the only
possible remaining case is for $n > 3$ and $r = 3$. In this case we
have $1/n - 1/6 = 1/E$, $n = 4$ or 5, whence $E = 12$ or 30 and $F = 6$
or 12 corresponding to the regular polyhedrons of 6 and 12 faces
respectively. We have used Euler's formula to prove that the only
possible regular polyhedra are those with 4, 6, 8, 12, or 20 faces; i.e.,
the only regular polyhedra are the tetrahedron, cube, octahedron,
dodecahedron, and icosahedron.

On any closed surface without boundary, Euler's Formula may be
expressed in the form

$$V - E + F = 2 - 2p,$$

where p is the genus of the surface (Section 9–4) and the number
$2 - 2p$ is called the *Euler characteristic* of the surface [3; 258–259].
The formula may also be expressed for n-dimensional spaces in
terms of Betti numbers (Section 9–4).

We have now considered several of the fundamental concepts of
topology — biunique correspondences, bicontinuous transformations,
topological transformations, homeomorphic figures, simple closed
curves, the Jordan Curve Theorem, genus, Betti number, connectivity, and Euler characteristic. Also we have used one of the important formulas of topology, Euler's Formula, to derive all possible
regular polyhedra in euclidean space. With this introduction to
the theories of topology, we now consider four topics based upon
problems that have attracted topologists and intrigued many non-
topologists for many years.

EXERCISES

1. Describe a polyhedron that is not a simple polyhedron.
2. Describe a surface with Euler characteristic
   (a) 2,   (b) 0.

9–6 Traversable networks. We define a *network* as a set of points and arcs of curves where the ends of the arcs are elements of the set of points. For example, the sides and vertices of any polygon form a network. The largest number of arcs that can be cut at interior points without separating the network is called the Betti number of the network (Exercise 4, Section 9–4). A connected network that does not contain any sets of points homeomorphic to a simple closed curve is called a *tree*.

A network is said to be *traversable* if there exists a point P of the network such that one may start at P and traverse (in one connected journey) the complete network by traversing each arc exactly once. The vertices may be traversed more than once. We shall consider the general criterion that was obtained by Euler for determining whether or not a network is traversable. A vertex that is an end point for an odd number of arcs is called an *odd vertex* of the network; a vertex that is an end for an even number of arcs is called an *even vertex* of the network. Since each arc has two ends, there must be an even number of odd vertices in any network. Euler found that the network was traversable if there were at most two odd vertices. If all the vertices are even, as in the case of a regular polygon, one may start at any vertex and traverse the network. In general, in order to traverse any network it is necessary to pass through each vertex that is on $2r$ arcs exactly r times. If there are two odd vertices, one must start at one of the odd vertices and will end up at the other odd vertex. If there are more than two odd vertices, then there are an even number, say $2k$, of odd vertices and it will require k journeys to traverse the graph.

One of the earliest and most famous topological problems — the Koenigsberg bridge problem — involved a traversable network. This problem was solved by Euler in 1735 when he devised the general theory discussed above. The problem is concerned with the bridges in the city of Koenigsberg. There was a river flowing through the city, two islands in the river, and seven bridges, as in Fig. 9–13. The people of Koenigsberg loved a Sunday stroll and thought that it would be nice to take a walk and to cross each bridge exactly once. But no matter where they started or what route they tried, they could not cross each bridge exactly once. This caused considerable discussion. Gradually it was observed that the basic problem was concerned with paths between the two sides of the river A, B and

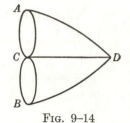

FIG. 9–13   FIG. 9–14

the two islands C, D, as in Fig. 9–14. With this geometric representation of the problem it was no longer necessary to discuss the problem in terms of walking across the bridges. Instead one could discuss whether or not the curve (Fig. 9–14) associated with the problem is traversable, i.e., whether or not one could start at some point of the curve and traverse each arc exactly once. The curve is often called the graph of the problem. In this form the problem could be considered by people who had never even been to Koenigsberg. The problem was solvable if and only if its graph was traversable.

The Koenigsberg bridge problem, like those in Section 9–3, has been attempted by many people. Euler used a network similar to that in Fig. 9–14 and found the underlying property of our geometry that rendered the problem impossible. The problem is impossible because the network has four odd vertices.

The Koenigsberg bridge problem led Euler to his general theory of traversable graphs and was one of the starting points of the whole subject of topology. Frequently we see in advanced mathematical theories only complicated manipulations and intricate statements involving precisely worded definitions and theorems. It is refreshing as well as enlightening to look back occasionally at the roots of the theory and see the problems that started great minds working for generalizations that have led to present theories. The Koenigsberg bridge problem, like all others in this chapter, is independent of the size and shape of the objects under consideration. It is a topological problem.

EXERCISES

1. Draw a traversable network.
2. Draw a network that is not traversable.
3. Find the Betti number of the network in Fig. 9–14.
4. Draw networks with Betti numbers

    (a) 0,   (b) 1,   (c) 2.   (d) 4.

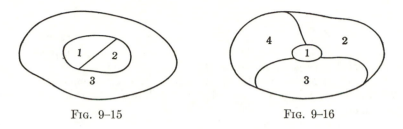

FIG. 9–15   FIG. 9–16

9–7 Four-color problem. The four-color problem is concerned with the number of colors that are required to color a map such that any two adjacent areas or countries have different colors. Two countries are considered to be adjacent if their boundaries have a common arc. They are not considered to be adjacent when their boundaries have only a single point in common or their common boundary consists only of isolated points. With these conventions it has been suggested that four colors are sufficient to color any map on a euclidean plane.

A preschool child will frequently color a whole sheet of paper a single color. It is possible to color a map consisting of a single country with one color. We may use two colors to color a green island in a blue ocean. Similarly, we may construct a map that may be colored with three colors. For example, we may consider an island divided into two countries, as in Fig. 9–15. We have seen that there exist maps that may be colored using one color, maps that require two colors, and maps that require three colors. Furthermore, if a country having more than one neighbor has an odd number of neighbors, as in Fig. 9–16, four colors are required. In the map of the United States, the state of Kentucky has seven neighbors. Thus four colors are required to color the map of the United States. The sufficiency of four colors has been established for maps having up to 35 countries but has not yet been proved in general.

The four-color problem is a topological problem independent of the size and shape of the countries. It has attracted considerable attention both amateur and professional. The reader may, as above, readily construct a map that may be colored using only three colors. He may also construct a map that requires four colors. His mathematical reputation is assured if he can prove in general that four colors are sufficient.

1. Describe a map covering a sphere that requires exactly one color.

2. Are all maps that may be obtained in Exercise 1 homeomorphic?

3. Repeat Exercise 1 for two colors.

4. Are all maps that may be obtained in Exercise 3 homeomorphic?

5. Repeat Exercise 1 for three colors. Are all maps obtained in this way homeomorphic?

6. Repeat Exercise 5 for four colors.

7. Discuss the possibility of constructing maps on a sphere such that at least five colors are needed to color the map according to the conventions of the four-color problem.

9–8 Fixed-point theorems. The fixed-point theorems are concerned with the number of points that are left fixed or invariant under topological transformations of a space into itself. For example, on a euclidean plane we may slide all points a fixed distance in a given direction and thereby move all points on the euclidean plane. However, when such a translation is considered on an affine plane, all points on the ideal line are invariant, or fixed points. Thus the fixed-point theorems depend upon the space that is under consideration.

Let us consider a disk (i.e., a circle and its interior) as a space. Does there exist a topological transformation of a disk into itself such that there are no fixed points? It is clear that the circular boundary of the disk can be transformed into itself without any fixed points simply by rotating the circle. However, this is not possible for the disk. If we rotate the disk and transform it into itself, we leave the center of the disk fixed. One of the fixed-point theorems implies that *every topological transformation of a disk into itself has at least one fixed point.* A very readable and intuitively evident proof of this theorem may be found in [3; 252–255]. We shall not attempt to prove it here. Some of the consequences of this fixed-point theorem and its generalizations may be readily visualized. If a dish of liquid is stirred so that the particles on the surface remain on the surface, then at any time t there is at least one particle that is in its original position although at different times the particular particles that are in their original positions may be different [3; 252].

Another fixed-point theorem implies that every orientation-preserving topological transformation of the surface of a sphere into itself has at least one fixed point. Accordingly, if we consider the winds as forces attempting a continuous transformation of the earth's surface into itself, there must always be at least one point on the earth's surface at which the wind is not blowing.

The fixed-point theorems provide another example of a problem having a representation that may be easily visualized and having a solution that has been generalized to provide fundamental results of an advanced mathematical theory.   The fixed-point theorems provide a basis for many existence proofs in mathematics.   For example, the *Fundamental Theorem of Algebra* (*every polynomial equation with complex coefficients has a complex number as a root*) may be proved using the first fixed-point theorem stated above.   In general, the fixed-point theorems, like the genus and many other topological concepts, have considerable significance in the applications of topology to other sciences as well as to other branches of mathematics.

9–9 Moebius strip.   The last topological problem that we shall consider is concerned with a surface that has several unusual properties. A fly walking on this page may obtain a new point of view by going over the edge of the page and walking on the back of it.   In other words, this page has two sides — front and back.   Similarly, a table top is a two-sided surface — top and bottom.   We are accustomed to surfaces that have two sides, and many readers will be surprised at some of the properties as well as the existence of one-sided surfaces. One popular writer has used the properties of a one-sided surface as a basis for a Paul Bunyan story [15].

The Paul Bunyan story is based upon the properties of a Moebius strip.   Since this one-sided surface is easily constructed and some of its properties are easily visualized using a model, let us consider it in some detail.   Consider a rectangular strip of paper (Fig. 9–17), paste the ends together such that the points A coincide and the points B coincide; i.e., give the paper a half twist before pasting the ends together.   This construction may be very easily performed

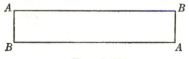

FIG. 9–17

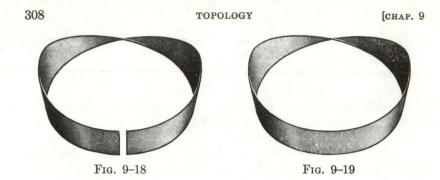

FIG. 9–18   FIG. 9–19

with a piece of gummed tape (gummed on one side).   Size is theoretically unimportant, but a strip an inch or two wide and about a
foot long is easy to handle.   We may construct a Moebius strip by
twisting the strip of gummed tape just enough to stick the gummed
edge of one end to the gummed edge of the other end (Fig. 9–18).
It is highly recommended that the reader construct a Moebius strip
and use it as a visual aid while considering some of the properties
of this unusual surface.

The Moebius strip is a one-sided surface.   A fly may start at a
point that appears to be on top of the surface and without crossing
an edge of the surface walk to a position that, when only a small
part of the surface is considered, appears to be on the "back" of the
surface.   In other words, the fly may start at a point, say P, on
the Moebius strip and walk around the strip until he appears to be
behind his previous position at P.   Note that in each case the fly
is at the point P since surfaces have only one layer of points.   However, unlike our usual two-sided surfaces, the Moebius strip has a
surface such that the fly could walk from one position at P to an
"opposite" position at P without crossing an edge of the surface.
This property of the Moebius strip is implied by calling it a one-
sided surface.

We next observe that the Jordan Curve Theorem does not hold
on a Moebius strip.   If one cuts along the dotted line in Fig. 9–19,
one gets two pieces.   However, if one cuts along the dotted line in
Fig. 9–20, one gets a single piece with two twists in it.   Even small
children enjoy this bit of paper cutting and are suitably impressed
with the result.   Now that we have cut a Moebius strip and found
only one piece, the question arises as to whether this can be done a

second time. Since most readers will be most vividly impressed by the result if they actually make a second cut of the type indicated in Fig. 9–20, we shall not answer the question here. Rather we leave all further cutting of Moebius strips to the leisurely enjoyment of the reader.

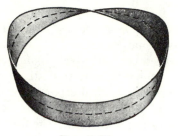

FIG. 9–20

EXERCISES

1. Make a Moebius strip.
2. Use a Moebius strip to demonstrate that
   (a) a Moebius strip is a one-sided surface,
   (b) a simple closed curve may separate a Moebius strip,
   (c) a simple closed curve need not separate a Moebius strip.
3. Make a single cut on a Moebius strip as indicated in Fig. 9–20.
4. Make two consecutive cuts of the type indicated in Fig. 9–20.
5. Make three consecutive cuts of the type indicated in Fig. 9–20.

9–10 Survey. In this chapter we have considered topology as a geometry based upon biunique and bicontinuous transformations. From the point of view of Klein, topology may be considered as a study of properties that are invariant under topological transformations, i.e., common properties of all figures that are homeomorphic to a given figure. In this regard topology is concerned with incidence relations (a point is on a curve, two figures have k points in common, . . .), genus, Betti number, connectivity, Euler characteristic, and other invariants. We have not attempted to discuss all topological invariants. Rather we have attempted to indicate what is meant by topology and to discuss a few of its properties in order to convince the reader that such a geometry exists and has mathematical significance. Several of the concepts that we have considered are discussed with excellent photographic illustrations in [12; 214–240].

Topology is concerned with points, curves, surfaces, and spaces of all sorts. Several common tricks used by magicians may be explained using topological concepts [7]. As in the case of the cutting of a Moebius strip, many topological properties may be visualized and demonstrated by children even though they involve difficult mathematical concepts. In this regard topological concepts may be used very effectively to impress upon both young people and adults that there is more to mathematics than formal algebraic manipulations and classical geometric constructions. In particular, all teachers of mathematics can use topological concepts to challenge and interest children of all ages. These concepts may also be used with superior students and mathematics clubs and for the general enrichment of one's teaching. They will help us encourage a genuine interest in fundamental principles of mathematics.

In conclusion, let us consider the hierarchy of geometries that we have discussed in our study of fundamental concepts of geometry. The array of transformations in Section 5–11 may now be extended and modified to obtain the following array of geometries:

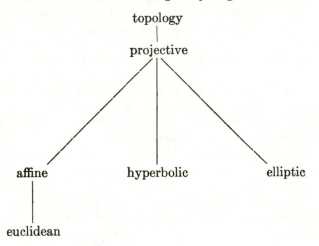

Topology is the most general geometry that we have considered. It is also the most recently developed geometry, since Euler's early work on the traversability of networks was done only about two centuries ago. Today, topology is still a very active and rapidly developing branch of mathematics.

BIBLIOGRAPHY

1. ARCHIBALD, R. C., Outline of the History of Mathematics, 6th ed., *American Mathematical Monthly*, **56**, Part II, 1949.

2. COOLIDGE, J. L., *A History of Geometrical Methods*. New York: Oxford University Press (Clarendon Press), 1940.

3. COURANT, R. and ROBBINS, H., *What is Mathematics?* New York: Oxford University Press, 1941.

4. COXETER, H. S. M., *The Real Projective Plane*. New York: McGraw-Hill Book Company, Inc., 1949.

5. DAVIS, D. R., *Modern College Geometry*. Cambridge, Mass.: Addison-Wesley Publishing Company, Inc., 1949.

6. EVES, HOWARD, *An Introduction to the History of Mathematics*. New York: Rinehart & Company, Inc., 1953.

7. GARDNER, MARTIN, Topology and Magic, *Scripta Mathematica*, **17**, 75–83, 1951.

8. HEATH, I. L., *The Thirteen Books of Euclid's Elements*. Vol. 1, 2nd ed. New York: Cambridge University Press, 1926; New York: Dover Publications. 1956.

9. ᴋᴀ̶ R, E. and NEWMAN, J., *Mathematics and the Imagination*. ... non and Schuster, Inc., 1940.

10. ᴛ, B. E., *Fundamental Concepts of Algebra*. Cambridge, Mass.: ...on-Wesley Publishing Company, Inc., 1953; New York: Dover Publications, 1982.

11. STABLER, E. R., *Introduction to Mathematical Thought*. Cambridge, Mass.: Addison-Wesley Publishing Company, Inc., 1953.

12. STEINHAUS, HUGO, *Mathematical Snapshots*. New York: Oxford University Press, 1950.

13. STRUIK, D. J., *A Concise History of Mathematics*. Vols. 1 and 2. New York: Dover Publications, 1948.

14. TUCKER, A. W. and BAILEY, H. S., JR., Topology, *Scientific American*, **182**, 18–24, 1950.

15. UPSON, W. H., Paul Bunyan versus the Conveyer Belt, *Ford Times*, **41**, 14–17, 1949.

16. VEBLEN, O. and YOUNG, J. W., *Projective Geometry*. Vol. 1. Boston: Ginn & Company, 1910.

17. VEBLEN, O. and YOUNG, J. W., *Projective Geometry*. Vol. 2. Boston: Ginn & Company, 1918.

18. WOLFE, H. E., *Introduction to Non-Euclidean Geometry*. New York: The Dryden Press, Inc., 1945.

19. YOUNG, J. W., *Fundamental Concepts of Algebra and Geometry*. New York: The Macmillan Company, 1911.

20. ZARISKI, OSCAR, The Fundamental Ideas of Abstract Algebraic

Geometry, *Proceedings of the International Congress of Mathematicians, Cambridge, Massachusetts, U.S.A., 1950.* Vol. II, 1952.

INDEX

A CATALOG OF SELECTED
DOVER BOOKS
IN SCIENCE AND MATHEMATICS

A CATALOG OF SELECTED
DOVER BOOKS
IN SCIENCE AND MATHEMATICS

QUALITATIVE THEORY OF DIFFERENTIAL EQUATIONS, V.V. Nemytskii and V.V. Stepanov. Classic graduate-level text by two prominent Soviet mathematicians covers classical differential equations as well as topological dynamics and ergodic theory. Bibliographies. 523pp. 5⅜ × 8½. 65954-2 Pa. $10.95

MATRICES AND LINEAR ALGEBRA, Hans Schneider and George Phillip Barker. Basic textbook covers theory of matrices and its applications to systems of linear equations and related topics such as determinants, eigenvalues and differential equations. Numerous exercises. 432pp. 5⅜ × 8½. 66014-1 Pa. $10.95

QUANTUM THEORY, David Bohm. This advanced undergraduate-level text presents the quantum theory in terms of qualitative and imaginative concepts, followed by specific applications worked out in mathematical detail. Preface. Index. 655pp. 5⅜ × 8½. 65969-0 Pa. $13.95

ATOMIC PHYSICS (8th edition), Max Born. Nobel laureate's lucid treatment of kinetic theory of gases, elementary particles, nuclear atom, wave-corpuscles, atomic structure and spectral lines, much more. Over 40 appendices, bibliography. 495pp. 5⅜ × 8½. 65984-4 Pa. $12.95

ELECTRONIC STRUCTURE AND THE PROPERTIES OF SOLIDS: The Physics of the Chemical Bond, Walter A. Harrison. Innovative text offers basic understanding of the electronic structure of covalent and ionic solids, simple metals, transition metals and their compounds. Problems. 1980 edition. 582pp. 6⅛ × 9¼. 66021-4 Pa. $15.95

BOUNDARY VALUE PROBLEMS OF HEAT CONDUCTION, M. Necati Özisik. Systematic, comprehensive treatment of modern mathematical methods of solving problems in heat conduction and diffusion. Numerous examples and problems. Selected references. Appendices. 505pp. 5⅜ × 8½. 65990-9 Pa. $12.95

A SHORT HISTORY OF CHEMISTRY (3rd edition), J.R. Partington. Classic exposition explores origins of chemistry, alchemy, early medical chemistry, nature of atmosphere, theory of valency, laws and structure of atomic theory, much more. 428pp. 5⅜ × 8½. (Available in U.S. only) 65977-1 Pa. $10.95

A HISTORY OF ASTRONOMY, A. Pannekoek. Well-balanced, carefully reasoned study covers such topics as Ptolemaic theory, work of Copernicus, Kepler, Newton, Eddington's work on stars, much more. Illustrated. References. 521pp. 5⅜ × 8½. 65994-1 Pa. $12.95

PRINCIPLES OF METEOROLOGICAL ANALYSIS, Walter J. Saucier. Highly respected, abundantly illustrated classic reviews atmospheric variables, hydrostatics, static stability, various analyses (scalar, cross-section, isobaric, isentropic, more). For intermediate meteorology students. 454pp. 6⅛ × 9¼. 65979-8 Pa. $14.95

RELATIVITY, THERMODYNAMICS AND COSMOLOGY, Richard C. Tolman. Landmark study extends thermodynamics to special, general relativity; also applications of relativistic mechanics, thermodynamics to cosmological models. 501pp. 5⅜ × 8½. 65383-8 Pa. $12.95

APPLIED ANALYSIS, Cornelius Lanczos. Classic work on analysis and design of finite processes for approximating solution of analytical problems. Algebraic equations, matrices, harmonic analysis, quadrature methods, much more. 559pp. 5⅜ × 8½. 65656-X Pa. $13.95

SPECIAL RELATIVITY FOR PHYSICISTS, G. Stephenson and C.W. Kilmister. Concise elegant account for nonspecialists. Lorentz transformation, optical and dynamical applications, more. Bibliography. 108pp. 5⅜ × 8½. 65519-9 Pa. $4.95

INTRODUCTION TO ANALYSIS, Maxwell Rosenlicht. Unusually clear, accessible coverage of set theory, real number system, metric spaces, continuous functions, Riemann integration, multiple integrals, more. Wide range of problems. Undergraduate level. Bibliography. 254pp. 5⅜ × 8½. 65038-3 Pa. $7.95

INTRODUCTION TO QUANTUM MECHANICS With Applications to Chemistry, Linus Pauling & E. Bright Wilson, Jr. Classic undergraduate text by Nobel Prize winner applies quantum mechanics to chemical and physical problems. Numerous tables and figures enhance the text. Chapter bibliographies. Appendices. Index. 468pp. 5⅜ × 8½. 64871-0 Pa. $11.95

ASYMPTOTIC EXPANSIONS OF INTEGRALS, Norman Bleistein & Richard A. Handelsman. Best introduction to important field with applications in a variety of scientific disciplines. New preface. Problems. Diagrams. Tables. Bibliography. Index. 448pp. 5⅜ × 8½. 65082-0 Pa. $12.95

MATHEMATICS APPLIED TO CONTINUUM MECHANICS, Lee A. Segel. Analyzes models of fluid flow and solid deformation. For upper-level math, science and engineering students. 608pp. 5⅜ × 8½. 65369-2 Pa. $13.95

ELEMENTS OF REAL ANALYSIS, David A. Sprecher. Classic text covers fundamental concepts, real number system, point sets, functions of a real variable, Fourier series, much more. Over 500 exercises. 352pp. 5⅜ × 8½. 65385-4 Pa. $10.95

PHYSICAL PRINCIPLES OF THE QUANTUM THEORY, Werner Heisenberg. Nobel Laureate discusses quantum theory, uncertainty, wave mechanics, work of Dirac, Schroedinger, Compton, Wilson, Einstein, etc. 184pp. 5⅜ × 8½. 60113-7 Pa. $5.95

INTRODUCTORY REAL ANALYSIS, A.N. Kolmogorov, S.V. Fomin. Translated by Richard A. Silverman. Self-contained, evenly paced introduction to real and functional analysis. Some 350 problems. 403pp. 5⅜ × 8½. 61226-0 Pa. $9.95

PROBLEMS AND SOLUTIONS IN QUANTUM CHEMISTRY AND PHYSICS, Charles S. Johnson, Jr. and Lee G. Pedersen. Unusually varied problems, detailed solutions in coverage of quantum mechanics, wave mechanics, angular momentum, molecular spectroscopy, scattering theory, more. 280 problems plus 139 supplementary exercises. 430pp. 6½ × 9¼. 65236-X Pa. $12.95

ASYMPTOTIC METHODS IN ANALYSIS, N.G. de Bruijn. An inexpensive, comprehensive guide to asymptotic methods—the pioneering work that teaches by explaining worked examples in detail. Index. 224pp. 5⅜ × 8½.   64221-6 Pa. $6.95

OPTICAL RESONANCE AND TWO-LEVEL ATOMS, L. Allen and J.H. Eberly. Clear, comprehensive introduction to basic principles behind all quantum optical resonance phenomena. 53 illustrations. Preface. Index. 256pp. 5⅜ × 8½.
65533-4 Pa. $7.95

COMPLEX VARIABLES, Francis J. Flanigan. Unusual approach, delaying complex algebra till harmonic functions have been analyzed from real variable viewpoint. Includes problems with answers. 364pp. 5⅜ × 8½.   61388-7 Pa. $8.95

ATOMIC SPECTRA AND ATOMIC STRUCTURE, Gerhard Herzberg. One of best introductions; especially for specialist in other fields. Treatment is physical rather than mathematical. 80 illustrations. 257pp. 5⅜ × 8½.   60115-3 Pa. $6.95

APPLIED COMPLEX VARIABLES, John W. Dettman. Step-by-step coverage of fundamentals of analytic function theory—plus lucid exposition of five important applications: Potential Theory; Ordinary Differential Equations; Fourier Transforms; Laplace Transforms; Asymptotic Expansions. 66 figures. Exercises at chapter ends. 512pp. 5⅜ × 8½.   64670-X Pa. $11.95

ULTRASONIC ABSORPTION: An Introduction to the Theory of Sound Absorption and Dispersion in Gases, Liquids and Solids, A.B. Bhatia. Standard reference in the field provides a clear, systematically organized introductory review of fundamental concepts for advanced graduate students, research workers. Numerous diagrams. Bibliography. 440pp. 5⅜ × 8½.   64917-2 Pa. $11.95

UNBOUNDED LINEAR OPERATORS: Theory and Applications, Seymour Goldberg. Classic presents systematic treatment of the theory of unbounded linear operators in normed linear spaces with applications to differential equations. Bibliography. 199pp. 5⅜ × 8½.   64830-3 Pa. $7.95

LIGHT SCATTERING BY SMALL PARTICLES, H.C. van de Hulst. Comprehensive treatment including full range of useful approximation methods for researchers in chemistry, meteorology and astronomy. 44 illustrations. 470pp. 5⅜ × 8½.   64228-3 Pa. $11.95

CONFORMAL MAPPING ON RIEMANN SURFACES, Harvey Cohn. Lucid, insightful book presents ideal coverage of subject. 334 exercises make book perfect for self-study. 55 figures. 352pp. 5⅜ × 8¼.   64025-6 Pa. $9.95

OPTICKS, Sir Isaac Newton. Newton's own experiments with spectroscopy, colors, lenses, reflection, refraction, etc., in language the layman can follow. Foreword by Albert Einstein. 532pp. 5⅜ × 8½.   60205-2 Pa. $9.95

GENERALIZED INTEGRAL TRANSFORMATIONS, A.H. Zemanian. Graduate-level study of recent generalizations of the Laplace, Mellin, Hankel, K. Weierstrass, convolution and other simple transformations. Bibliography. 320pp. 5⅜ × 8½.   65375-7 Pa. $8.95

THE ELECTROMAGNETIC FIELD, Albert Shadowitz. Comprehensive undergraduate text covers basics of electric and magnetic fields, builds up to electromagnetic theory. Also related topics, including relativity. Over 900 problems. 768pp. 5⅜ × 8¼. 65660-8 Pa. $18.95

FOURIER SERIES, Georgi P. Tolstov. Translated by Richard A. Silverman. A valuable addition to the literature on the subject, moving clearly from subject to subject and theorem to theorem. 107 problems, answers. 336pp. 5⅜ × 8½. 63317-9 Pa. $8.95

THEORY OF ELECTROMAGNETIC WAVE PROPAGATION, Charles Herach Papas. Graduate-level study discusses the Maxwell field equations, radiation from wire antennas, the Doppler effect and more. xiii + 244pp. 5⅜ × 8½. 65678-0 Pa. $6.95

DISTRIBUTION THEORY AND TRANSFORM ANALYSIS: An Introduction to Generalized Functions, with Applications, A.H. Zemanian. Provides basics of distribution theory, describes generalized Fourier and Laplace transformations. Numerous problems. 384pp. 5⅜ × 8½. 65479-6 Pa. $9.95

THE PHYSICS OF WAVES, William C. Elmore and Mark A. Heald. Unique overview of classical wave theory. Acoustics, optics, electromagnetic radiation, more. Ideal as classroom text or for self-study. Problems. 477pp. 5⅜ × 8½. 64926-1 Pa. $12.95

CALCULUS OF VARIATIONS WITH APPLICATIONS, George M. Ewing. Applications-oriented introduction to variational theory develops insight and promotes understanding of specialized books, research papers. Suitable for advanced undergraduate/graduate students as primary, supplementary text. 352pp. 5⅜ × 8½. 64856-7 Pa. $8.95

A TREATISE ON ELECTRICITY AND MAGNETISM, James Clerk Maxwell. Important foundation work of modern physics. Brings to final form Maxwell's theory of electromagnetism and rigorously derives his general equations of field theory. 1,084pp. 5⅜ × 8½. 60636-8, 60637-6 Pa., Two-vol. set $21.90

AN INTRODUCTION TO THE CALCULUS OF VARIATIONS, Charles Fox. Graduate-level text covers variations of an integral, isoperimetrical problems, least action, special relativity, approximations, more. References. 279pp. 5⅜ × 8½. 65499-0 Pa. $7.95

HYDRODYNAMIC AND HYDROMAGNETIC STABILITY, S. Chandrasekhar. Lucid examination of the Rayleigh-Benard problem; clear coverage of the theory of instabilities causing convection. 704pp. 5⅜ × 8¼. 64071-X Pa. $14.95

CALCULUS OF VARIATIONS, Robert Weinstock. Basic introduction covering isoperimetric problems, theory of elasticity, quantum mechanics, electrostatics, etc. Exercises throughout. 326pp. 5⅜ × 8½. 63069-2 Pa. $8.95

DYNAMICS OF FLUIDS IN POROUS MEDIA, Jacob Bear. For advanced students of ground water hydrology, soil mechanics and physics, drainage and irrigation engineering and more. 335 illustrations. Exercises, with answers. 784pp. 6⅛ × 9¼. 65675-6 Pa. $19.95

NUMERICAL METHODS FOR SCIENTISTS AND ENGINEERS, Richard Hamming. Classic text stresses frequency approach in coverage of algorithms, polynomial approximation, Fourier approximation, exponential approximation, other topics. Revised and enlarged 2nd edition. 721pp. 5⅜ × 8½.

65241-6 Pa. $14.95

THEORETICAL SOLID STATE PHYSICS, Vol. I: Perfect Lattices in Equilibrium; Vol. II: Non-Equilibrium and Disorder, William Jones and Norman H. March. Monumental reference work covers fundamental theory of equilibrium properties of perfect crystalline solids, non-equilibrium properties, defects and disordered systems. Appendices. Problems. Preface. Diagrams. Index. Bibliography. Total of 1,301pp. 5⅜ × 8½. Two volumes. Vol. I 65015-4 Pa. $14.95
Vol. II 65016-2 Pa. $14.95

OPTIMIZATION THEORY WITH APPLICATIONS, Donald A. Pierre. Broad-spectrum approach to important topic. Classical theory of minima and maxima, calculus of variations, simplex technique and linear programming, more. Many problems, examples. 640pp. 5⅜ × 8½. 65205-X Pa. $14.95

THE CONTINUUM: A Critical Examination of the Foundation of Analysis, Hermann Weyl. Classic of 20th-century foundational research deals with the conceptual problem posed by the continuum. 156pp. 5⅜ × 8½. 67982-9 Pa. $5.95

ESSAYS ON THE THEORY OF NUMBERS, Richard Dedekind. Two classic essays by great German mathematician: on the theory of irrational numbers; and on transfinite numbers and properties of natural numbers. 115pp. 5⅜ × 8½.

21010-3 Pa. $4.95

THE FUNCTIONS OF MATHEMATICAL PHYSICS, Harry Hochstadt. Comprehensive treatment of orthogonal polynomials, hypergeometric functions, Hill's equation, much more. Bibliography. Index. 322pp. 5⅜ × 8½. 65214-9 Pa. $9.95

NUMBER THEORY AND ITS HISTORY, Oystein Ore. Unusually clear, accessible introduction covers counting, properties of numbers, prime numbers, much more. Bibliography. 380pp. 5⅜ × 8½. 65620-9 Pa. $9.95

THE VARIATIONAL PRINCIPLES OF MECHANICS, Cornelius Lanczos. Graduate level coverage of calculus of variations, equations of motion, relativistic mechanics, more. First inexpensive paperbound edition of classic treatise. Index. Bibliography. 418pp. 5⅜ × 8½. 65067-7 Pa. $11.95

MATHEMATICAL TABLES AND FORMULAS, Robert D. Carmichael and Edwin R. Smith. Logarithms, sines, tangents, trig functions, powers, roots, reciprocals, exponential and hyperbolic functions, formulas and theorems. 269pp. 5⅜ × 8½. 60111-0 Pa. $6.95

THEORETICAL PHYSICS, Georg Joos, with Ira M. Freeman. Classic overview covers essential math, mechanics, electromagnetic theory, thermodynamics, quantum mechanics, nuclear physics, other topics. First paperback edition. xxiii + 885pp. 5⅜ × 8½. 65227-0 Pa. $19.95

HANDBOOK OF MATHEMATICAL FUNCTIONS WITH FORMULAS, GRAPHS, AND MATHEMATICAL TABLES, edited by Milton Abramowitz and Irene A. Stegun. Vast compendium: 29 sets of tables, some to as high as 20 places. 1,046pp. 8 × 10½.   61272-4 Pa. $24.95

MATHEMATICAL METHODS IN PHYSICS AND ENGINEERING, John W. Dettman. Algebraically based approach to vectors, mapping, diffraction, other topics in applied math. Also generalized functions, analytic function theory, more. Exercises. 448pp. 5⅜ × 8¼.   65649-7 Pa. $9.95

A SURVEY OF NUMERICAL MATHEMATICS, David M. Young and Robert Todd Gregory. Broad self-contained coverage of computer-oriented numerical algorithms for solving various types of mathematical problems in linear algebra, ordinary and partial, differential equations, much more. Exercises. Total of 1,248pp. 5⅜ × 8½. Two volumes.   Vol. I 65691-8 Pa. $14.95
                                                               Vol. II 65692-6 Pa. $14.95

TENSOR ANALYSIS FOR PHYSICISTS, J.A. Schouten. Concise exposition of the mathematical basis of tensor analysis, integrated with well-chosen physical examples of the theory. Exercises. Index. Bibliography. 289pp. 5⅜ × 8½.
                                                               65582-2 Pa. $8.95

INTRODUCTION TO NUMERICAL ANALYSIS (2nd Edition), F.B. Hildebrand. Classic, fundamental treatment covers computation, approximation, interpolation, numerical differentiation and integration, other topics. 150 new problems. 669pp. 5⅜ × 8½.   65363-3 Pa. $15.95

INVESTIGATIONS ON THE THEORY OF THE BROWNIAN MOVEMENT, Albert Einstein. Five papers (1905–8) investigating dynamics of Brownian motion and evolving elementary theory. Notes by R. Fürth. 122pp. 5⅜ × 8½.
                                                               60304-0 Pa. $4.95

CATASTROPHE THEORY FOR SCIENTISTS AND ENGINEERS, Robert Gilmore. Advanced-level treatment describes mathematics of theory grounded in the work of Poincaré, R. Thom, other mathematicians. Also important applications to problems in mathematics, physics, chemistry and engineering. 1981 edition. References. 28 tables. 397 black-and-white illustrations. xvii + 666pp. 6⅛ × 9¼.
                                                               67539-4 Pa. $16.95

AN INTRODUCTION TO STATISTICAL THERMODYNAMICS, Terrell L. Hill. Excellent basic text offers wide-ranging coverage of quantum statistical mechanics, systems of interacting molecules, quantum statistics, more. 523pp. 5⅜ × 8½.   65242-4 Pa. $12.95

ELEMENTARY DIFFERENTIAL EQUATIONS, William Ted Martin and Eric Reissner. Exceptionally clear, comprehensive introduction at undergraduate level. Nature and origin of differential equations, differential equations of first, second and higher orders. Picard's Theorem, much more. Problems with solutions. 331pp. 5⅜ × 8½.   65024-3 Pa. $8.95

STATISTICAL PHYSICS, Gregory H. Wannier. Classic text combines thermodynamics, statistical mechanics and kinetic theory in one unified presentation of thermal physics. Problems with solutions. Bibliography. 532pp. 5⅜ × 8½.
                                                               65401-X Pa. $12.95

ORDINARY DIFFERENTIAL EQUATIONS, Morris Tenenbaum and Harry Pollard. Exhaustive survey of ordinary differential equations for undergraduates in mathematics, engineering, science. Thorough analysis of theorems. Diagrams. Bibliography. Index. 818pp. 5⅜ × 8½. 64940-7 Pa. $16.95

STATISTICAL MECHANICS: Principles and Applications, Terrell L. Hill. Standard text covers fundamentals of statistical mechanics, applications to fluctuation theory, imperfect gases, distribution functions, more. 448pp. 5⅜ × 8½. 65390-0 Pa. $11.95

ORDINARY DIFFERENTIAL EQUATIONS AND STABILITY THEORY: An Introduction, David A. Sánchez. Brief, modern treatment. Linear equation, stability theory for autonomous and nonautonomous systems, etc. 164pp. 5⅜ × 8¼. 63828-6 Pa. $5.95

THIRTY YEARS THAT SHOOK PHYSICS: The Story of Quantum Theory, George Gamow. Lucid, accessible introduction to influential theory of energy and matter. Careful explanations of Dirac's anti-particles, Bohr's model of the atom, much more. 12 plates. Numerous drawings. 240pp. 5⅜ × 8½. 24895-X Pa. $6.95

THEORY OF MATRICES, Sam Perlis. Outstanding text covering rank, nonsingularity and inverses in connection with the development of canonical matrices under the relation of equivalence, and without the intervention of determinants. Includes exercises. 237pp. 5⅜ × 8½. 66810-X Pa. $7.95

GREAT EXPERIMENTS IN PHYSICS: Firsthand Accounts from Galileo to Einstein, edited by Morris H. Shamos. 25 crucial discoveries: Newton's laws of motion, Chadwick's study of the neutron, Hertz on electromagnetic waves, more. Original accounts clearly annotated. 370pp. 5⅜ × 8½. 25346-5 Pa. $10.95

INTRODUCTION TO PARTIAL DIFFERENTIAL EQUATIONS WITH APPLICATIONS, E.C. Zachmanoglou and Dale W. Thoe. Essentials of partial differential equations applied to common problems in engineering and the physical sciences. Problems and answers. 416pp. 5⅜ × 8½. 65251-3 Pa. $10.95

BURNHAM'S CELESTIAL HANDBOOK, Robert Burnham, Jr. Thorough guide to the stars beyond our solar system. Exhaustive treatment. Alphabetical by constellation: Andromeda to Cetus in Vol. 1; Chamaeleon to Orion in Vol. 2; and Pavo to Vulpecula in Vol. 3. Hundreds of illustrations. Index in Vol. 3. 2,000pp. 6⅛ × 9¼. 23567-X, 23568-8, 23673-0 Pa., Three-vol. set $41.85

CHEMICAL MAGIC, Leonard A. Ford. Second Edition, Revised by E. Winston Grundmeier. Over 100 unusual stunts demonstrating cold fire, dust explosions, much more. Text explains scientific principles and stresses safety precautions. 128pp. 5⅜ × 8½. 67628-5 Pa. $5.95

AMATEUR ASTRONOMER'S HANDBOOK, J.B. Sidgwick. Timeless, comprehensive coverage of telescopes, mirrors, lenses, mountings, telescope drives, micrometers, spectroscopes, more. 189 illustrations. 576pp. 5⅜ × 8¼. (Available in U.S. only) 24034-7 Pa. $9.95

SPECIAL FUNCTIONS, N.N. Lebedev. Translated by Richard Silverman. Famous Russian work treating more important special functions, with applications to specific problems of physics and engineering. 38 figures. 308pp. 5⅜ × 8½.
60624-4 Pa. $8.95

OBSERVATIONAL ASTRONOMY FOR AMATEURS, J.B. Sidgwick. Mine of useful data for observation of sun, moon, planets, asteroids, aurorae, meteors, comets, variables, binaries, etc. 39 illustrations. 384pp. 5⅜ × 8¼. (Available in U.S. only)
24033-9 Pa. $8.95

INTEGRAL EQUATIONS, F.G. Tricomi. Authoritative, well-written treatment of extremely useful mathematical tool with wide applications. Volterra Equations, Fredholm Equations, much more. Advanced undergraduate to graduate level. Exercises. Bibliography. 238pp. 5⅜ × 8½.
64828-1 Pa. $7.95

POPULAR LECTURES ON MATHEMATICAL LOGIC, Hao Wang. Noted logician's lucid treatment of historical developments, set theory, model theory, recursion theory and constructivism, proof theory, more. 3 appendixes. Bibliography. 1981 edition. ix + 283pp. 5⅜ × 8½.
67632-3 Pa. $8.95

MODERN NONLINEAR EQUATIONS, Thomas L. Saaty. Emphasizes practical solution of problems; covers seven types of equations. ". . . a welcome contribution to the existing literature. . . ."—*Math Reviews*. 490pp. 5⅜ × 8½. 64232-1 Pa. $11.95

FUNDAMENTALS OF ASTRODYNAMICS, Roger Bate et al. Modern approach developed by U.S. Air Force Academy. Designed as a first course. Problems, exercises. Numerous illustrations. 455pp. 5⅜ × 8½.
60061-0 Pa. $9.95

INTRODUCTION TO LINEAR ALGEBRA AND DIFFERENTIAL EQUATIONS, John W. Dettman. Excellent text covers complex numbers, determinants, orthonormal bases, Laplace transforms, much more. Exercises with solutions. Undergraduate level. 416pp. 5⅜ × 8½.
65191-6 Pa. $10.95

INCOMPRESSIBLE AERODYNAMICS, edited by Bryan Thwaites. Covers theoretical and experimental treatment of the uniform flow of air and viscous fluids past two-dimensional aerofoils and three-dimensional wings; many other topics. 654pp. 5⅜ × 8½.
65465-6 Pa. $16.95

INTRODUCTION TO DIFFERENCE EQUATIONS, Samuel Goldberg. Exceptionally clear exposition of important discipline with applications to sociology, psychology, economics. Many illustrative examples; over 250 problems. 260pp. 5⅜ × 8½.
65084-7 Pa. $7.95

LAMINAR BOUNDARY LAYERS, edited by L. Rosenhead. Engineering classic covers steady boundary layers in two- and three-dimensional flow, unsteady boundary layers, stability, observational techniques, much more. 708pp. 5⅜ × 8½.
65646-2 Pa. $18.95

LECTURES ON CLASSICAL DIFFERENTIAL GEOMETRY, Second Edition, Dirk J. Struik. Excellent brief introduction covers curves, theory of surfaces, fundamental equations, geometry on a surface, conformal mapping, other topics. Problems. 240pp. 5⅜ × 8½.
65609-8 Pa. $8.95

ROTARY-WING AERODYNAMICS, W.Z. Stepniewski. Clear, concise text covers aerodynamic phenomena of the rotor and offers guidelines for helicopter performance evaluation. Originally prepared for NASA. 537 figures. 640pp. 6⅛ × 9¼.
64647-5 Pa. $15.95

DIFFERENTIAL GEOMETRY, Heinrich W. Guggenheimer. Local differential geometry as an application of advanced calculus and linear algebra. Curvature, transformation groups, surfaces, more. Exercises. 62 figures. 378pp. 5⅜ × 8½.
63433-7 Pa. $8.95

INTRODUCTION TO SPACE DYNAMICS, William Tyrrell Thomson. Comprehensive, classic introduction to space-flight engineering for advanced undergraduate and graduate students. Includes vector algebra, kinematics, transformation of coordinates. Bibliography. Index. 352pp. 5⅜ × 8½. 65113-4 Pa. $8.95

A SURVEY OF MINIMAL SURFACES, Robert Osserman. Up-to-date, in-depth discussion of the field for advanced students. Corrected and enlarged edition covers new developments. Includes numerous problems. 192pp. 5⅜ × 8½.
64998-9 Pa. $8.95

ANALYTICAL MECHANICS OF GEARS, Earle Buckingham. Indispensable reference for modern gear manufacture covers conjugate gear-tooth action, gear-tooth profiles of various gears, many other topics. 263 figures. 102 tables. 546pp. 5⅜ × 8½. 65712-4 Pa. $14.95

SET THEORY AND LOGIC, Robert R. Stoll. Lucid introduction to unified theory of mathematical concepts. Set theory and logic seen as tools for conceptual understanding of real number system. 496pp. 5⅜ × 8¼. 63829-4 Pa. $12.95

A HISTORY OF MECHANICS, René Dugas. Monumental study of mechanical principles from antiquity to quantum mechanics. Contributions of ancient Greeks, Galileo, Leonardo, Kepler, Lagrange, many others. 671pp. 5⅜ × 8½.
65632-2 Pa. $14.95

FAMOUS PROBLEMS OF GEOMETRY AND HOW TO SOLVE THEM, Benjamin Bold. Squaring the circle, trisecting the angle, duplicating the cube: learn their history, why they are impossible to solve, then solve them yourself. 128pp. 5⅜ × 8½. 24297-8 Pa. $4.95

MECHANICAL VIBRATIONS, J.P. Den Hartog. Classic textbook offers lucid explanations and illustrative models, applying theories of vibrations to a variety of practical industrial engineering problems. Numerous figures. 233 problems, solutions. Appendix. Index. Preface. 436pp. 5⅜ × 8½. 64785-4 Pa. $10.95

CURVATURE AND HOMOLOGY, Samuel I. Goldberg. Thorough treatment of specialized branch of differential geometry. Covers Riemannian manifolds, topology of differentiable manifolds, compact Lie groups, other topics. Exercises. 315pp. 5⅜ × 8½. 64314-X Pa. $9.95

HISTORY OF STRENGTH OF MATERIALS, Stephen P. Timoshenko. Excellent historical survey of the strength of materials with many references to the theories of elasticity and structure. 245 figures. 452pp. 5⅜ × 8½. 61187-6 Pa. $11.95

GEOMETRY OF COMPLEX NUMBERS, Hans Schwerdtfeger. Illuminating, widely praised book on analytic geometry of circles, the Moebius transformation, and two-dimensional non-Euclidean geometries. 200pp. 5⅜ × 8¼.
63830-8 Pa. $8.95

MECHANICS, J.P. Den Hartog. A classic introductory text or refresher. Hundreds of applications and design problems illuminate fundamentals of trusses, loaded beams and cables, etc. 334 answered problems. 462pp. 5⅜ × 8½. 60754-2 Pa. $9.95

TOPOLOGY, John G. Hocking and Gail S. Young. Superb one-year course in classical topology. Topological spaces and functions, point-set topology, much more. Examples and problems. Bibliography. Index. 384pp. 5⅜ × 8¼.
65676-4 Pa. $9.95

STRENGTH OF MATERIALS, J.P. Den Hartog. Full, clear treatment of basic material (tension, torsion, bending, etc.) plus advanced material on engineering methods, applications. 350 answered problems. 323pp. 5⅜ × 8½. 60755-0 Pa. $8.95

ELEMENTARY CONCEPTS OF TOPOLOGY, Paul Alexandroff. Elegant, intuitive approach to topology from set-theoretic topology to Betti groups; how concepts of topology are useful in math and physics. 25 figures. 57pp. 5⅜ × 8½.
60747-X Pa. $3.50

ADVANCED STRENGTH OF MATERIALS, J.P. Den Hartog. Superbly written advanced text covers torsion, rotating disks, membrane stresses in shells, much more. Many problems and answers. 388pp. 5⅜ × 8½. 65407-9 Pa. $9.95

COMPUTABILITY AND UNSOLVABILITY, Martin Davis. Classic graduate-level introduction to theory of computability, usually referred to as theory of recurrent functions. New preface and appendix. 288pp. 5⅜ × 8½. 61471-9 Pa. $7.95

GENERAL CHEMISTRY, Linus Pauling. Revised 3rd edition of classic first-year text by Nobel laureate. Atomic and molecular structure, quantum mechanics, statistical mechanics, thermodynamics correlated with descriptive chemistry. Problems. 992pp. 5⅜ × 8½. 65622-5 Pa. $19.95

AN INTRODUCTION TO MATRICES, SETS AND GROUPS FOR SCIENCE STUDENTS, G. Stephenson. Concise, readable text introduces sets, groups, and most importantly, matrices to undergraduate students of physics, chemistry, and engineering. Problems. 164pp. 5⅜ × 8½. 65077-4 Pa. $6.95

THE HISTORICAL BACKGROUND OF CHEMISTRY, Henry M. Leicester. Evolution of ideas, not individual biography. Concentrates on formulation of a coherent set of chemical laws. 260pp. 5⅜ × 8½. 61053-5 Pa. $6.95

THE PHILOSOPHY OF MATHEMATICS: An Introductory Essay, Stephan Körner. Surveys the views of Plato, Aristotle, Leibniz & Kant concerning propositions and theories of applied and pure mathematics. Introduction. Two appendices. Index. 198pp. 5⅜ × 8½. 25048-2 Pa. $7.95

THE DEVELOPMENT OF MODERN CHEMISTRY, Aaron J. Ihde. Authoritative history of chemistry from ancient Greek theory to 20th-century innovation. Covers major chemists and their discoveries. 209 illustrations. 14 tables. Bibliographies. Indices. Appendices. 851pp. 5⅜ × 8½. 64235-6 Pa. $18.95

DE RE METALLICA, Georgius Agricola. The famous Hoover translation of greatest treatise on technological chemistry, engineering, geology, mining of early modern times (1556). All 289 original woodcuts. 638pp. 6¾ × 11.
60006-8 Pa. $18.95

SOME THEORY OF SAMPLING, William Edwards Deming. Analysis of the problems, theory and design of sampling techniques for social scientists, industrial managers and others who find statistics increasingly important in their work. 61 tables. 90 figures. xvii + 602pp. 5⅜ × 8½.
64684-X Pa. $15.95

THE VARIOUS AND INGENIOUS MACHINES OF AGOSTINO RAMELLI: A Classic Sixteenth-Century Illustrated Treatise on Technology, Agostino Ramelli. One of the most widely known and copied works on machinery in the 16th century. 194 detailed plates of water pumps, grain mills, cranes, more. 608pp. 9 × 12.
28180-9 Pa. $24.95

LINEAR PROGRAMMING AND ECONOMIC ANALYSIS, Robert Dorfman, Paul A. Samuelson and Robert M. Solow. First comprehensive treatment of linear programming in standard economic analysis. Game theory, modern welfare economics, Leontief input-output, more. 525pp. 5⅜ × 8½.
65491-5 Pa. $14.95

ELEMENTARY DECISION THEORY, Herman Chernoff and Lincoln E. Moses. Clear introduction to statistics and statistical theory covers data processing, probability and random variables, testing hypotheses, much more. Exercises. 364pp. 5⅜ × 8½.
65218-1 Pa. $9.95

THE COMPLEAT STRATEGYST: Being a Primer on the Theory of Games of Strategy, J.D. Williams. Highly entertaining classic describes, with many illustrated examples, how to select best strategies in conflict situations. Prefaces. Appendices. 268pp. 5⅜ × 8½.
25101-2 Pa. $7.95

MATHEMATICAL METHODS OF OPERATIONS RESEARCH, Thomas L. Saaty. Classic graduate-level text covers historical background, classical methods of forming models, optimization, game theory, probability, queueing theory, much more. Exercises. Bibliography. 448pp. 5⅜ × 8¼.
65703-5 Pa. $12.95

CONSTRUCTIONS AND COMBINATORIAL PROBLEMS IN DESIGN OF EXPERIMENTS, Damaraju Raghavarao. In-depth reference work examines orthogonal Latin squares, incomplete block designs, tactical configuration, partial geometry, much more. Abundant explanations, examples. 416pp. 5⅜ × 8¼.
65685-3 Pa. $10.95

THE ABSOLUTE DIFFERENTIAL CALCULUS (CALCULUS OF TENSORS), Tullio Levi-Civita. Great 20th-century mathematician's classic work on material necessary for mathematical grasp of theory of relativity. 452pp. 5⅜ × 8½.
63401-9 Pa. $9.95

VECTOR AND TENSOR ANALYSIS WITH APPLICATIONS, A.I. Borisenko and I.E. Tarapov. Concise introduction. Worked-out problems, solutions, exercises. 257pp. 5⅜ × 8¼.
63833-2 Pa. $7.95

THE FOUR-COLOR PROBLEM: Assaults and Conquest, Thomas L. Saaty and Paul G. Kainen. Engrossing, comprehensive account of the century-old combinatorial topological problem, its history and solution. Bibliographies. Index. 110 figures. 228pp. 5⅜ × 8½. 65092-8 Pa. $6.95

CATALYSIS IN CHEMISTRY AND ENZYMOLOGY, William P. Jencks. Exceptionally clear coverage of mechanisms for catalysis, forces in aqueous solution, carbonyl- and acyl-group reactions, practical kinetics, more. 864pp. 5⅜ × 8½. 65460-5 Pa. $19.95

PROBABILITY: An Introduction, Samuel Goldberg. Excellent basic text covers set theory, probability theory for finite sample spaces, binomial theorem, much more. 360 problems. Bibliographies. 322pp. 5⅜ × 8½. 65252-1 Pa. $8.95

LIGHTNING, Martin A. Uman. Revised, updated edition of classic work on the physics of lightning. Phenomena, terminology, measurement, photography, spectroscopy, thunder, more. Reviews recent research. Bibliography. Indices. 320pp. 5⅜ × 8¼. 64575-4 Pa. $8.95

PROBABILITY THEORY: A Concise Course, Y.A. Rozanov. Highly readable, self-contained introduction covers combination of events, dependent events, Bernoulli trials, etc. Translation by Richard Silverman. 148pp. 5⅜ × 8¼. 63544-9 Pa. $5.95

AN INTRODUCTION TO HAMILTONIAN OPTICS, H. A. Buchdahl. Detailed account of the Hamiltonian treatment of aberration theory in geometrical optics. Many classes of optical systems defined in terms of the symmetries they possess. Problems with detailed solutions. 1970 edition. xv + 360pp. 5⅜ × 8½. 67597-1 Pa. $10.95

STATISTICS MANUAL, Edwin L. Crow, et al. Comprehensive, practical collection of classical and modern methods prepared by U.S. Naval Ordnance Test Station. Stress on use. Basics of statistics assumed. 288pp. 5⅜ × 8½. 60599-X Pa. $6.95

DICTIONARY/OUTLINE OF BASIC STATISTICS, John E. Freund and Frank J. Williams. A clear concise dictionary of over 1,000 statistical terms and an outline of statistical formulas covering probability, nonparametric tests, much more. 208pp. 5⅜ × 8½. 66796-0 Pa. $6.95

STATISTICAL METHOD FROM THE VIEWPOINT OF QUALITY CONTROL, Walter A. Shewhart. Important text explains regulation of variables, uses of statistical control to achieve quality control in industry, agriculture, other areas. 192pp. 5⅜ × 8½. 65232-7 Pa. $7.95

THE INTERPRETATION OF GEOLOGICAL PHASE DIAGRAMS, Ernest G. Ehlers. Clear, concise text emphasizes diagrams of systems under fluid or containing pressure; also coverage of complex binary systems, hydrothermal melting, more. 288pp. 6½ × 9¼. 65389-7 Pa. $10.95

STATISTICAL ADJUSTMENT OF DATA, W. Edwards Deming. Introduction to basic concepts of statistics, curve fitting, least squares solution, conditions without parameter, conditions containing parameters. 26 exercises worked out. 271pp. 5⅜ × 8½. 64685-8 Pa. $8.95

TENSOR CALCULUS, J.L. Synge and A. Schild. Widely used introductory text covers spaces and tensors, basic operations in Riemannian space, non-Riemannian spaces, etc. 324pp. 5⅜ × 8¼. 63612-7 Pa. $8.95

A CONCISE HISTORY OF MATHEMATICS, Dirk J. Struik. The best brief history of mathematics. Stresses origins and covers every major figure from ancient Near East to 19th century. 41 illustrations. 195pp. 5⅜ × 8½. 60255-9 Pa. $7.95

A SHORT ACCOUNT OF THE HISTORY OF MATHEMATICS, W.W. Rouse Ball. One of clearest, most authoritative surveys from the Egyptians and Phoenicians through 19th-century figures such as Grassman, Galois, Riemann. Fourth edition. 522pp. 5⅜ × 8½. 20630-0 Pa. $10.95

HISTORY OF MATHEMATICS, David E. Smith. Nontechnical survey from ancient Greece and Orient to late 19th century; evolution of arithmetic, geometry, trigonometry, calculating devices, algebra, the calculus. 362 illustrations. 1,355pp. 5⅜ × 8½. 20429-4, 20430-8 Pa., Two-vol. set $23.90

THE GEOMETRY OF RENÉ DESCARTES, René Descartes. The great work founded analytical geometry. Original French text, Descartes' own diagrams, together with definitive Smith-Latham translation. 244pp. 5⅜ × 8½. 60068-8 Pa. $7.95

THE ORIGINS OF THE INFINITESIMAL CALCULUS, Margaret E. Baron. Only fully detailed and documented account of crucial discipline: origins; development by Galileo, Kepler, Cavalieri; contributions of Newton, Leibniz, more. 304pp. 5⅜ × 8½. (Available in U.S. and Canada only) 65371-4 Pa. $9.95

THE HISTORY OF THE CALCULUS AND ITS CONCEPTUAL DEVELOPMENT, Carl B. Boyer. Origins in antiquity, medieval contributions, work of Newton, Leibniz, rigorous formulation. Treatment is verbal. 346pp. 5⅜ × 8½. 60509-4 Pa. $8.95

THE THIRTEEN BOOKS OF EUCLID'S ELEMENTS, translated with introduction and commentary by Sir Thomas L. Heath. Definitive edition. Textual and linguistic notes, mathematical analysis. 2,500 years of critical commentary. Not abridged. 1,414pp. 5⅜ × 8½. 60088-2, 60089-0, 60090-4 Pa., Three-vol. set $29.85

GAMES AND DECISIONS: Introduction and Critical Survey, R. Duncan Luce and Howard Raiffa. Superb nontechnical introduction to game theory, primarily applied to social sciences. Utility theory, zero-sum games, n-person games, decision-making, much more. Bibliography. 509pp. 5⅜ × 8½. 65943-7 Pa. $12.95

THE HISTORICAL ROOTS OF ELEMENTARY MATHEMATICS, Lucas N.H. Bunt, Phillip S. Jones, and Jack D. Bedient. Fundamental underpinnings of modern arithmetic, algebra, geometry and number systems derived from ancient civilizations. 320pp. 5⅜ × 8½. 25563-8 Pa. $8.95

CALCULUS REFRESHER FOR TECHNICAL PEOPLE, A. Albert Klaf. Covers important aspects of integral and differential calculus via 756 questions. 566 problems, most answered. 431pp. 5⅜ × 8½. 20370-0 Pa. $8.95

CHALLENGING MATHEMATICAL PROBLEMS WITH ELEMENTARY SOLUTIONS, A.M. Yaglom and I.M. Yaglom. Over 170 challenging problems on probability theory, combinatorial analysis, points and lines, topology, convex polygons, many other topics. Solutions. Total of 445pp. 5⅜ × 8½. Two-vol. set.
Vol. I 65536-9 Pa. $7.95
Vol. II 65537-7 Pa. $6.95

FIFTY CHALLENGING PROBLEMS IN PROBABILITY WITH SOLUTIONS, Frederick Mosteller. Remarkable puzzlers, graded in difficulty, illustrate elementary and advanced aspects of probability. Detailed solutions. 88pp. 5⅜ × 8½.
65355-2 Pa. $4.95

EXPERIMENTS IN TOPOLOGY, Stephen Barr. Classic, lively explanation of one of the byways of mathematics. Klein bottles, Moebius strips, projective planes, map coloring, problem of the Koenigsberg bridges, much more, described with clarity and wit. 43 figures. 210pp. 5⅜ × 8½.
25933-1 Pa. $5.95

RELATIVITY IN ILLUSTRATIONS, Jacob T. Schwartz. Clear nontechnical treatment makes relativity more accessible than ever before. Over 60 drawings illustrate concepts more clearly than text alone. Only high school geometry needed. Bibliography. 128pp. 6⅛ × 9¼.
25965-X Pa. $6.95

AN INTRODUCTION TO ORDINARY DIFFERENTIAL EQUATIONS, Earl A. Coddington. A thorough and systematic first course in elementary differential equations for undergraduates in mathematics and science, with many exercises and problems (with answers). Index. 304pp. 5⅜ × 8½.
65942-9 Pa. $8.95

FOURIER SERIES AND ORTHOGONAL FUNCTIONS, Harry F. Davis. An incisive text combining theory and practical example to introduce Fourier series, orthogonal functions and applications of the Fourier method to boundary-value problems. 570 exercises. Answers and notes. 416pp. 5⅜ × 8½.
65973-9 Pa. $9.95

THE THEORY OF BRANCHING PROCESSES, Theodore E. Harris. First systematic, comprehensive treatment of branching (i.e. multiplicative) processes and their applications. Galton-Watson model, Markov branching processes, electron-photon cascade, many other topics. Rigorous proofs. Bibliography. 240pp. 5⅜ × 8½.
65952-6 Pa. $6.95

AN INTRODUCTION TO ALGEBRAIC STRUCTURES, Joseph Landin. Superb self-contained text covers "abstract algebra": sets and numbers, theory of groups, theory of rings, much more. Numerous well-chosen examples, exercises. 247pp. 5⅜ × 8½.
65940-2 Pa. $7.95
